INSTANCE SELECTION AND CONSTRUCTION FOR DATA MINING

T0189265

The Kluwer International Series
in Engineering and Computer Science

INSTANCE SELECTION AND CONSTRUCTION FOR DATA MINING

edited by

Huan Liu
Arizona State University, U.S.A.

Hiroshi Motoda
Osaka University, Japan

KLUWER ACADEMIC PUBLISHERS
Boston / Dordrecht / London

Distributors for North, Central and South America:
Kluwer Academic Publishers
101 Philip Drive
Assinippi Park
Norwell, Massachusetts 02061 USA
Telephone (781) 871-6600
Fax (781) 681-9045
E-Mail <kluwer@wkap.com>

Distributors for all other countries:
Kluwer Academic Publishers Group
Distribution Centre
Post Office Box 322
3300 AH Dordrecht, THE NETHERLANDS
Telephone 31 78 6392 392
Fax 31 78 6392 254
E-Mail <services@wkap.nl>

 Electronic Services <http://www.wkap.nl>

Library of Congress Cataloging-in-Publication Data

Instance selection and construction for data mining / edited by Huan Liu and Hiroshi Motoda.
    p. cm. -- (Kluwer international series in engineering and computer science ; SECS 608)
    Includes bibliographical references and index.

    1. Data mining. I. Liu, Huan, 1958- II. Motoda, Hiroshi. III. Series.

QA76.9.D343 I57 2001 ISBN 978-1-4419-4861-8
006.3--dc21   00-067106

Printed on acid-free paper.

Printed in the United States of America

The Publisher offers discounts on this book for course use and bulk purchases. For further information, send email to <scott.delman@wkap.com>.

Contents

Foreword

When Huan Liu and Hiroshi Motoda asked me to write a foreword for this book, I was intrigued as to why they have chosen me for this honor. They reminded me that in the mid-seventies of the last century (this sounds impossibly long ago!), I developed a method called "outstanding representatives," specifically addressing the problem of selecting best representative examples from a database for the purpose of machine learning. Indeed, it was an early method concerned with instance selection and construction, the topic of their new book. My subsequent research has taken different directions and I completely forgot about my "outstanding representative" method. However, I have always felt that instance selection and construction is an important area for investigation, and now, with the rapid growth of interests in data mining, its importance is even greater. I am, therefore, thrilled to see such a large volume of excellent and diverse contributions in this area.

The editors of this volume authored an earlier book "Feature Selection for Knowledge Discovery and Data Mining," and edited a book "Feature Extraction, Construction and Selection." This third book "Instance Selection and Construction for Data Mining" is a beautifully logical complement to the previous books and closes an important gap in the literature that existed in this area. The above three books jointly address a general and exceedingly important field of data selection, preparation and improvement for data mining and knowledge discovery.

The new book deals with the heretofore largely neglected topic of selecting instances from large datasets in order to facilitate the tasks of detecting important patterns and deriving useful knowledge from the data. Clearly, if the dataset to analyze is overwhelmingly large and noisy, which is a frequent case today, it is crucial for the success of the data mining task to be able to work with a high quality representative subset of the data.

The book provides a comprehensive and state-of-the-art coverage of the area of instance selection and construction, including such topics as basics of data sampling and an explanation of the issues in this area, a presentation of diverse methods for instance selection and construction, and illustrative examples of their application. The authors of individual chapters did an excellent job in presenting various specialized topics. This book is very well written and logically organized, giving the reader a modern and highly readable review of this increasingly important field.

In conclusion, I am delighted to have the opportunity to write a fore-word to this important and valuable collection. I recommend it whole-heartedly to any reader interested in topics relevant to data mining and knowledge discovery, machine learning, and the emerging general field of data science. The editors and authors should be congratulated on producing a fine book, the first one to take on this topic.

RYSZARD S. MICHALSKI

Preface

The most raw form of information is data. Data is accumulating increasingly fast as the use of computers widens. The storage of data serves many purposes, but it alone does not bring about knowledge that can be used for further actions to our advantages. The continuum from the past to the future makes it possible for us to seek clues from the past and to attempt for a better future. That is, we wish to guide our pursuit with data (records of the past). Data mining is an emerging technique that extracts implicit, previously unknown, and potentially useful information (or patters) from data. As with any burgeoning new technology that enjoys intense attention, the use of data mining is greatly simplified and trivialized. One of the challenges to effective data mining is how to handle vast amounts of data. A previously working data mining algorithm can falter with inundation of data. Some often encountered difficulties are that (1) data mining can become intolerably slow, (2) if we can get results, they are simply too overwhelming to be useful, and (3) a query can take very long time to respond. One solution is to reduce data for mining and keep those most relevant data for time critical tasks such as on-line analytic processing.

This book is a joint effort from leading and active researchers with a theme about instance selection and construction. It provides a forum for a wide variety of research work to be presented ranging from sampling, boundary hunting, to data squashing, to name a few. It is a timely report on the forefront of data preprocessing for data mining. It offers a contemporary overview of modern solutions with real-world applications, shares hard-learned experiences to avoid tricky pitfalls, and sheds light on future development of instance selection and construction.

Instance selection and construction are a set of techniques that reduce quantity of data by selecting a subset of data and/or constructing a reduced set of data that resembles the original data. Sampling is a commonly used, conventional technique. It relies on a random mechanism to form a subset of data, or a sample, for data mining tasks. Another technique of instance selection has recourse to search. Based on what kind of data mining algorithm is used, a set of relevant data is obtained in forms of critical points, boundary points, prototypes, etc. These various types of selected instances boil down to the basic questions whether we wish to best separate groups of data, or best represent each group. Instance construction takes another approach to data reduction by creating pseudo data points from original data with similar sufficient statistics. We can see in this book reviews of the current development and recent successful applications of instance selection for data mining.

This collection evolved from a project on instance selection spanning over two years. It consists of 6 parts. Part I presents some background knowledge and overview of the field. Parts II & III introduce instance selection methods and use of sampling methods, respectively. Part IV consists of chapters on unconventional methods of instance selection. Part V is dedicated to model combination using instance selection. Part VI concludes a full description of instance selection with some successful applications. The topics covered in 22 chapters include sampling, focusing, forcasting, feature (attribute) selection, discretization, filtering, active learning, instance based learning, squashing, boosting, footprinting, approximation, classification, clustering, and many more.

This book is intended for a wide audience, from *graduate students* who wish to learn basic concepts and principles of instance selection and construction to seasoned *practitioners and researchers* who want to take advantage of the state-of-the-art development for data mining. The book can be used as a reference to find recent techniques and their applications, as a starting point to find other related research topics on data preparation and data mining, or as a stepping stone to develop novel theories and techniques meeting the exciting challenges ahead of us.

HUAN LIU AND HIROSHI MOTODA

Acknowledgments

As the field of data mining advances, the interest in preprocessing data for effective data mining intensifies. A lot of research ensues. This book project resulted from the initiatives such as feature selection and data reduction. We received many constructive suggestions and support from researchers in machine learning, data mining and database communities from the very begining of the project. The completion of this book is particularly due to the contributors from all over the world, their ardent and creative research work, and successful applications in data preparation and data mining. The diligent reviewers have kindly provided their detailed and constructive comments and suggestions to help clarify terms, concepts, and writing in this truly multi-disciplinary collection. Their unsung efforts are greatly acknowledged.

We are also grateful to the editorial staff of Kluwer Academic Publishers, especially Scott Delmann and Melissa Fearon for their swift and timely help in bringing this book to a successful conclusion.

During the process of this book development, we were generously supported by our colleagues and friends at Osaka University, National University of Singapore, and Arizona State University. Our special thanks go to Melissa Fearon for her innovative cover design for this book.

Contributing Authors

Khaled AlSabti is Assistant Professor in the computer science department at King Saud University, Saudi Arabia. He received his M.S. and Ph.D. in computer science from Syracuse University, in 1994 and 1998, respectively. His research interest include knowledge discovery and data mining, databases and information retrieval, high performance computing and parallel algorithms.

Henry Brighton is a PhD student at the Language Evolution and Computation Research Unit in the Department of Linguistics, the University of Edinburgh. Previously he worked for Sharp Laboratories of Europe in the field of Machine Translation. His research interests include all forms of machine learning, but specifically, how different learning biases impact on the cultural evolution of language.

Daniel Bush is a recent Masters graduate from Coventry University. He is now working in London for a data-mining company in the financial sector.

Jean-Hugues Chauchat is Associate Professor of Applied Statistics at University Lumiére of Lyon, France. His present research interest lies in survey and data mining. He is directing a postgraduate program on the gathering, organization, and processing of data with a view to improve the decision-making process in the firm. He is consultant to both government and private industry.

Winton Davies is a graduate student at the Department of Computer Science, University of Aberdeen, Scotland, and a Principal Software Engineer with GOTO.COM, Inc. He was with Digital Equipment Corporation from 1985 to 1991. He is currently completing a PhD entitled

"Communication of Inductive Inference". His research interests include machine learning, software agents and multi-agent systems.

Carlos Domingo obtained his PhD in Computer Science at the Technical University of Catalonia (UPC), Barcelona, Spain in 1999. He is currently a European Union Science and Technology Fellow at the Department of Mathematical and Computing Sciences of the Tokyo Institute of Technology, Tokyo, Japan. His research interests are machine learning and data mining.

William DuMouchel received the Ph.D. in Statistics from Yale University in 1971 and has held a number of positions in academia and industry since then. He is currently a Technology Consultant at the AT&T Shannon Laboratory in Florham Park, New Jersey, conducting research on data mining, Bayesian modeling and other statistical methods.

Pete Edwards is Lecturer at the Department of Computer Science University of Aberdeen. He obtained his PhD from the University of Leeds for work on blackboard systems and chemistry. His current research interests are in the area of machine learning; specifically, the development of learning techniques for intelligent software agents, and data mining.

Dragan Gamberger is Research Associate at the Rudjer Bošković Institute in Zagreb, Croatia. His research interests include the development of techniques and the application of machine learning, knowledge discovery in databases, Occam's razor, and predicate invention. He received a PhD in computer science from the University of Zagreb.

Ricard Gavaldà received his Ph.D. in Computer Science at the Technical University of Catalonia (UPC) in 1992. He is Associate Professor at the Department of Software, UPC, since 1993. He worked mostly in computational complexity theory for several years. More recent research interests include also algorithmic aspects of machine learning and data mining.

Baohua Gu is a PhD student in the department of computer science of National University of Singapore. He got his BSc in Xi'an Jiaotong University (Xi'an, China) with major in Applied Mathematics & Engineering Software. His current study is mainly on statistical methods and data mining.

Julia Hodges is Professor and Department Head in the Department of Computer Science at Mississippi State University. She received her Ph.D. in computer science from the University of Southwestern Louisiana in 1985. Her research interests include knowledge discovery, data mining, and document understanding.

Feifang Hu is Assistant Professor at the Department of Statistics and Applied Probability, National University of Singapore. He obtained his Ph. D. Degree from University of British Columbia in statistics. His current research interests are in the area of biostatistics, bootstrap method and data mining.

Hisao Ishibuchi is Professor of Department of Industrial Engineering, Osaka Prefecture University. His research interests include linguistic knowledge extraction, fuzzy rule-based systems, fuzzy neural networks, genetic algorithms, and multi-objective optimization.

David Jensen is Research Assistant Professor of Computer Science at the University of Massachusetts, Amherst. His research focuses on the intersection of knowledge discovery, statistics, and decision making.

Chi-Kin Keung is a M. Phil. student in the Department of Systems Engineering and Engineering Management, the Chinese University of Hong Kong. He received the Bachelor degree in the same department in 1998. His research interests are machine learning and artificial intelligence, especially for instance-based learning and classification.

Wai Lam received a Ph.D. in Computer Science from the University of Waterloo, Canada in 1994. Currently he is Assistant Professor at Department of Systems Engineering and Engineering Management in the Chinese University of Hong Kong. His current interests include data mining, intelligent information retrieval, machine learning, reasoning under uncertainty, and digital library.

Nada Lavrač is Head of Intelligent Data Analysis and Computational Linguistics Group at the Department of Intelligent Systems, J. Stefan

Institute, Ljubljana, Slovenia. Her research interests include machine learning, inductive logic programming and intelligent data analysis in medicine. She is (co)author of numerous publications, including *Inductive Logic Programming: Techniques and Applications*, Ellis Horwood 1994.

Charles X. Ling obtained his BSc in Computer Science at Shanghai JiaoTong University in 1985, and his Msc and PhD in 1987 and 1989 respectively, from Computer Science at University of Pennsylvania. Since then he has been a faculty member at Univ of Western Ontario (UWO), Canada. He is currently Associate Professor, Adjunct Professor at University of Waterloo, and Director of Data Mining and E-commerce Laboratory.

Huan Liu is Associate Professor in the department of Computer Science and Engineering at Arizona State University. He earned a Ph.D. in Computer Science at University of Southern California (USC) in 1989. He held positions at Telecom Australia Research Laboratories (Telstra), and at School of Computing, National University of Singapore. His major research interests include machine learning, data mining, data preprocessing, and real-world applications.

David Madigan is Director of Dialogue Mining at Soliloquy, Inc., an internet company based in New York City. Previously he held positions at AT&T Shannon Labs and at the University of Washington. He was program co-chair for KDD-99 and serves on the editorial board of the Journal of Data Mining and Knowledge Discovery. He received a Ph.D. in Statistics from Trinity College Dublin, Ireland, in 1990.

Elizabeth McKenna received her BSc from the Department of Computer Science at University College Dublin and is currently completing her PhD in the area of Case-Based Reasoning and competence modelling. She has published her work in numerous international conferences and journals.

Chris Mellish is Reader in the Division of Informatics at the University of Edinburgh. He received his PhD in Artificial Intelligence from Edinburgh. He has worked mainly in the fields of natural language processing and logic programming. He is especially interested in natural language generation. He has co-authored textbooks on Prolog and natural language processing.

Ryszard S. Michalski is Planning Research Corporation Chaired Professor of Computational Sciences and Information Technology and Director of Machine Learning and Inference Laboratory at George Mason University, and Affiliate Professor at the Institute of Computer Science, Polish Academy of Sciences. He has authored over 350 publications in machine learning, plausible inference, data mining, computer vision and computational intelligence.

Hiroshi Motoda is Professor of the Division of Intelligent Systems Science at the Institute of Sicentific and Industrial Research of Osaka University. He was with Hitachi, Ltd. from 1967 to 1995. He obtained his Ph.D from University of Tokyo. His research interests include machine learning, knowledge acquisition, scientific knowledge discovery, data mining, qualitative reasoning, and knowledge based systems.

Tomoharu Nakashima is Research Associate of Department of Industrial Engineering at Osaka Prefecture University. He received the Ph.D. degree from Osaka Prefecture University in 2000. His current research interests include fuzzy systems, machine learning, evolutionary algorithms, reinforcement learning and game theory.

Martha Nason is a graduate student at the Department of Biostatistics, University of Washington. She is also a Research Scientist at Talaria, Inc. in Seattle. Her research interests include statistical modeling of massive datasets and computational statisics.

Manabu Nii is Ph.D student of Department of Industrial Engineering at Osaka Prefecture University. His current research interests include fuzzy neural networks, linguistic knowledge extraction, fuzzy rule-based systems, and genetic algorithms.

Partha Niyogi received his B.Tech from the Indian Institute of Technology, New Delhi, and his Ph.D. in electrical engineering and computer science from the Massachusetts Institute of Technology, USA. His research interests are in pattern recognition, learning theory, and their applications to problems in speech and language processing. He is currently with Lucent Technologies, Bell Laboratories, Murray Hill, NJ.

Richard Nock is Assistant Professor of Computer Science at the Université Antilles-Guyane, France. He obtained his PhD from the Université de Montpellier II, France in 1998. He holds also an Agronomical Engineering degree from the Ecole Nationale Supérieure Agronomique de Montpellier (ENSA.M), France. His research interests include machine learning, data mining, computational complexity and image processing.

Tim Oates is a researcher in the Department of Electrical Engineering and Computer Science at the Massachusetts Institute of Technology. His research focuses on learning and discovery.

D. Stott Parker is Professor of Computer Science at UCLA. He received a M.S. and Ph.D. at the University of Illinois, and since 1979 has been at UCLA. His interests center on developing more effective information models, and computer systems that manage them for applications in data mining, decision support, and scientific data management.

Chang-Shing Perng is Research Staff Member the Performance Management Group at IBM Thomas J. Watson Rsearch center. He received a Ph.D. at University of California, Los Angeles in 2000. His research interests includes querying and mining time series patterns, mining sequential data and sequential data modeling.

Christian Posse is Research Scientist at Talaria Inc, Seattle. He was Visiting Assistant Professor at University of Washington, Seattle, from 1995 to 1996, then Assistant Professor at University of Minnesota, Minneapolis between 1996 and 1998. He got his PhD from the Swiss Federal Institute of Technology at Lausanne, Switzerland. His research interests include graphical exploratory data analysis, clustering techniques, data mining and Bayesian computations.

Foster Provost is on the faculty of the Stern School of Business at New York University, where he studies and teaches information technology, knowledge discovery, and systems for intelligent commerce.

Nandini Raghavan is Senior Statistician at DoubleClick, Inc. in New York, New York. She was with AT&T Shannon Labs from 1999-2000 and Assistant Professor of Statistics at the Ohio State University from 1993-1999. Her research interests include data mining, statistical modeling of massive datasets and Bayesian inference.

Ricco Rakotomalala is Associate Professor at University Lumiére of Lyon, France. He developed SIPINA, a decision tree based data mining software, which is used for teaching and research in numerous universities. He is also coauthor of the book "Graphes d'Induction - Apprentissage et Data Mining (Hermes, Paris, 2000)" which makes a wide state of the art about methods and techniques implemented in induction of decision trees.

Sanjay Ranka is Professor in the department of computer science at University of Florida, Gainesville. He was Associate Professor at Syracuse University from 1988 to 1995. He received his B.Tech in computer science and engineering from IIT, Kanpur, in India in 1985 and his Ph.D. in computer and information science from University of Minnesota, Minneapolis, in 1988. Professor Ranka has coauthored over 120 papers, at least 40 of which have appeared in archival journals.

Colin Reeves is Senior Lecturer in Statistics in the School of Mathematical and Information Sciences at Coventry University. His research interests centre around heuristic methods, especially genetic algorithms, and their applications. He has authored over 80 papers on these topics. He is also the editor and primary author of the book *Modern Heuristic Techniques for Combinatorial Problems.*

Thomas Reinartz is currently Senior Researcher in the data mining group at DaimlerChrysler Research and Technology in Ulm, Germany. He received his Ph.D. in Computer Science from the University of Kaiserslautern. His research interests include knowledge acquisition, machine learning, data mining, and case-based reasoning.

Greg Ridgeway is Associate Statistician in the RAND Statistics Group in Santa Monica, California. He earned a Ph.D. in Statistics at the University of Washington in 1999. His research interests include computational statistics, boosting and bagging for non-linear prediction, and inference in massive datasets.

Marc Sebban is Assistant Professor in Computer Science of the TRIVIA team at the French West Indies and Guiana University. He obtained his Ph.D from University of Lyon 2 in 1996. His research interests include machine learning, feature and prototype selection, knowledge discovery, data mining, computational geometry

David Skalak is Senior Data Mining Analyst with IBM. A former Fulbright Fellow and a graduate of the Harvard Law School, he obtained his Ph.D. in 1997 from the University of Massachusetts at Amherst. His research interests include instance-based learning, classifier combination, artificial intelligence and law, and computational finance.

Barry Smyth is Lecturer in Artificial Intelligence in the Department of Computer Science at University College Dublin. Dr. Smyth specialises in the areas of Case-Based Reasoning and intelligent Internet Systems. He has published over 70 research papers, has won numerous international awards for his research, and has recently co-founded ChangingWorlds.com.

Shinsuke Sugaya is currently a Master course student at the graduate school of Yokohama National University, Japan. His research interests include machine learning, data mining, web analysis and computer vision.

Kah-Kay Sung received his Ph.D. degree in electrical engineering and computer science from the Massachusetts Institute of Technology, 1995. He is currently Assistant Professor at the Department of Computer Science, National University of Singapore. His research interests include computer vision and machine learning.

Einoshin Suzuki is Associate Professor of the Division of Electrical and Computer Engineering, Faculty of Engineering, Yokohama National University. He obtained his B. E., M. E. and Dr. Eng. degrees from the University of Tokyo. His research interests include artificial intelligence, machine learning, and data mining.

Shusaku Tsumoto received his Ph.D from Tokyo Institute of Technology and now is Professor at Department of Medical Informatics, Shimane Medical University. His interests include approximate reasoning, data mining, fuzzy sets, granular computing, non classical logic, knowledge acquisition, and rough sets (alphabetical order).

Hui Wang is Lecturer in Faculty of Informatics, University of Ulster. He earned his BSc and MSc in Jilin University, P. R. China, and his Ph.D in University of Ulster, N. Ireland. His research interests include statistical and algebraic machine learning, data mining, neural networks, and financial applications.

Osamu Watanabe is Professor of the Department of Mathematical and Computing Sciences, Tokyo Institute of Technology. He obtained Dr. of Engineering from Tokyo Institute of Technology on 1986. His research interests include computational complexity theory, design and analysis of algorithms, machine learning, and data mining.

Peggy Wright has been employed at the Information Technology Laboratory, Engineer Research & Development Center, U S Army Corps of Engineers for 15 years. She received her Ph.D. in computer science from Mississippi State University in May 2000. Her research interests include knowledge discovery, data mining, and knowledge management.

Hankil Yoon received his M.S. in computer engineering from University of California at Irvine in 1993, and his Ph.D. in computer science from University of Florida in 2000. He is currently a data mining engineer at Oracle Corporation in the U.S., and was formerly with Information Technology Laboratories of LG Electronics in Korea from 1988 to 1993. His research interest includes data mining on large data, data warehousing.

Sylvia R. Zhang is Software Engineer at Candle Corporation. She obtained her M.S. at University of California, Los Angeles on 1996.

I

BACKGROUND AND FOUNDATION

Chapter 1

DATA REDUCTION VIA INSTANCE SELECTION

Huan Liu
Department of Computer Science & Engineering
Arizona State University
Tempe, AZ 85287-5406, USA
hliu@asu.edu

Hiroshi Motoda
Institute of Scientific & Industrial Research
Osaka University
Ibaraki, Osaka 567-0047, Japan
motoda@sanken.osaka-u.ac.jp

Abstract   Selection pressures are pervasive. As data grows, the demand for data reduction increases for effective data mining. Instance selection is one of effective means to data reduction. This chapter expounds basic concepts of instance selection, its context, necessity and functionality. It briefly introduces the state-of-the-art methods for instance selection, and presents an overview of the field as well as a summary of contributing chapters in this collection. Its coverage also includes evaluation issues, related work, and future directions.

Keywords: Selection pressure, data reduction, instance selection, data mining.

1.1.   BACKGROUND

The digital technologies and computer advances with the booming internet uses have led to massive data collection (corporate data, data warehouses, webs, just to name a few) and information (or misinformation) explosion. In (Szalay and Gray, 1999), they described this phenomenon as "drowning in data" and reported that each year the detectors at the CERN particle collider in Switzerland record 1 petabyte of data; and researchers in areas of science from astronomy to the human genome are facing the same problems and choking on information. A natural question is "now that we have gathered so much data, what do we do with it?" Raw data is rarely of direct use and manual analysis simply cannot keep pace with the fast growth of data. Knowledge discovery and data mining (KDD), an emerging field comprising disciplines such as databases, statistics, machine learning, comes to the rescue. KDD aims to turn raw data into nuggets and create special edges in this ever competitive world for science discovery and business intelligence.

The KDD process is defined in (Fayyad et al., 1996) as *the nontrivial process of identifying valid, novel, potentially useful, and ultimately understandable patterns in data.* Data Mining processes include data selection, preprocessing, data mining, interpretation and evaluation. The first two processes (data selection and preprocessing) play a pivotal role in successful data mining. Facing the mounting challenges of enormous amounts of data, much of the current research concerns itself with scaling up data mining algorithms (Provost and Kolluri, 1999). Researchers have also worked on scaling down the data - an alternative to the algorithm scaling-up. The major issue of scaling down data is to select the relevant data and then present it to a data mining algorithm. This line of work is in parallel with the work on algorithm scaling-up and the combination of the two is a two edged sword in mining nuggets from massive data sets.

Selection pressures are present everywhere. They can be found in scientific enterprises, organizations, business environments, human being or even bacteria evolution. Making a decision is also about selecting among choices. Selection seems a necessity in the world surrounding us. It stems from the sheer fact of limited resources. No exception for data mining. Many factors give rise to data selection. First, data is not purely collected for data mining or for one particular application. Second, there are missing data, redundant data, and errors during collecting and recording. Third, data can be too overwhelming to handle. Instance selection is one avenue to the empire of data selection. Data is stored in a *flat file* and described by terms called *attributes* or *features*. Each line in the file consists of attribute-values and forms an *instance*, also named as a *record, tuple*, or a *data point* in a multi-dimensional space defined by the attributes. Data reduction can be achieved in many ways (Liu and Motoda, 1998; Blum and Langley, 1997). By selecting features, we reduce the number of columns in a data set; by discretizing feature-values, we reduce the number of possible values of discretized features; and by selecting instances, we reduce the number of rows in a data set. It is the last on which we focus in this collection. Instance selection is to choose a subset of data to achieve the original purpose of a data mining application as if the whole data is used (Michalski, 1975). Many variants of instance selection exist such as squashed data, critical points, prototype construction (or instance averaging), in addition to many forms of sampling.

The ideal outcome of instance selection is a model independent, minimum sample of data that can accomplish tasks with little or no performance deterioration, i.e., for a given data mining algorithm M, its performance $\mathcal{P}$ on a sample s of selected instances and the whole data w is roughly

$$\mathcal{P}(M_s) \doteq \mathcal{P}(M_w).$$

By model independence, we mean that for any two data mining algorithms M_i and M_j, let $\Delta\mathcal{P}$ be the performance difference in using data s with respect to using data w,

$$\Delta\mathcal{P}(M_i) \doteq \Delta\mathcal{P}(M_j).$$

Performance issues of instance selection will be elaborated in Section 1.3.

Although selection pressures always exist, the surge of data mining applications and wide availability of data mining algorithms/products give rise to the pressing need for data reduction. Among many, the well studied topic is feature selection, extraction and construction (Liu and Motoda, 1998). Instance selection, as another topic for data reduction, is recently getting more and more attention from researchers and practitioners. There are many reasons for this new trend: first, instance selection concerns some aspects of data reduction that feature selection cannot blanket; second, it is possible to attempt it now with advanced statistics and accumulated experience; and third, doing so can result in many advantages in data mining applications.

The issue of instance selection comes to the spot light because of the vast amounts of data and increasing needs of preparing data for data mining applications. More often than not, we have to perform instance selection in order to obtain meaningful results. Instance selection has the following prominent functions:

Enabling Instance selection renders the impossible possible. As we know, every data mining algorithm is somehow limited by its capability in handling data in terms of sizes, types, formats. When a data set is too huge, it may not be possible to run a data mining algorithm or the data mining task cannot be effectively carried out without data reduction. Instance selection reduces data and enables a data mining algorithm to function and work effectively with huge data.

Focusing   The data includes almost everything in a domain (recall that data is not solely collected for data mining), but one application is normally only about one aspect of the domain. It is natural and sensible to focus on the relevant part of the data for the application so that search is more focused and the mining is more efficient.

Cleaning   The GIGO (garbage-in-garbage-out) principle applies to almost all, if not all, data mining algorithms. It is therefore paramount to clean data, if possible, before mining. By selecting relevant instances, we can usually remove irrelevant ones as well as noise and/or redundant data. The high quality data will lead to high quality results and reduced costs for data mining.

The above three main functions of instance selection may intertwine. E.g., cleaning can sometimes be a by-product of the first two. Focusing can also serve a function of enabling under certain circumstances.

From its data aspect, instance selection takes the whole data w as input and churns out a sample s of the data as output. Although benefits reaped from this exercise are many, instance selection does require resources (machine, labor, and time). Therefore, we need to consider the gain vs. the loss due to instance selection. This problem is tightly associated with another two issues: (1) sample size and (2) mining quality. Usually, the more (relevant) data, the better the mining quality. In the context of data reduction, we are required to reduce data but maintain the mining quality. Hence, in selecting instances, we often face an issue of trade-off between the sample size and the mining quality. In this sense, instance selection is an optimization problem that attempts to maintain the mining quality while minimizing the sample size. The issue of measuring the mining quality is associated with the sample quality issue in which we aim to achieve a true representative sample with the minimum size.

Clearly, instance selection is not a simple matter. Many difficult issues related to instance selection in KDD remain to be solved. In the next section, we will briefly review major lines of research and development in instance selection. The first common reaction on instance selection is *random sampling*, we shall see that there are works other than random sampling. In the end of Section 1.2, we attempt to gain insights from the existing methods in our search of the next generation of instance selection methods. In Section 1.3, we are concerned about evaluation issues of instance selection. Section 1.4 is about the work related to data reduction and/or performance improvement. Section 1.5 is a brief introduction of the chapters in this collection with emphasis on their distinctive contributions. In the last section, we summarize the achievements and project further research, development, and application of instance selection.

1.2. MAJOR LINES OF WORK

A spontaneous response to the challenge of instance selection is, without fail, some form of sampling. Although it is an important part of instance selection, there are other approaches that do not rely on sampling, but resort to search or take advantage of data mining algorithms. As we can select the best feature to build a decision tree (stump), for example, we can also select the most representative data points for a decision tree induction algorithm to efficiently divide the data space. An overview on sampling methods is presented in Chapter 2. We discuss here instance selection methods associated with data mining tasks such as classification and clustering.

1.2.1. METHODS OF CLASSIFICATION

One key data mining application is classification. It is to discover patterns from data with two purposes: (1) classification - grouping the data points in terms of common properties, and (2) prediction - predicting the class of an unseen instance. The data for this type of application is usually labeled with class values. Instance selection in the context of classification has been attempted by researchers according to the classifiers being built. Instance selection here clearly serves the purposes of enabling, focusing, and cleaning. Besides the explicitness shown in ways of selecting instances, sample size can be automatically determined in the process. We include below four types of selected instances.

Critical points They are the points that matter the most to a classifier. The issue was originated from the learning method of Nearest Neighbor (NN) (Cover and Thomas, 1991). Instead of generalizing the data, NN does nothing in terms of learning; only when it is required to classify a new instance does NN search the data to find the nearest neighbor in the data for the new instance and use the nearest neighbor's class value to predict it. During the latter phase, NN could be very slow if the data is large and be extremely sensitive to noise. Therefore, many suggestions have been made to keep only the critical points so that noisy data points are removed as well as the data set is reduced. Here we use

Instance-Based Learning (IBL) (Aha et al., 1991) as an example of NN to illustrate the concept of critical points. IB1 is a modified version of NN with normalized feature values. IB2 is similar to IB1 but saves only misclassified instances (critical points) so that the storage requirements for IB2 can be significantly smaller than IB1. The idea is that instances near the concept boundaries are possible to wrongly classified. IB2 is prone to noise as noisy instances are almost always misclassified. An improve version of IB2 is IB3 that can better handle noise.

Boundary points   They are the instances that lie on borders between classes. Support vector machines (SVM) provide a principled way of finding these points through minimizing structural risk (refer to a tutorial in (Burges, 1998)). Using a non-linear function ϕ to map data points to a high-dimensional feature space, a non-linearly separable data set becomes linearly separable. Data points on the boundaries, which maximize the margin band, are the support vectors. Support vectors are instances in the original data sets, and contain all the information a given classifier needs for constructing the decision function. Boundary points and critical points are different in the ways how they are found.

Prototypes   They are representatives of groups of instances via averaging (Chang, 1974). A prototype that represents the typicality of a class is used in characterizing a class, instead of describing the differences between classes. Therefore they are different from critical points or boundary points.

Tree based sampling   Decision trees are a commonly used classification tool in data mining and machine learning. An attribute is chosen to split the data forming branches when we grow a tree from its root. It is a recursive process and stops when some stopping criterion is satisfied (for example, a fixed number of layers has been reached or all instances are classified). In order to build a compact yet accurate tree, some attribute selection criteria have been used, e.g., Gini (Breiman et al., 1984) and Information Gain (Quinlan, 1993). Instance selection can be done via the decision tree built. In (Breiman and Friedman, 1984), they propose *delegate sampling*. The basic idea is to construct a decision tree such that instances at the leaves of the tree are approximately uniformly distributed. Delegate sampling then samples instances from the leaves in inverse proportion to the density at the leaf and assigns weights to the sampled points that are proportional to the leaf density.

1.2.2.   METHODS OF CLUSTERING

When data is unlabeled, methods associated classification algorithms cannot be directly applied to instance selection. The widespread use of computers results in huge amounts of data stored without labels (web pages, transaction data, newspaper articles, email messages) (Baeza-Yates and Ribeiro-Neto, 1999). Clustering is one approach to finding regularities from unlabeled data. Being constrained by limited resources, we cannot just simply dump all the data to a computer (no matter how powerful the computer is) and let it crunch. Besides, in many applications, the experience from practice suggests that not all instances in a data set are of equal importance to the description of clustering. In the following, we discuss three types of selected instances in clustering.

Prototypes   They are pseudo data points generated from the formed clusters. The basic idea is that after the clusters are found in a space of high dimension, one may just keep the prototypes of the clusters and discard the rest. The k-means clustering algorithm is a good example of this sort. Given a data set and a constant k, the k-means clustering algorithm is to partition the data into k subsets such that instances in each subset are similar under some measure. The k means are iteratively updated until a stopping criterion is satisfied. The prototypes in this case are the k means.

Prototypes & sufficient statistics   In (Bradley et al., 1998), they extend the k-means algorithm to perform clustering in one scan of the data. By keeping some points that defy compression plus some sufficient statistics, they demonstrate a scalable k-means algorithm. From the viewpoint of instance selection, it is a method of representing a cluster using both defiant points and pseudo points that can be reconstructed from sufficient statistics, instead of keeping only the k means.

Data description in a hierarchy   When the clustering produces a hierarchy (concepts of different details or taxonomy, it is obvious that the prototype approach to instance selection will not work as there are many ways of choosing (defining) appropriate clusters. In other words, it is an issue about which layer in the hierarchy is the most representative description (or basic concepts that are neither too general nor too specific). This is because at the leaf level, each instance forms a cluster, and at the root level, all instances are under one cluster. In COBWEB (Fisher, 1987), for example, when basic concepts are determined, pseudo data points are the descriptions of basic concepts in probability distributions over the space of possible attribute values.

Squashed data They are some pseudo data points generated from the original data. In this aspect, they are similar to prototypes as both may or may not be in the original data set. Squashed data are different from prototypes in that each pseudo data point has a weight and the sum of the weights is equal to the number of instances in the original data set. Presently two ways of obtaining squashed data are (1) model free and (2) model dependent (or likelihood based). In (DuMouchel et al., 1999), the approach to squashing is model-free and relies on moment-matching to ensure that the original data and the squashed data are sufficiently similar. In other words, the squashed data set is attempted to replicate the moments of the original data. The model dependent approach is explained in Chapter 13 of this collection.

1.2.3.   INSTANCE LABELING

In real world applications, classification models are extremely useful analytic techniques. Although large amounts of data are potentially available, the majority of data are not labeled as we mentioned earlier. Manually labeling the data is a labor intensive and costly process. Researchers investigate whether experts are asked to only label a small portion of the data that means the most to the task if it is too expensive and time consuming to label all data. This is equivalent to a problem of selecting which unlabeled data for labeling. Usually an expert can be engaged to label a small portion of the selected data at various stages. So we wish to select as little data as possible at each stage, and use an adaptive algorithm to guess what else should be selected for labeling in the next stage. Clustering techniques are often used to group unlabeled data in the beginning. Instance labeling is closely associated with adaptive sampling (Chapter 2), clustering, active learning, and boosting. The last two topics will be discussed in Section 1.4.

1.2.4.   TOWARDS THE NEXT STEP

As shown above, instance selection has been studied and employed in various tasks (sampling, classification, clustering, and instance labeling). Life would become much simpler if we could have a universal model of instance selection. The most important of all is that we could have a component of instance selection available as a library function for applications (and it is transparent to the user). The tasks shown above are, however, very unique in themselves as each task has different information, requirements, and underlying principles. It is clear that a universal model of instance selection is out of the question. However, is it possible for us to generalize what we have discussed so far, explore their properties to invent new and better instance selection methods? In (Syed et al., 1999b), for example, they investigated if support vector machines could be used as a general model for instance selection in classification. The reason to choose SVMs is the nice features exhibited by SVMs and the statistical learning theory under SVMs - it aims to minimize both empirical risk and structural risk. The learning algorithms included in the limited scope of investigation are C4.5, SVMs, Multi-layered Perceptron, IBL (instance based learning). Since the instances selected by SVMs were empirically shown to be worse than the sample selected by random sampling for some other learning algorithms, the conclusion is that SVMs cannot serve as a universal model for instance selection with respect to classification. An instant derivation of this finding is that different groups of learning algorithms need different instance selectors in order to suit their learning/search bias well. This negative results should not deter our efforts in searching for a specific universal model of instance selection. It is hoped that this collection of works by various researchers can lead to more concerted study and development of new methods for instance selection. This is because even modest achievements along this line will harvest great profits in data mining applications: this would allow the selected instances to be useful for a group of mining algorithms in solving real world problems.

1.3.   EVALUATION ISSUES

Naturally, other things being equal, an instance selection algorithm would be the best if it achieves the largest reduction. However, we are not content with sheer data reduction. The goal of instance selection is multi-fold. Therefore, evaluation of instance selection is also related to tasks for which instance selection is performed. Some examples are (1) *Sampling* - performance is of sufficient statistics and can depend on the data distribution. For example, if it is of normal distribution, means and variances are the two major measures, higher moments are also used to check if sufficient statistics are preserved. (2) *Classification* - performance is more about predictive accuracy as well as aspects such as comprehensibility and simplicity, when applicable. (3) *Clustering* - performance is naturally about clusters: inter- and intra-cluster similarity, shapes of clusters, number of clusters, etc.

In practice, researchers often conduct experimental studies to evaluate performance. One can divide evaluation methods into two categories: *direct* and *indirect* measures. Direct measures work only with data and indirect measures involve a data mining algorithm. The basic idea for direct measures is to keep as much resemblance as possible between the

selected data and the original data so that the data selected can be used
by any data mining algorithms. Some examples of direct measures are
entropy, moments, and histograms.

Ironically, it would be more direct to see the effect of instance selection using indirect measures. For example, a classifier can be used to
check whether instance selection results in better, equal, or worse predictive accuracy. The more sophisticated version of measuring predictive
accuracy is to engage multiple runs of cross validation (Weiss and Indurkhya, 1998). In many cases, labeled data sets are used to measure
the quality of clusters. The class labels are not used during clustering.
After clustering, for each class, we can define precision and recall to
evaluate the clusters: how many data points in a cluster are correctly
assigned over the total data points in the cluster - the percentage of the
two is *precision*, and how many correctly assigned data points are in a
cluster with respect to the total data points that should belong to the
cluster - the percentage of the two is *recall*. One can observe the effect
of instance selection by comparing the two sets of precision and recall
before and after instance selection. In the case that class labels are not
known, one can examine how different the standard clustering measures
(inter- and intra-cluster similarity, number of clusters, etc.) are before
and after instance selection. In short, conventional evaluation methods
in sampling, classification, and clustering can be used in assessing the
performance of instance selection.

Since instance selection is usually a preprocessing step between data
and mining, one should also be aware of two practical and important
points in instance selection: *platform compatibility* and *database protec-
tion*. The former suggests that the subset of selected instances should
be compatible with the mining algorithm used in application; the latter
requires that under any circumstances, the original data should be kept
intact. For example, if a data mining algorithm cannot take weighted
instances, then it does not make sense to have squashed data as selected
instances. It is also wise to compare any new instance selection algorithm
with a basic sampling method (say random sampling) because the latter
is simple and fast, and check which maintains the better compatibility
between the original data and the selected instances.

The bottom line performance measures are *time complexity* and *space
complexity*. The former is about the time required for instance selection
to be performed. The latter is about the space needed during instance
selection. Time spent on instance selection should, in general, be no
more than what a data mining algorithm requires. The space requirement for instance selection should not be as demanding as data mining.
Instance selection stems from the need that we have to reduce the data

as it is just too much to handle in some applications. Nevertheless, we can sometimes relax the requirements on time and space a little because instance selection is normally performed once in a while. In other words, even if it requires similar time or space complexities as a data mining algorithm does, it may still be acceptable.

1.4. RELATED WORK

Instance selection is just one of many data reduction techniques. We briefly review some related techniques that can be used either directly or indirectly to reduce data or to identify distinguish instances or attributes. Considering data as a flat file as is in instance selection, we can transpose an instance selection problem to a problem of attribute selection. An updated collection on feature selection, extraction and construction (Liu and Motoda, 1998) provides a comprehensive range of techniques in this regard. The basic link of this work to instance selection is that we can reduce duplicates of instances by removing irrelevant attributes. Take an example of feature selection. Many evaluation measures have been designed for feature selection. In most experimental cases, the number of features can be reduced drastically without performance deterioration, or sometimes with performance improvement. This type of work, however, is not directly designed for instance selection, but can indirectly reduce instances by removing duplicates.

Boosting (Schapire, 1990) is, in a general sense, of instance selection though there is no data reduction involved. It is a general method which attempts to boost the accuracy of any given learning algorithm. One of the main ideas of the AdaBoost algorithm (Freund and Schapire, 1997) is to maintain a distribution or set of weights over the instances. It repeatedly updates the weights after generating weak hypotheses. The wrongly classified instances in many iterations will have larger weights, otherwise the instances will have smaller weights. At the termination of the algorithm, the difficult instances for the chosen learner will have higher weights.

Active learning also concerns instance selection. It assumes some control over the subset of the input space and is an approach to improving learning performance by identifying a specific set of data during learning instead of employing all data available. The key issue is to choose instances most needed for the learning at a particular learning time. One way is to measure uncertainty of the estimated predictions - uncertainty sampling (Lewis and Gale, 1994). The degree of uncertainty influences the amount of data as well as which subset of data. Query-by-committee is another way of choosing instances for learning (Seung et al., 1992).

The disagreement on a certain query among the committee will cause more data related to the query to be selected.

Incremental learning uses the data whenever it is available to maintain a "best-so-far" concept. The subsequently available data is used to revise and improve the concept. However, not every classifier is capable of learning incrementally. Incremental learning can be extended to a relaxed version - keeping the core data instead of maintaining a concept. Instance selection plays a role in selecting the core data to alleviate the storage problem. In (Syed et al., 1999a), they examine this type of incremental learning for support vector machines and show that support vectors can be used as the core data for learning in the next phase.

In the work of on-line analytic processing (OLAP), there is one task that is related to instance construction, i.e., aggregated or summarized data in a data warehouse. It's really about focusing and efficient online analytic processing. Aggregation strategies rely on the fact that most common queries will analyze either a subset or an aggregation of the detailed data. Aggregation consists of grouping the data according to some criterion and totaling, averaging, or applying some other statistical methods to the resultant set of data. The definition of a specific aggregation is driven by the business need, rather than by the perceived information need. The appropriate aggregation will substantially reduce the processing time required to run a query, at the cost of preprocessing and storing the intermediate results.

1.5.   DISTINCTIVE CONTRIBUTIONS

The chapters have been categorized into six basic parts that focus on: Background and Foundation, Instance Selection Methods, Use of Sampling Methods, Unconventional Methods, Instance Selection in Model Combination, and Applications of Instance Selection.

Part I introduces basic concepts of instance selection. Gu *et al* (Chapter 2) survey sampling methods, which, with a solid foundation in statistics, can be profitably used to estimate characteristics of a population of interest with less cost, faster speed, greater scope and even greater accuracy compared to a complete enumeration. They summarize the basic ideas, assumptions, considerations and advantages as well as limitations, categorize representative sampling methods by their features, provide a preliminary guideline on how to choose suitable sampling methods. Reinartz (Chapter 3) views the problem of instance selection as a special case of a more general problem - *focusing* and tries to provide a unifying framework for various approaches of instance selection. Sampling, clustering, and prototyping are the three basic generic components of the unified framework, and each of them can be instantiated to individual algorithms. The most important merit of this unified view is that it allows for systematic comparisons of various approaches along different aspects of interest and provide a basis to find out which instantiation of instance selection is best suited in a given context.

Part II spotlights some representative methods of instance selection. Smyth *et al* (Chapter 4) describe an instance selection problem in CBR systems. They introduce a notion of competence group and show that every case-base is organized into a unique set of competence group each of which makes its own contribution to competence. They devise a number of strategies to select footprint set (union of a highly competent subset of cases in each group) and empirically evaluated their effectiveness. Brighton *et al* (Chapter 5) focus on the instance selection problem of nearest neighbor (NN) classifiers, use a similar idea to the footprint and conduct an extensive study to compare their approach with other state-of-art techniques. They argue that there is no free lunch in that a single instance selection method can work for all data sets, and emphasize that whether the class distribution is homogeneous or heterogeneous in the feature space is a decisive factor. Both Smyth *et al* and Brighton *et al* view "selection as removal". Ishibuchi *et al* (Chapter 6) use a genetic algorithm to select both instances and features for a nearest neighbor classifier. Its fitness function reflects three terms: classification accuracy, number of instances and number of features. They introduce a biased mutation to accelerate reducing the number of instances. They also confirm that the selected instances and features give good performance when applied to a neural network classifier. The data sets used are small, though. Perng *et al* (Chapter 7) deal with time series data and propose the Landmark Model which integrates similarity measures, data representation and smoothing techniques in a unified framework. With this model the number of time points can be reduced to two ordered of magnitude with a little loss of accuracy. Their method can define features from the landmarks in a time series segment and find segments that are similar to each other under six basic transformations, which is achieved by comparing features invariant under these transformations.

Part III draws attention to sampling methods in instance selection. Domingo *et al* (Chapter 8) propose an adaptive on-line sampling method in which sampling size is not fixed a priori but adaptively decided based on the samples so far seen. Their problem is to select a nearly optimal hypothesis from a finite known hypothesis space by using a fraction of data without going through the whole data. Their method is meant to be used as a tool to solve some sub-tasks that fit in the framework. Provost *et al* (Chapter 9) consider progressive sampling in which sampling size is updated and the data are sampled randomly, starting with a small sample and using progressively larger ones until model accuracy no longer improves. They prove that geometric progressive sampling is optimal in an asymptotic sense, and show that it compares well with the optimal progressive sampling that can be computed by dynamic programming with some assumptions. They also propose a method to detect convergence efficiently. Chauchat *et al* (Chapter 10) address the problem of discretizing numeric attributes in building decision trees, in which sorting becomes problematic when the data set is very large. They employ sequential sampling at each node with the sampling size determined by statistical testing. Theoretical and empirical evidence shows that the strategy reduces substantially the learning time without sacrificing accuracy. Yoon *et al* (Chapter 11) present a method to incrementally construct classifiers as a dataset grows over time. A data set is divided into a series of epochs (subsets) from each of which an intermediate decision tree is constructed to generate weighted samples using a clustering technique at each leaf node (tree-based sampling). At each epoch the weighted samples are combined with the previously generated samples to obtain an up-to-date classifier. The empirical results show that the method is independent of data distribution and constructs a series of up-to-date classifiers in an efficient, incremental manner.

Part IV presents unconventional methods in instance selection. Madigan *et al* (Chapter 12) consider likelihood-based data squashing (LDS) to construct new instances. Squashing is a lossy data compression technique that preserves statistical information. Squashing can usually achieve significant data reduction. LDS uses a statistical model to squash the data. What is more interesting is that their results show that LDS provides excellent squashing performance even when the target statistical analysis departs from the model used to squash the data. Lam *et al* (Chapter 13) explore the important issue of applying prototype generation and filtering to instance-based learning. The motivation is to reduce its high data retention rate and sensitivity to noise. The combined approach is insensitive to the order of presentation of instances as well as provides noise tolerance. Their approach is shown to be effective in data reduction while keeping or improving classification accuracy in an extensive empirical study. Wang (Chapter 14) proposes a procedure for instance selection based on hypertuples which is a generalization of database tuples. The procedure is of two steps: creating hypertuples to form a model, and generating representative instances from the model. The two key criteria are identified: preserving classification structures and maximizing density of hypertuples. Experiments suggest that it can help improve the performance of an NN classifier in comparison with C5.

Wright and Hodges (Chapter 15) tell us how to use knowledge about attributes to select instances. Knowledge is provided by a domain expert on missing values and relative importance of attributes. The knowledge is then used to determine if an instance should be kept or discarded by the potential value of the instance based on the significance a feature has to the goal. Their work shows the significance and possibility in using knowledge for instance selection.

Part V demonstrates the role of instance selection in model combination. Skalak (Chapter 16) extends previous work on prototype selection for one nearest neighbor classifier and introduces an algorithm that combines an NN classifier with another small, coarse-hypothesis NN classifier to realize paired boosting. The results are encouraging. Nock and Sebban (Chapter 17) adapts boosting to prototype selection with a weighting scheme. They examine this new approach with reference to other existing ones and obtain favorable results. In addition, they employ visualization to display the relevance of selected prototypes. Davies and Edwards (Chapter 18) investigate a novel instance selection method for combining multiple learned models. A special feature of the method is that it results in a single comprehensible model. The comprehensibility is usually lost in a voting scheme or a committee approach. In their treatment, it is achieved by selecting instances using multiple local models and building a single model from the selected instances.

Part VI brings home real-world experience in instance selection. Reeves and Bush (Chapter 19) describe the problem of generalization in the application of neural networks to classification and regression problems. In particular, they investigate an application in which a fairly small reduction in error could lead to substantial reductions in cost. Genetic algorithms are used for training data selection in RBF networks. Sung and Niyogi (Chapter 20) discuss an active learning formulation for instance selection with applications to object detection. Different from a passive learner, an active learner explicitly seeks for new training instances of high utility. They propose a Bayesian formulation within a classical function approximation learning framework. A real-world learning scenario on object (face) detection is presented to show the effectiveness of active learning. Gamberger and Lavrac (Chapter 21) report on methods for noise and outliers detection that can be incorporated into sampling as filters for data cleaning. It is one of the most important but non-glamorous components of data mining. They investigate various types of filters in this regard. The noise filtering methods are put into test in a medical application - diagnosing coronary artery disease. The evaluation shows that the detected instances are indeed noisy or non-typical class representatives as judged by a domain expert. Hence, it is, in fact, instance removal - the opposite case of instance selection. Sugaya and Suzuki (Chapter 22) point out the application of instance selection in another medical domain - meningoencephalitis. They apply support vector machines to selecting instances as well as features. The comparison of SVM is conducted with a physician and with Fisher's linear discriminant. What is particularly interesting is that their findings have led the physician to interesting discoveries - one of ideal objectives of data mining.

1.6.   CONCLUSION AND FUTURE WORK

With the constraints imposed by computer memory and mining algorithms, we experience selection pressures more than ever. The central point of instance selection is *approximation*. Our task is to achieve as good mining results as possible by approximating the whole data with the selected instances and hope to do better in data mining with instance selection (that is, *less is more*) as it is possible to remove noisy and irrelevant data in the process.

There are many ways of achieving approximation of the original data via instance selection. Therefore, it would be nice if there were a single general purpose instance selection method that is guaranteed of good performance in any situation. Unfortunately, the best we have in line with this hope is random sampling, though it obviously ignores the nature of data mining tasks, mining goals, and data characteristics. We have presented an initial attempt to categorize the methods of instance selection in terms of sampling, classification, clustering, and instance labeling. We addressed fundamental issues of instance selection, outlined representative sampling methods and task-dependent instance selection methods, and discussed evaluation issues associated with instance selection. As each method has its strengths and weaknesses, identifying what to gain and what to lose is crucial in designing an instance selection method that matches the user's need. One of the immediate tasks we are facing is how to develop some theory that can guide the use of instance selection and help a user to apply it in data mining applications.

Much work still remains to be done. Instance selection deals with scaling down data. When we understand better instance selection, it is natural to investigate if this work can be combined with other lines of research in overcoming the problem of huge amounts of data, such as algorithm scaling-up, feature selection and construction. It is a big challenge to integrate these different techniques in achieving the common goal - effective and efficient data mining.

References

Aha, D. W., Kibler, D., and Albert, M. K. (1991). Instance-based learning algorithms. *Machine Learning*, 6:37–66.

Baeza-Yates, R. and Ribeiro-Neto, B. (1999). *Morden Information Retrieval*. Addison Wesley and ACM Press.

Blum, A. and Langley, P. (1997). Selection of relevant features and examples in machine learning. *Artificial Intelligence*, 97:245–271.

Bradley, P., Fayyad, U., and Reina, C. (1998). Scaling clustering algorithms to large databases. In *Proceedings of the Fourth International Conference on Knowledge Discovery & Data Mining*, pages 9 – 15. AAAI PRESS, California.

Breiman, L. and Friedman, J. (1984). Tool for large data set analysis. In Wegman, E. and Smith, J., editors, *Statistical Signal Processing*, pages 191 – 197. New York: M. Dekker.

Breiman, L., Friedman, J., Olshen, R., and Stone, C. (1984). *Classification and Regression Trees*. Wadsworth & Brooks/Cole Advanced Books & Software.

Burges, C. (1998). A tutorial on support vector machines. *Journal of Data Mining and Knowledge Discovery*, 2.

Chang, C. (1974). Finding prototypes for nearest neighbor classifiers. *IEEE Transactions on Computers*, C-23.

Cover, T. M. and Thomas, J. A. (1991). *Elements of Information Theory*. Wiley.

DuMouchel, W., Volinsky, C., Johnson, T., Cortes, C., and Pregibon, D. (1999). Squashing flat files flatter. In *Proceedings of the 5th ACM Conference on Knowlededge Discovery and Data Mining*. AIII/MIT Press.

Fayyad, U., Piatetsky-Shapiro, G., Smyth, P., and Uthurusamy, R., editors (1996). *Advances in Knowledge Discovery and Data Mining*. AAAI Press / The MIT Press.

Fisher, D. (1987). Knowledge acquisition via incremental conceptual clustering. *Machine Learning*, 2:139–172.

Freund, Y. and Schapire, R. (1997). A decision-theoretic generalization of on-line learning and an application to boosting. *Journal of Computer Systems and Science*, 55(1):119 – 139.

Lewis, D. and Gale, W. (1994). A sequential algorithm for training text classifiers. In *Proceedings of the Seventeenth Annual ACM-SIGR Conference on Research and Development in Information Retrieval*, pages 3 – 12.

Liu, H. and Motoda, H., editors (1998). *Feature Extraction, Construction and Selection: A Data Mining Perspective.* Boston: Kluwer Academic Publishers.

Michalski, R. (1975). On the selection of representative samples from large relational tables for inductive inference. *Report No. M.D.C. 1.1.9, Department of Engineering, University of Illinois at Chicago Circle.*

Provost, F. and Kolluri, V. (1999). A survey fo methods for scaling up inductive algorithms. *Journal of Data Mining and Knowledge Discovery,* 3:131 – 169.

Quinlan, J. (1993). *C4.5: Programs for Machine Learning.* Morgan Kaufmann.

Schapire, R. (1990). The strength of weak learnability. *Machine Learning,* 5(2):197 – 227.

Seung, H., Opper, M., and Sompolinsky, H. (1992). Query by committee. In *Proceedings of the Fifth Annual Workshop on Computational Learning Theory,* pages 287–294, Pittsburgh, PA. ACM Press, New York.

Syed, N., Liu, H., and Sung, K. (1999a). Handling concept drifts in incremental learning with support vector machines. In Chaudhuri, S. and Madigan, D., editors, *Proceedings of ACM SIGKDD, International Conference on Knowledge Discovery and Data Mining,* pages 317 – 321, New York, NY. ACM.

Syed, N., Liu, H., and Sung, K. (1999b). A study of support vectors on model independent example selection. In Chaudhuri, S. and Madigan, D., editors, *Proceedings of ACM SIGKDD, International Conference on Knowledge Discovery and Data Mining,* pages 272 – 276, New York, NY. ACM.

Szalay, A. and Gray, J. (1999). Drowning in data. *Scientific American,* page www.sciam.com/explorations/1999/.

Weiss, S. and Indurkhya, N. (1998). *Predictive Data Mining.* Morgan Kaufmann Publishers, San Francisco, California.

Chapter 2

SAMPLING: KNOWING WHOLE FROM ITS PART

Baohua Gu

Department of Computer Science, National University of Singapore

gubh@comp.nus.edu.sg

Feifang Hu

Department of Statistics and Applied Probability, National University of Singapore

stahuff@nus.edu.sg

Huan Liu

Department of Computer Science and Engineering, Arizona State University

hliu@asu.edu or liuh@comp.nus.edu.sg

Abstract   Sampling is a well-established statistical technique that selects a part from a whole to make inferences about the whole. It can be employed to overcome problems caused by high dimensionality of attributes as well as large volumes of data in data mining. This chapter summarizes the basic ideas, assumptions, considerations and advantages as well as limitations of sampling, categorizes representative sampling methods by their features, provides a preliminary guideline on how to choose suitable sampling methods. We hope this can help users build a big picture of sampling methods and apply them in data mining.

Keywords: sampling, sampling methods.

2.1.   INTRODUCTION

When studying the characteristics of a population, generally there are two approaches. The first one is to study every unit of the population, and such a process is statistically called a **complete enumeration** or a **census**. The other approach is to study the characteristics of the population by examining only a part of it, and such a technique is known as **sampling**. Theoretically a complete census is more desirable, however, if the population is very large, this approach would be time-consuming or expensive thus often infeasible. Practically, sampling can demonstrate remarkable advantages such as reduced cost, reduced time, greater scope and even greater accuracy over a complete census (Cochran, 1977). Therefore, sampling has been widely applied in many fields (Hedayat, 1991) including data mining (refer to (Gu et al., 2000) for more applications of sampling).

Plenty of sampling methods have been invented and it is widely accepted that the selection procedure is the first step of a sampling and also a main criterion to differentiate sampling methods. However, there is few literature dedicated to comparing different existing sampling methods by their selection procedures. A comprehensive comparison of existing sampling methods based on their selection procedures may make it easier for users to grasp the technique as a whole. With this motivation in mind, this chapter summarizes the basics of sampling, categorizes existing sampling methods based on their features in forming samples, provides a guideline of choosing a suitable sampling method. The main contribution of this paper is summarizing and categorizing many existing sampling methods.

The content of this chapter is organized as follows. We summarize basic terminology, underlying assumptions of sampling in Section 2.2, discuss general considerations of sampling in Section 2.3, categorize traditional statistical sampling methods along some dimensions in Section 2.4, and propose a guideline for choosing a suitable sampling method in Section 2.5 We conclude this chapter in Section 2.6

2.2.   BASICS OF SAMPLING

To acquire information about an unknown population, we can perform sampling on it. Statistically, a **population** is a set of elements about which we want to study. An **element** is a unit for which information is sought. **Sampling** is to select a subset, so called a **sample**, of a studied population and estimate some interested characteristics about the population. Before selecting a sample, we need to do two things. First, the population must be divided into collections of elements, which are called **sampling units** and must cover the whole of the population without overlapping, in the sense that every element in the population belongs to one and only one unit. The construction of a list of sampling units is called a **sampling frame**. Second, we make a *scheme* or *design* to select elements into a sample, i.e., a mechanism that can determine whether an element will be accepted or rejected as a member of a sample. The number of elements to be selected in the sample, called **sample size**, usually also need to be considered within a scheme. Such a scheme or design will be called **sampling method** thereafter in this paper, but it may have other names elsewhere perhaps with slightly different meaning, for example, **sampling design**.

After the sample is obtained, it is used to provide some statistical information of the population, which is often called **estimator** and expressed in constructed **statistic**, which is a function of the elements in a sample. A particular value taken by an estimator for a given sample is called an **estimate**. The estimation procedure usually needs techniques of **statistical inference**.

There are some basic assumptions and/or ideas underlying sampling. (1) The interested characteristics of a population are usually not available to us or hard to obtain, whereas the interested characteristics of its sample are much easier to obtain. This is the basic motivation that we need sampling. (2) A sample is always a subset selected from a population. However, an arbitrary selection of a subset from a population is hardly accepted as a "sampling method" by statisticians. In other words, an acceptable sampling method should be well-designed to give better estimation. This is why sampling is called a technique. (3) With a careful design of sampling and using a suitable estimation approach, one can obtain estimates that are "unbiased" for population quantities, such as mean value or total value of a target characteristic of the population, without relying on any assumptions about the population itself. This is why sampling can often work. (4) If the sample size were expanded until all units of the population were included in the sample, then the population characteristic of interest would be known exactly. Any uncertainty in estimates obtained by sampling thus stems from the fact that only part of the population is observed. This is why there are errors inherent to sampling. (5) With the population characteristics remain fixed, the estimate of them depends on which sample is selected and what estimation method is used. This is why different sampling methods may produce different estimations about the same population and why we should adopt a suitable one for a specific purpose.

2.3.   GENERAL CONSIDERATIONS

How large a sample should be? A suitable sample size is determined by taking into account cost, accuracy and some other considerations.

Generally, a sample size can be determined so that the estimates of the true value of a population does not differ by more than a stated margin error in more than δ of the cases. By setting up a probability inequality: $P(|e - e_0| \geq \epsilon) \leq \delta$, we solve for the sample size n for a given value ϵ and δ, where e stands for an estimate from the sample, which is generally a function of the sample size n, e_0 stands for the true value of the population, ϵ is called *confidence limit*, and $1 - \delta$ *confidence level*.

However, e_0 is usually unknown either. In this case, a practical way to determine the required sample size can be done as follow (Singh and Mangat, 1996): in the first step, we select a small preliminary sample of size m. Observations made on the units selected in this sample, will be used to estimate e_0 involved in the above expression. After replacing e_0 by the estimate just obtained from the preliminary sample, the equation is then solved for n, the required sample size. If $n \geq m$, then $n - m$ additional units are selected in the final sample. If $n \leq m$, then no more units are selected and preliminary sample is taken as the final sample.

How good an estimate will be? The ultimate objective of any sampling is to make inferences about a population of interest. The goodness of any estimate made from a sample depends both on the sampling method by which the sample is obtained and on the method by which the estimate is calculated from the sample data. A good estimator is expected to have two main properties. One is known as *unbiasedness*. The other is small *sampling variance*. An estimator e is called to be *unbiased* for the actual but unknown value e_0, if the expected value of e is equal to e_0, i.e., $E(e) = e_0$. Thus unbiasedness of an estimator ensures that on the average, e will take a value equal to e_0, although for most of the samples, the values taken by e will be more or less than e_0. Sampling variance is defined to measure the divergence of the estimator e from its expected value e_0, i.e., $V(e) = E[e - E(e)]^2$. With small sampling variance, we are ensured that the estimators differ little from sample to sample as well as from the true value e_0. To enhance the usefulness of estimation, we can also calculate a specific region $(e_0 - \epsilon, e_0 + \epsilon)$, known as *confidence interval*, in which the true value of interested parameter e_0 lies with a probability $1 - \delta$ (i.e., the *confidence level*), where the δ satisfies $P(|e - e_0| \geq \epsilon) = \delta$.

What types of errors may involve? In sampling theory, it is usually assumed that the variable of interest is measured on every unit in the sample without error, so that errors in the estimates occur only because just part of the population is included in the sample. Such errors are called **sampling errors**. But in fact, **non-sampling errors** may also arise due to defective sampling procedures, ambiguity in definitions, faulty measurement techniques and so forth.

Sampling error is inherent and unavoidable in every sampling scheme. This error, however, will decrease when sample size increases and shall theoretically become nonexistent in case of complete enumeration. In many situations, the decrease is inversely proportional to the square root of sample size (Singh and Mangat, 1996). Unlike sampling errors, the non-sampling errors are likely to increase with increase in sample size. It is quite possible that non-sampling errors in a complete enumeration survey are greater than both the sampling and non-sampling errors taken together in a sample survey. One should, therefore, be careful in evaluating and checking the processing of the sample data right from its collection to its analysis in order to minimize the occurrence of non-sampling errors.

Any useful auxiliary information or relationship? In many cases, there is additional information about the population elements that can be exploited. One use of auxiliary information is to divide the population into groups, for example, to make "clusters" in *cluster sampling*, and to make "strata" in *stratified sampling*. Auxiliary information opens the door to a variety of different sampling methods (Tryfos, 1996). Besides used in sampling design, auxiliary information may also be used in estimation.

Sometimes there exists a relationship between a unknown characteristic Y and another known characteristic X. It is practically better to exploit the relationship in order to estimate the Y rather than ignore it. Ratio and regression estimators are examples of the use of auxiliary information in estimation (Thompson, 1992).

What are the pros and cons of sampling? Compared to a complete enumeration, many practical samplings possess one or more of the following advantages: reduced cost, greater speed, greater scope, or great accuracy. With a small number of observations in sampling, it is possible to provide results much faster but with much less cost than a complete census. Sampling also has a greater scope than a complete enumeration regarding the variety of information by virtue of its flexibility and adaptability and the possibility of studying the interrelations of various factors. Moreover, under suitable conditions, more accurate data can be provided by a sample than by a complete enumeration (Cochran, 1977; Singh and Mangat, 1996; Som, 1995).

However, it is a misunderstanding that sampling can reveal the true characteristics of a population. In other words, it is wrong to believe that there is some way of selecting a part of a population so that the resulting estimates are guaranteed to be correct. There is no known methods of sample selection and estimation which ensure with certainty that the sample estimates will be equal to the unknown population characteristics. It should be understood that relying on a sample nearly always involves a risk of reaching incorrect conclusions. Sampling theory can assist in reducing that risk, but a certain risk is always present in every sampling (Tryfos, 1996).

2.4. CATEGORIES OF SAMPLING METHODS

Statisticians have developed many sampling methods and each of them has its own characters. Some sampling methods are applied to **general purpose**, while some others are only used for **specific domain**. Some methods treat each unit of a population with **equal probability**, while some with **varying probability**. Some samplings are done by selecting elements from a population **with replacement**, while some are done **without replacement**. Some sampling can be done in **one stage**, while some must be done in **multi-stage**. Many sampling are based on probability theory, while a few are not. Some sampling procedures work in a **non-adaptive** way (i.e, its scheme including sample size be determined before sampling and does not change during sampling), while some others use **adaptive** strategy by changing its scheme according to the result obtained so far from the sampling. Appropriate categorization of all these sampling methods may help to grasp their basic ideas. In this section, we categorize these sampling methods by the above-mentioned dimensions.

2.4.1. General vs. Domain-Specific

General-purpose sampling methods

This group of sampling methods refers to those that are widely used for general purpose, including random sampling, stratified sampling, cluster sampling, systematic sampling, double sampling, network sampling, and inverse sampling.

Random Sampling: This is a sampling method by which every population unit has the same chance of being selected in the sample. It has two variants: random sampling without replacement and random sampling with replacement. *Random sampling without replacement* is the most popular sampling design in which n distinct units are selected from the N units in the population in such a way that every possible combination of n units is equally likely to be selected. The selections are more commonly made by using a random number table or a computer "pseudo random number" generator. The most prominent advantage of using random sampling is that it does not introduce any bias in selecting the sample. Another advantage is that it is easy to operate (Fan, 1967).

Random sampling with replacement is a sampling design in which a sample of n units is selected from a population of N units by such a procedure of a series of draws that at any draw, all N units of the population are given an equal chance of being draw, no matter how often they already been drawn, i.e., any of the units may be selected more than once. One of its practical advantages is that, in some situations, it is an important convenience not to have to determine whether any unit in the population is included more than once. However, for a given sample size, simple random sampling with replacement is inherently less efficient than simple random sampling without replacement (Thompson, 1992).

Both two strategies are widely accepted, however, they may not apply when the objective is to sample some rare units in a population, because a random sample of the population may not include any information about the objective units.

Stratified Sampling: When the population is homogeneous (i.e., all the elements are identical), it is obvious that a sample of just one element is sufficient to provide correct estimates. Naturally, when a population consists of a number of approximately homogeneous groups, it will be convenient and effective to select a small number of elements from each of such groups. This is the basic idea of stratified sampling.

When making a stratified sample, we first divide a population into some non-overlapping sub-populations or strata. Then small samples could be selected from these different strata independently of one another. At last the total sample is formed by combining all the small samples. The strata are generally purposively formed by utilizing available relevant auxiliary information. In order to obtain high precision of estimates, the units within a stratum are required to be as similar as possible whereas differ as much as possible between strata. Then, even though one stratum may differ markedly from another, a stratified sample will tend to be "representative" of the population as a whole. A commonly used design is called *Stratified Random Sampling* in which the design within each stratum is *simple random sampling*.

The main advantages that stratified sampling over unstratified sampling are: (1) it could be an alternative way apart from increasing the sample size to reach higher precision of estimates and in fact it normally provides more efficient estimates than unstratified sampling; (2) it enables effective utilization of the available auxiliary information; (3) by grouping extreme values in a separate stratum, it can reduce the variability within other strata thus it can deal with "outliers". Its main disadvantage is that sometimes it is difficult in terms of time and cost to construct ideal strata (Singh and Mangat, 1996).

Cluster Sampling: When a population consists of a number of groups, each of which is a "miniature" of the entire population, it is possible to estimate correctly the population characteristics by selecting the smallest group and all its elements. This is the basic idea of cluster sampling. In cluster sampling, the population is firstly separated into mutually exclusive sub-populations called "clusters". Then a sample (often a random sample) of these clusters is chosen. All units in the selected clusters are selected to form the sample. Unlike stratified sampling, the units in each cluster are desired to be as much heterogeneous as possible and all clusters are similar to each other to obtain estimators of low variance. An ideal cluster contains the full diversity of the population and hence is "representative".

The advantage of cluster sampling is that it is often less costly to sample a collection of clusters than to sample an equal number of units from the population. However, it may be hard to ensure the "representativeness" of the sample since it almost inevitably introduces biasness when forming clusters (Yamane, 1967).

Systematic Sampling: One strategy of selecting n units from a population of N units is to select first a random number between 1 and k and then select the unit with this serial number and every kth unit afterwards. This procedure is known as *systematic sampling* and ensures that each unit has the same chance of being included in the sample, the constant k is known as the "sampling interval" and generally taken as the integer nearest to N/n, the inverse of the sampling fraction. There are several variants of systematic sampling (Krishnaiah and Rao, 1988).

This strategy has two main advantages. First, it is operationally convenient. The selection of the first unit determines the whole sample. Second, systematic samples are well spread over the population, thus there is less risk that any large contiguous part of the population will be left unrepresented. Therefore, this strategy can sometimes provide more precise estimates than a simple random sampling and a stratified random sampling. The main disadvantages of the strategy are: first, variance estimators cannot be obtained unbiasedly from a single systematic sample; second, a bad arrangement of the units (such as periodicity of the units in the population) may produce a very inefficient even misleading sample (Stuart, 1976).

Double Sampling: It is also called two-phase sampling (i.e., sampling is done in two phases) referring to a design in which initially a sample of units is selected from a large first-phase sample for obtaining auxiliary information only, and then a second-phase sample is selected in which the studied variable is observed in addition to the auxiliary information. The second-phase sample is often selected as a subsample of the first (Scheaffer et al., 1996).

The purpose of double sampling is to obtain better estimators by using the relationship between auxiliary variables and the studied variable. When the sampling procedure is completed in three or more phases, it is termed *multi-phase sampling*.

Network Sampling: In network sampling (sometimes called *multiplicity sampling*), a simple random selection or stratified random selection of units is made, and all units which are linked to any of the selected units are also included in the sample. The *multiplicity* of a unit is the number of selection units to which a unit is linked. Using network (or relation) among elements may provide a convenient way to select elements but it may also result in complicate estimation procedure.

Inverse Sampling: When an attribute under consideration occurs with rare frequency, even a large sample may not give enough information to estimate the attribute. A proper sampling method for such a situation is *inverse sampling*, in which sampling is continued until some specified conditions are satisfied, which often depend on the results of the sampling, have been fulfilled. By this strategy, we can obtain enough information about rare-occurring attribute by setting a suitable sample size. However, it may also be difficult to know how much is enough for estimating the unknown attribute. Therefore, this sampling method may work better with some auxiliary information available.

Domain-specific sampling methods

Sampling methods in this category are highly related to their respective domains, i.e., they are purposively designed for certain specific domain. Generally, these strategies may not be applied in other domains. However, their ideas may be useful to other domains.

Distance Sampling: It is used to estimate density or abundance of biological populations. *Line transect* and *point transect* are two main approaches to estimation of density. In distance sampling, a set of randomly placed lines or points is established and distances are measured to those objects detected by traveling along the lines or surveying the points (Buckland et al., 1993).

Spatial Sampling: The theory of spatial sampling is concerned with two-dimensional populations such as fields or groups of contiguous quadrants. It has been widely used in geostatistics, where it is frequently desired to predict the amount of ore or fossil fuel that will be found at a site. The prediction may be based on values observed at other sites in the region, and these other sites may be irregularly spaced in the region. This prediction problem and its solution–termed *kriging* in geostatistics–are essentially sampling with auxiliary information, where one can use the information in a covariance function to determine which sample sites will give the best prediction, that is, which sampling scheme is most effective (Thompson, 1992; Krishnaiah and Rao, 1988).

Capture-Recapture Sampling: It is used to estimate the total number of individuals in a population (Krishnaiah and Rao, 1988). An initial sample is obtained and the individuals in that sample are marked, then released. A second sample is independently obtained, and it is noted how many of the individuals in that sample are marked before. If the second sample is representative of the population as a whole, then the sample proportion of marked individuals should be about the same as the population proportion of marked individuals. From this relationship, the total number of individuals in the population can be estimated.

Line-Intercept Sampling: It is also used to estimate elusive populations such as rare animals. A sample is formed by the way that a sample of lines is selected in a study area and whenever an object of the population is intersected by one or more of the sample lines, a variable of interest associated with that object is recorded. With the design, a large bush has a higher probability of inclusion in the sample than does a small bush. Unbiased estimation of the population quantities depends on the ways to determine these probabilities (Thompson, 1992).

Composite Sampling: It is used in ecological studies. A sample is formed by taking a number of individual samples and then physically mixing them to form a composite sample. The goal is to obtain the desired information in the original samples but at reduced cost or effort. It can be used for detection of a certain trait, compliance, reduction of variance in the presence of inexact analytical procedures and so on (Krishnaiah and Rao, 1988).

Panel Sampling: This strategy may be used in social surveys. Sampling inquires are often carried out at successive intervals of time on a continued basis covering the same population. When the same sample of respondents in a mail, personal interview is selected to provide information on more than one occasion, the sampling plan is termed panel sampling (Som, 1995).

Monte Carlo Strategies: These series of strategies are successfully used in Bayesian Inference. The aims of Monte Carlo strategies are to solve one or both of the following problems: (1) to generate samples from a given probability distribution $P(x)$; (2) to estimate expectations of functions under this distribution. Common Monte Carlo strategies are: Importance Sampling, Rejection Sampling, Metropolis strategy, and Gibbs Sampling. *Importance Sampling* is basically not for generating samples from $P(x)$ but for estimating the expectation of a function. *Rejection Sampling* generate a sample of the complicated object distribution by using a simpler proposal distribution $Q(x)$. It randomly generates v first, and then decides if v is accepted into the sample set by comparing two values related to $P(v)$ and $Q(v)$. *Metropolis strategy* makes use of a proposal density $Q(x)$ which depend on the current state $x(t)$. By computing a function value b of $P(x)$ and $Q(x)$ at each new state x', it can make a decision by comparing b with 1. If $b \geq 1$, then the new state x' is accepted. Otherwise, the new state is accepted with probability b. *Gibbs Sampling* is a strategy for sampling from distributions over at least two dimensions. It can be viewed as a Metropolis strategy in which the proposal distribution Q is defined in terms of the conditional distributions of the joint distribution $P(x)$. It is assumed that $P(x)$ is too complex to draw samples from directly, its conditional distributions are tractable to work with (Mackay, 1998; Gamerman, 1997).

Shannon Sampling: It is named after the famous Shannon sampling theory and used in signal process. Given a sampling rate, Shannon's theorem gives the cutoff frequency of the pre-filter h. The original signal $x(t)$ changes into band-limited signal by computing $x(t) * h$ and then is transformed into $x(n)$ by a discretization process in which impulse train is made. After the Shannon sampling, the original signal continuous $x(t)$ is sampled as discrete $x(n)$ which can be passed through an ideal low-pass filter and then can be reconstructed after the transfer (Higgins, 1996).

2.4.2. Non-Probability Sampling

Statistically, a sampling is called a **probability sampling** when its selection procedure is based on the theory of probability. However, sampling can also be done without the theory of probability, i.e., **non-probability sampling**. The difference between non-probabilistic and probabilistic sampling does not necessarily mean that non-probabilistic samples are not representative of the population. But it does mean that non-probabilistic samples do not depend upon the rationale of probability theory. With a probabilistic sample, we know that the probabilities that we have represent the population well. We are able to estimate confidence intervals for a statistic. With non-probabilistic samples, it will often be hard for us to know how well we have done so. In general, researchers prefer probabilistic strategies to non-probabilistic ones, and consider them to be more accurate and rigorous. However, especially in social researches, when it is not feasible or theoretically sensible to do probability sampling, a wide range of non-probabilistic alternatives listed below may work better (Trochim, 1999).

Accidental Sampling: It is also called *Haphazard* or *Convenience Sampling*. This strategy refers to making a sample by accidental encountering or selection. For instance, accidentally interviewing a people on the street to get a quick reading of public opinion. Clearly, the problem with this sampling is that we have no evidence that they are representative of the populations we are interested in.

Purposive Sampling: In purposive sampling, we usually would have one or more specific predefined groups we are seeking. Purposive sampling can be very useful for situations where you need to reach a targeted sample quickly and where sampling for proportionality is not the primary concern. With a purposive sample, you are likely to get the opinions of your target population, but you are also likely to overweight subgroups in your population that are more readily accessible. All the strategies that follow in this subsection can be considered sub-categories of purposive sampling.

Modal Instance Sampling: In statistics, a *mode* is a value most frequently occurring in a distribution. In sampling, when we do a modal instance sample, we are sampling the most frequent cases, or the typical cases. Basically, modal instance sampling is only sensible for informal sampling contexts.

Expert Sampling: It involves the assembling of a group of persons who expertise in some area and making conclusions only based on information that they provide. The advantage of doing this is that we have some acknowledged experts to back us. The disadvantage is that even the experts can be wrong.

Quota Sampling: In quota sampling, we select people non-randomly according to some fixed quota. There are two types of quota sampling: proportional and non proportional. In proportional quota sampling we want to represent the major characteristics of the population by sampling a proportional amount of each. In non-proportional, we specify the minimum number of sampled units we want in each category. This strategy is the non-probabilistic analogue of stratified random sampling in that it is typically used to assure that smaller groups are adequately represented in our sample.

Heterogeneity Sampling: We sample for heterogeneity or diversity when we want to include all opinions or views, and we are not concerned about representing these views proportionately. In order to get all of certain diversity, and especially the "outlier" or unusual ones, we have to include a broad and diverse range of participants. Heterogeneity sampling is, in this sense, almost the opposite of *modal instance sampling*.

Snowball Sampling: It can be seen as the counterpart of network sampling in probability sampling. In snowball sampling, we begin by identifying someone who meets the criteria for inclusion in our study. We then ask them to recommend others who they may know who also meet the criteria. Although this strategy would hardly lead to representative samples, snowball sampling is especially useful when we are trying to reach populations that are inaccessible or hard to find.

It should be pointed out that, under the right conditions, any of these non-probability methods can give useful results. They are not, however, amenable to the development of the sampling theory, since no element of random selection is involved. The only way to examine how good the results are may be to find a situation in which the true results are known, either for the whole population or for a probability sample, and make comparisons (Cochran, 1977).

2.4.3.   One-Stage vs. Multi-Stage

Many sampling methods can be done in one-stage. Sometimes, however, we can combine the one-stage strategies in different ways, called **multi-stage sampling**, that help us address our sampling needs in a more efficient and effective manner (Yamane, 1967). In multi-stage sampling, the sample is selected in stages: the population is divided into a number of first-stage (or primary) units, which are sampled; then the selected first-stage units are sub-divided into a number of smaller second-stage (or secondary) units, which are again sampled: the process is continued until the ultimate sampling units are reached.

Multi-stage sampling differs from the multi-phase sampling in that the sampling unit in multi-phase sampling remains the same at each phase of sampling, whereas it changes in multi-stage sampling. Multi-stage sampling is used in a number of situations: (1) when it is extremely laborious and expensive to prepare such a complete frame; and (2) when a multi-stage sampling plan may be more convenient than a one-stage sampling.

Some commonly used multi-stage sampling methods are: (1) *Multi-stage Simple Random Sampling*, which makes simple random sampling at each stage; (2) *Multi-stage Varying Probability Sampling*, in which varying probability sampling is used at some stage; and (3) *Stratified Multi-stage Sampling*, in which a population is at first divided into some strata, then sampling is conducted in stages separately within each stratum.

2.4.4.   Equal vs. Varying Probability

With **equal probability sampling**, every units in a population have the same probabilities to be selected, while with **varying (or unequal) probabilities**, different units in the population have different probabilities of being included in the sample. The differing inclusion probabilities may result from some inherent features of the sampling procedure, or they may be deliberately imposed in order to obtain better estimates by including "more important" units with higher probability.

Such a strategy can be useful, for example, when units vary considerably in size, simple random sampling does not seem to be an appropriate procedure, since it does not take into account the possible importance of the size of the unit. Under such circumstances, selection of units with unequal probabilities may provide more efficient estimators than equal probability sampling. When the units are selected with probability proportional to a given measure of size, this type of sampling is known as *probability proportional to size* (PPS) sampling.

2.4.5.   With or Without Replacement

The distinction between the two applies regardless of the sampling method used. It can be better understood if it is imagined that the elements that will form the sample are step by step selected with one at each step. If previously selected elements are not eligible for selection at any other step, we say that sampling is **without replacement**. If, on the other hand, all the population elements are eligible for selection at any step of the selection process, we say that sampling is **with replacement**. Intuitively, sampling with replacement seems rather wasteful: once an element is selected and inspected, it appears to yield all the information it could possibly provide. Replacing it in the population, from which it may be selected again, does not seem to serve a useful purpose. In fact, although sampling without replacement is more popular, sampling with replacement has its own uses. For example, sampling with replacement is, in effect, equivalent to drawing elements from an infinitely large population. This property is sometimes very useful when handling sample of small size. For example, the famous "Bootstrap" method (Efron and Tibshirani, 1993) re-uses the original dataset to obtain new datasets by re-sampling with replacement.

2.4.6.   Adaptive vs. Non-Adaptive

For *non-adaptive* sampling methods, the selection procedure does not depend in any way on observations made during the sampling, so the sample size can be set before sampling. *Adaptive sampling* refers to sampling method in which the procedure for selecting sites or units to be included in the sample may depend on values of the variable of interest observed during the sampling.

The primary purpose of adaptive sampling is to take advantage of population characteristics to obtain more precise estimates than using non-adaptive strategies. Adaptive sampling methods offer the potential to give large increase in efficiency for some populations. However, adaptive selection procedures may introduce biases into conventional estimators. Whether an adaptive sampling is more efficient or less efficient than a non-adaptive sampling such as simple random sampling depends on the type of population being sampled (Thompson, 1992).

Adaptive Cluster Sampling: It refers to strategies in which an initial set of units is selected by some probability sampling procedure, and, whenever the variable of interest of a selected unit satisfies a given condition, additional units in the neighborhood of that unit are added to the sample. The initial sample can be systematic, striped or stratified, and thus respectively results in three variants: *systematic adaptive cluster sampling*, *strip adaptive cluster sampling*, and *stratified adaptive cluster sampling*.

2.4.7.   Summary of Sampling Methods

So far, we have described many sampling methods. Although they are not exhaustive, the sampling methods described in this chapter, from basic to special, one-stage to multi-stage, with replacement to without replacement, probability to non-probability, and adaptive to non-adaptive, constitute representatives of existing sampling methods.

To illustrate the relationship among these sampling methods, we draw a "sampling tree" in Figure 2.1. In the figure, each leaf node is a sampling method, each path from the root to a leaf shows a procedure to classify a given sampling method, while each internal node on such a path shows one criterion used to classify a sampling method. It should be pointed out that our categorization of sampling methods is not the only one even by the used standards. The main motivation for this categorization is to

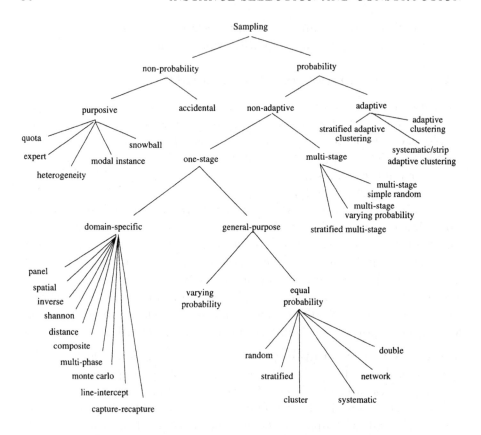

Figure 2.1   A Categorization Tree of Sampling Methods

provide a big picture and help users understand the relationships among these sampling methods as well as a preliminary guideline to use them.

Finally, it should be understood that having a sample in hand is not the end of a complete sampling procedure. To get estimations of the interested population is the very final objective of sampling. Due to length limitation, we deliberately omit the estimation procedures. We wish this would not give readers wrong impression that only a selection procedure is important to sampling.

2.5.   CHOOSING SAMPLING METHODS

Does there exist an optimal sampling that will produce an unbiased estimator of certain population characteristic with the lowest possible variance? The answer is that it depends (Thompson and Seber, 1996). With the fixed population approach, where no model is assumed about the population, with the population values being considered fixed constants and probability entering only through the sampling design, no optimal strategy exists. On the other hand, when a model of the population is known exactly, i.e., the joint probability distribution of the population values is known, the optimal strategy is known in general to be adaptive. Between the two extremes of no model and an exact statistical model assumed for the population are the cases where partial knowledge is assumed about the probability distribution for the population. For such cases, the optimal strategy is only possible for such selection procedures that do not depend on any observations of the variable of interest. Fortunately, these selection procedures cover a majority of common sampling methods, like random sampling, systematic sampling, stratified random sampling, cluster sampling, and multistage sampling.

However, in practice, the above conclusions may require unrealistic amounts of prior knowledge about the population, and may be difficult to implement and computationally complex to analyze. The greatest potential of the optimal strategies may be in suggesting procedures which, though not optimal, are efficient yet practical and simple.

The determination of an suitable sampling method is not straightforward and it should also take into account various technical, operational and cost considerations. Our categorization and the "sampling tree" can be used as a rough guideline to choose feasible sampling methods from many candidate strategies. For example, random sampling due to its ease and unbiasedness, should be in the first choices. If we just want this kind of ease, systematic sampling may also be a good choice. If we care less about precision of estimates as computed in probability, we can choose non-probability strategies. If sampling in one stage involves many factors which may be complex and lead to high cost or long sampling time and sampling in stages may help, multi-stage sampling or multi-phase sampling should be considered. If the research is very domain-specific, then besides common sampling methods, existing sampling methods for that specific domain may work better.

2.6.   CONCLUSION

Sampling methodology, with a solid foundation in statistics, can be profitably used to estimate characteristics of a population of interest with less cost, faster speed, greater scope and even greater accuracy compared to a complete enumeration. We discuss basic ideas, underlying assumptions, and general considerations of sampling. By surveying sampling methods, we collect many sampling methods that have been applied in different domains of real world applications. We categorize them according to their relations and characteristics. We provide a rough guideline of determining a suitable sampling method. We try to enlist as many as possible existing sampling methods to make a comprehensive comparison mainly on their ideas of selection procedure. We hope this can help users to understand sampling better.

References

Buckland, S., Anderson, D., Burnham, K., and Laake, J. (1993). *Distance Sampling: Estimating Abundance of Biological Populations.* Chapman – Hall, Inc.

Cochran, W. (1977). *Sampling Techniques.* Wiley, third edition.

Efron, B. and Tibshirani, R. (1993). *An Introduction to the Bootstrap.* Chapman & Hall, Inc.

Fan, S. (1967). *Basic Sampling Methods.* University of Singapore.

Gamerman, D. (1997). *Markov Chain Monte Carlo: Stochastic Simulation For Bayesian Inference.* Chapman – Hall.

Gu, B., Hu, F., and Liu, H. (2000). Sampling and its application in data mining. Technical Report http://techrep.comp.nus.edu.sg/techreports/2000/TRA6-00.asp, Department of Computer Science, National University of Singapore.

Hedayat, A. (1991). *Design and Inference in Finite Population Sampling.* NY: John Wiley – Sons, Inc.

Higgins, J. (1996). *Sampling Theory in Fourier and Signal Analysis.* Oxford Science Publications.

Krishnaiah, P. and Rao, C. (1988). *Handbook of Statistics 6: Sampling.* North-Holland.

Mackay, D. (1998). *Introduction to Monte Carlo Methods, in Learning in Graphical Models.* Kluwer Academic Publishers.

Scheaffer, R., Mendenhall, W., and L., O. (1996). *Elementary Survey Sampling.* Duxbury Press, fifth edition.

Singh, R. and Mangat, N. (1996). *Elements of Survey Sampling.* Kluwer Academic Publishers.

Som, R. (1995). *Practical Sampling Techniques.* M. Dekker, second edition.

Stuart, A. (1976). *Basic Ideas of Scientific Sampling.* London: Griffin.

Thompson, S. (1992). *Sampling.* Wiley.

Thompson, S. and Seber, G. A. F. (1996). *Adaptive Sampling.* Wiley.

Trochim, W. (1999). *Research Methods in Knowledge Base.* Cornell Custom Publishing, second edition.

Tryfos, P. (1996). *Sampling Methods for Applied Research: Text and Cases.* John Wiley & Sons, Inc.

Yamane, T. (1967). *Elementary Sampling Theory.* Prentice Hall.

Chapter 3

A UNIFYING VIEW ON INSTANCE SELECTION

Thomas Reinartz

DaimlerChrysler AG, Research & Technology, FT3/AD

P.O. Box 2360, 89013 Ulm, Germany

thomas.reinartz@daimlerchrysler.com

Abstract   In this chapter, we consider instance selection as a focusing task in the data preparation phase of knowledge discovery and data mining. Focusing covers all issues related to data reduction. First, we define a broader perspective on focusing tasks, choose instance selection as one particular focusing task, and outline the specification of evaluation criteria to measure success of instance selection approaches. Thereafter, we present a unifying framework that covers existing approaches for instance selection as instantiations. We describe a specific example instantiation of this framework and discuss its strengths and weaknesses. Then, we propose an enhanced framework for instance selection, generic sampling, and summarize evaluation results for several instantiations of its implementation. Finally, we conclude with open issues and research challenges for instance selection as well as focusing in general.

Keywords: Focusing, prototyping, leader sampling, generic sampling.

3.1.   INTRODUCTION

Knowledge discovery in databases (KDD) and data mining is increasingly important as the number and size of existing data sources grow at phenomenal rates, and all industry segments rely on analysis of data to compete. KDD is defined as a complex, iterative, and interactive process (e.g., (Fayyad et al., 1996)). This process contains several phases, each consisting of various tasks and activities. The CRoss-Industry Standard Process for Data Mining (CRISP-DM) suggests to separate the KDD process into six different phases: Business understanding, data understanding, data preparation, modeling, evaluation, and deployment.

CRISP-DM also describes tasks, activities and results of single steps within each phase in detail (Chapman et al., 1999).

As more experience exists in the field, it is clear that data preparation is one of the most important and time consuming phases in KDD. Preparation tasks such as data selection, data cleaning, data construction, data integration, and data formatting often determine the success of data mining engagements.

In this chapter, we emphasize the importance of data selection since the size of today's databases often exceeds the size of data which current data mining algorithms handle properly. Hence, we argue to use data reduction to shrink the data before data mining, and then apply data mining algorithms to the reduced data set.

From our perspective, data reduction covers a broad spectrum of different tasks varying in the data that is reduced, in the context in which data reduction is performed, and in the specific way of evaluation. As a broader term for all issues arising in data reduction, we use the term *focusing* throughout this chapter (Matheus et al., 1993). As we will show, instance selection is one particular focusing task, still varying in different manners.

This chapter is organized as follows. In the next section, we define a broader perspective on focusing tasks and describe instance selection as a subtask in focusing. Thereafter, we outline different evaluation strategies to measure success of instance selection approaches and define an example criterion in more detail. Then, we specify a unifying framework which covers existing instance selection methods as instantiations. We present an example instantiation of this framework and discuss its strengths and weaknesses. Thereafter, we propose an enhanced framework, generic sampling, and summarize evaluation results for several instantiations of its implementation. Finally, we conclude with open issues and remaining research challenges in the area of instance selection and focusing in general.

3.2.   FOCUSING TASKS

The first issue in dealing with focusing is to define focusing tasks in detail. This includes the specification of evaluation criteria to measure success of approaches solving these tasks in the context of data mining. In this section, we propose a framework for the definition of focusing tasks and respective evaluation criteria.

3.2.1.   Focusing Specification

In the following, we assume data in form of a *table*. A table is a pair of a set of *attributes* and a set of *tuples*. Each *attribute* is characterized by its *name*, its specific *type*, and its *domain* of values. Each *tuple* (or instance) contains a sequence of attribute values (see definition 1).

Definition 1 (Attribute, Tuple, and Table). We define an *attribute* a_j by a unique *name*, an attribute *type*, and a *domain $dom(a_j) = \{a_{j1},$ $a_{j2}, \ldots, a_{jk}, \ldots, a_{jN_j}\}$ or $dom(a_j) \subseteq R$. We specify a *tuple* t_i as a sequence of values $t_i = (t_{i1}, t_{i2}, \ldots, t_{ij}, \ldots, t_{iN})$. Each value t_{ij} is an element of attribute domain $dom(a_j)$, i.e., $\forall t_{ij}, 1 \leq j \leq N : t_{ij} \in dom(a_j)$.

Assume a set of attributes $A = \{a_1, a_2, \ldots, a_j, \ldots, a_N\}$, and a (multi-)set of tuples $T = \{t_1, t_2, \ldots, t_i, \ldots, t_M\}$. We define a *table* as a pair (A, T). We denote the *power set*, i.e., the set of all subsets, of A and T as $A^{\subseteq}$ and $T^{\subseteq}$, respectively.

We start the definition of focusing tasks by specifications of input, output, and the relation between them.

Definition 2 (Focusing Specification). Assume a table (A, T). A *focusing input* f_{in} is either a *set of tuples* $(f_{in} \subseteq \mathrm{T})$, a *set of attributes* $(f_{in} \subseteq \mathrm{A})$, or a *set of values* $(f_{in} \subseteq dom(a_j), a_j \in A)$. A *focusing output* f_{out} is either a simple *subset* $(f_{out} \subseteq f_{in})$, or a *constrained subset* $(f_{out} \subset f_{in}$, and $\mathrm{P}(f_{in}, f_{out}))$, or a *constructed set* of artificial entities $(f_{out} \not\subseteq f_{in}$, and $\mathrm{P}(f_{in}, f_{out}))$. P is a predicate or a function that relates f_{in} to f_{out}. A *focusing criterion* $f(f_{in}, f_{out})$ is either *relevance* or *representativeness*. A *focusing specification* is a tuple $(f_{in}, f_{out}, f(f_{in}, f_{out}))$ with a focusing input f_{in}, a focusing output f_{out}, and a focusing criterion $f(f_{in}, f_{out})$.

3.2.1.1   Focusing Input.   The original *focusing input* corresponds to a table. Each table is composed of a set of attributes and a set of tuples, and each attribute has its own domain of values. These three components result in three different inputs of focusing specifications (see definition 2). Hence, the focusing input specifies which component of a table is manipulated. Each type of input results in different tasks which need different solutions.

3.2.1.2   Focusing Output.   The overall *focusing output* still represents all components of a table. However, the output differs depending on the input and type of operations performed during focusing. If the input is a set of tuples, the output is also a set of tuples. Similarly, if the input is a set of attributes or a set of values, the output is a set of attributes or a set of values, respectively.

We further distinguish between the following three types of outputs (see definition 2). In case of *simple subsets*, we do not define any constraints on the output except that it is a subset of the input. Whereas the first type allows any arbitrary subset of the input, *constrained subsets* restrict the output to specific subsets that meet desired constraints. Finally, the output is a *constructed set*, if we allow the output to include artificial representatives of the input. This type no longer corresponds to subsets of the input but constructions of new entities.

Note, in case of constrained subsets and constructed sets, we assume predicates P or functions P. A predicate P defines the set of constraints which the output meets. A function P specifies the construction principles how to generate the output from the input.

3.2.1.3   Focusing Criterion.   The *focusing criterion* determines the relation between input and output. At an abstract level, we distinguish between *relevance* and *representativeness*. Relevance means that focusing restricts the available information in the input to a subset. In contrast, representativeness ensures that the output still represents the entire information in the input.

Specific focusing criteria are again either predicates or functions. In case of predicates, focusing criteria evaluate to true, if the relation between input and output meets the criterion. In order to get more fine-grained criteria, we also consider functions which return positive real values. These functions enable more detailed comparisons of different pairs of inputs and outputs.

3.2.2.   Focusing Context

Besides the core components of focusing tasks described previously, additional aspects affect focusing tasks and their solutions since focusing depends on phases before and after data preparation in the knowledge discovery and data mining process, and the results of other phases influence focusing. Here, we concentrate the *focusing context* on the most important aspects: Data mining goal, data characteristics, and data mining algorithm.

3.2.2.1   Data Mining Goal.   The *data mining goal* describes the primary purpose of an engagement. Data description and summarization, segmentation, concept description, deviation detection, dependency analysis, classification, and prediction are typical data mining goals (Chapman et al., 1999). Each of these goals has its own characteristics and hence its own influence on focusing tasks.

3.2.2.2   Data Characteristics.   Similarly, *data characteristics* affect the definition of focusing tasks, too. Data characteristics summarize information on the shape of data. Typical characteristics range from simple statistics up to information on the data quality.

3.2.2.3   Data Mining Algorithm.   Finally, the *data mining algorithm*, which we want to apply to achieve the data mining goal, is relevant for the definition of focusing tasks. Even for a fixed data mining goal and fixed data characteristics there exist numerous alternative algorithms, and the specific choice of approach is important for defining the focusing task.

3.2.3. Instance Selection as A Focusing Task

In summary, this section defined prerequisites for formal specifications of focusing tasks. Each aspect of specification and context influences the definition of focusing tasks as well as the development or selection of appropriate solutions. In conclusion, *focusing tasks* consist of focusing specification and focusing context, and evaluation criteria implement focusing criteria that enable comparisons of solutions.

Instance selection is a particular focusing task where the input is a set of tuples and the output is a (simple or constrained) subset of the input. From now on, we will restrict the focus of this chapter on this instance selection task, and start with an outline of specific evaluation strategies for instance selection.

3.3. EVALUATION CRITERIA FOR INSTANCE SELECTION

The definition of focusing criteria in section 3.2.1 is too general for evaluation of specific instance selection results. In this section, we argue to consider the focusing context in order to define specific evaluation criteria as implementations of relevance and representativeness to measure success of instance selection.

3.3.1. Different Evaluation Strategies

We distinguish different evaluation strategies along three separate dimensions.

3.3.1.1 Filter and Wrapper Evaluation. First of all, we distinguish between *filter* and *wrapper* approaches (John et al., 1994). Filter evaluation only considers data reduction but does not take into account any data mining activities. In contrast, wrapper approaches explicitly emphasize the data mining aspect and evaluate results by using the specific data mining algorithm to trigger instance selection.

In this chapter, we suggest four different components for each of filter and wrapper evaluation. For filter evaluation, we propose to consider typical statistical characteristics to measure the appropriateness of instance selection. For example, we expect a sample to represent the same or similar characteristics for *modus* and *mean, variance, distribution*, and *joint distribution* within attribute domains of interest.

For wrapper evaluation, we suggest aspects such as *execution time, storage requirements, accuracy,* and *complexity* (of data mining results).

For example, if we assume classification as the data mining goal, accuracy is often described in terms of the number of correct classifications given a classifier and a separate test set. If we use top down induction of decision trees to build the classifier, the number of nodes in the resulting decision tree represents the complexity of the result, for instance.

3.3.1.2   Isolated and Comparative Evaluation.   The second differentiation in evaluation distinguishes between *isolated* and *comparative* approaches. An evaluation strategy is isolated if it only takes into account a single instance selection approach and its results. In contrast, comparative approaches compare different solutions and their results to each other. For example, if an approach does not yield optimal results but all competitive approaches are worse, then, this approach is still good in terms of comparative evaluation.

3.3.1.3   Separated and Combined Evaluation.   For each aspect, we can define a single *separated* evaluation criterion. However, in some cases we want to consider more than only a single aspect for evaluation. Therefore, we can also define *combined* criteria that either cumulate different aspects or that also combine isolated and comparative evaluations. In combined evaluation, we can add relevance weights to different aspects and stress the most important factor given a specific data mining context depending on user requirements.

3.3.2.   An Example Evaluation Criterion

An example for an isolated separated wrapper evaluation criterion takes into account execution time of instance selection and data mining. We expect that for appropriate solutions the sum of execution time of instance selection and data mining on the output is smaller than execution time of data mining on the entire input. Consequently, we define the evaluation criterion which takes into account execution time as a relation between these two durations (see definition 3).

$E_T^{\leftrightarrow}$ returns values smaller than 1, if the sum of execution time of instance selection and data mining on the output is smaller than execution time of data mining on the input. It reveals values larger than 1, if the inverse is true. Hence, more appropriate solutions with respect to execution time yield smaller values of $E_T^{\leftrightarrow}$.

Definition 3 (Evaluation Criterion $E_T^{\leftrightarrow}$). Assume a table (A, T), a focusing input $f_{in} \subseteq T$, an instance selection approach $F : T^{\subseteq} \longrightarrow T^{\subseteq}$, a focusing output $f_{out} = F(f_{in}) \subseteq f_{in}$, if F is deterministic, or a set of focusing outputs F_{out}, $\forall f_{out} \in F_{out} : f_{out} = F(f_{in}) \subseteq f_{in}$, if F is non-deterministic, a data mining algorithm D, a function t that measures execution time of F and D, an additional value $\sigma(t) \geq 0$ that depicts the sum of t and its standard deviation in case of non-deterministic instance selection and their averaged application, and $0 \leq \eta_\sigma \leq 0.5$. We define the evaluation criterion $E_T^{\leftrightarrow}$ as a function $E_T^{\leftrightarrow} : F \times D \times T^{\subseteq} \times T^{\subseteq} \longrightarrow]0; \infty[$,

$$
\begin{aligned}
E_T^{\leftrightarrow}(F, D, f_{in}, f_{out}) := \quad & (1 - \eta_\sigma) \cdot \left(\frac{t(F, f_{in}) + t(D, f_{out})}{t(D, f_{in})} \right) \\
& + \eta_\sigma \cdot \left(\frac{\sigma\left(t(F, f_{in})\right) + \sigma\left(t(D, f_{out})\right)}{t(D, f_{in})} \right).
\end{aligned}
$$

At this point, we emphasize the difference between deterministic and non-deterministic instance selection approaches. *Deterministic* methods always generate the same output given the same input, whereas *non-deterministic* solutions possibly create different outputs although the input is the same. Systematic sampling for a fixed input order and a pre-specified start position is an example for a deterministic approach, whereas random sampling is a typical non-deterministic solution.

Evaluation of non-deterministic approaches has to consider the potential of randomly generating the best or worst possible result in a single try. Thus, a single application of a non-deterministic approach is not sufficient for an appropriate instance selection and its evaluation. We consider this aspect in definition 3 for non-deterministic solutions, by referring to average applications. We run non-deterministic approaches several times and average their results.[1]

In addition, definiton 3 punishes the worst possible case as an extra amount of $E_T^{\leftrightarrow}$ by adding the standard deviation of t across multiple experiments. Note, in definition 3 function σ covers both the average value plus the standard deviation.

3.4. A UNIFYING FRAMEWORK FOR INSTANCE SELECTION

Up to now, we specified instance selection as a focusing task and outlined evaluation strategies to measure success of instance selection approaches. In this section, we present a unifying framework which covers individual state of the art approaches related to instance selection.

3.4.1. Sampling, Clustering, and Prototyping

The unifying framework consists of three steps (see figure 3.1): Sampling, clustering, and prototyping. The general assumption of the framework is either to utilize (statistical) sampling techniques, or to search for sets of sufficiently similar tuples and to replace each set by a single or a few prototypes. At any intermediate point, instance selection approaches possibly apply an evaluation step in order to judge whether the output already meets the pre-defined evaluation criterion.

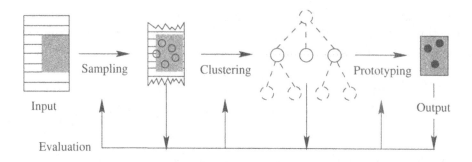

Figure 3.1   A Unifying Framework for Instance Selection

For instance, an example instantiation works as follows. First, a statistical sampling technique draws an initial sample. In the next step, a clustering technique groups the initial sample into subsets of similar tuples. For each of these subsets, the prototyping step selects or constructs a smaller set of representative prototypes. The set of prototypes then constitutes the final output of instance selection.

The order of steps in this framework is not strict. Instantiations are able to apply the basic steps in any order, or skip some steps completely. Similarly, instantiations sometimes skip one of the basic steps, perform other steps first, and then return to the previously ignored step. And finally, some approaches combine more than a single step into a compound step.

3.4.1.1   Sampling.   The first step of the unifying framework uses (statistical) *sampling* techniques (Cochran, 1977; Scheaffer et al., 1996). Originally, sampling techniques supply surveys in human populations. Starting with an initial question which requires measurements of variables in the population, sampling leads to approximate answers by drawing samples from the population, measuring variables of interest in these samples, and then concluding from results on samples to the entire population. Typically, variables of interest are numeric values such as averages, sums, or proportions.

As an advantage of sampling, we recognize that most sampling techniques are efficient in terms of execution time, and from KDD perspectives costs for drawing samples from tables and calculating statistical values are low. The theory of sampling is also well understood, although it only applies to estimations of a few numeric statistical values, and a similar theory for the appropriateness of sampling in KDD does not exist. We also notice that statistical sampling techniques generally do not take into account the focusing context and represent non-deterministic solutions. Hence, single experiments using sampling techniques are not sufficient for KDD purposes.

Due to space limitations, we can not describe specific sampling approaches in detail but typical instantiations of the unifying framework apply *simple random sampling, systematic sampling,* or *stratified sampling.*

3.4.1.2   Clustering.   The second step of the unifying framework uses *clustering* techniques to identify subsets of similar tuples within the input. In statistics as well as in machine learning exist many efforts to identify and to describe groups of similar objects (see Hartigan (1975) and Genari et al. (1989) for example). The idea to utilize clustering techniques for instance selection suggests to select a few representatives for each cluster. This set of representatives then forms the output of instance selection.

Again, detailed descriptions of clustering methods are beyond the scope of this chapter. All approaches mainly differ in their way to build clusters (coverage, separation, and structure) and whether they only create clusters without descriptions or whether they also provide characterizations of each cluster.

3.4.1.3   Prototyping.   The third step in the unifying framework is *prototyping*. In general, prototypes are more condensed descriptions of sets of tuples. Prototyping assumes that a single tuple is able to represent information of an entire subset of tuples, and we distinguish between the following types of approaches:

- *Prototype Selection* results in a subset of the original set of tuples and is often based on notions of *prototypicality*. For examples of prototype selection approaches, we refer to Zhang (1992), Skalak (1993), Skalak (1994), and Smyth & Keane (1995).

- *Prototype Construction* generates prototypes which do not necessarily refer to existing tuples. Therefore, this type of prototyping uses a specific mechanism to explicitly build new tuples that represent information of an entire subset of tuples. For examples of prototype construction methods, see Linde et al. (1980), Barreis (1989), Sen & Knight (1995), and Datta & Kibler (1995).

3.4.2.   An Example Instantiation of The Unifying Framework

In this section, we discuss an example instantiation of the unifying framework for instance selection. We focus our attention on a particularly promising approach that is both efficient in computation and appropriate for data mining. Further example instantiations are described in (Reinartz, 1999).

3.4.2.1   Leader Sampling.

Leader sampling (LEASAM) is a derivative of leader clustering which adopts the clustering algorithm for instance selection without explicitly constructing the clusters. Leader clustering constructs exhaustive, non-overlapping, non-hierarchical clusterings of the input (Hartigan, 1975). Therefore, we assume a similarity measure Sim to compute similarities between pairs of tuples, and a similarity threshold δ. The algorithm generates a partition of all tuples into clusters and selects a leader tuple for each cluster, such that each tuple in a cluster is within a distance of $1 - \delta$ from the leader tuple.

Leader sampling uses the same approach to perform instance selection. It only needs one pass through the input and assigns each tuple to the cluster of the first leader tuple which is sufficiently similar (in terms of δ) to this tuple. It selects new leader tuples for tuples which are not close enough to any existing leader tuple. As the final output of instance selection, LEASAM returns all leader tuples as representative prototypes.

3.4.2.2   Prototype Selection in Leader Sampling.

In terms of the framework for the definition of focusing tasks, leader sampling refers to constrained subsets as the output. Definition 4 states the prototype selection criterion in LEASAM. This criterion evaluates to true, if for all tuples in the input, at least one tuple in the output exists which exceeds similarity threshold δ in comparison to this tuple, and for all pairs of distinct tuples in the output, similarities between these tuples are smaller than δ.

Definition 4 (Prototype Selection in LEASAM). Assume a table (A, T), a focusing input $f_{in} \subseteq T$, a focusing output $f_{out} \subseteq f_{in}$, a similarity measure Sim, and $0 < \delta \leq 1$. Predicate $P(f_{in}, f_{out})$ in LEASAM is

$$P(f_{in}, f_{out}) \quad :\Longleftrightarrow \quad \forall t_i \in f_{in} \, \exists t_{i'} \in f_{out} : Sim(t_i, t_{i'}) \geq \delta$$
$$\wedge \quad \forall t_i, t_{i'} \in f_{out} : t_i \neq t_{i'} \implies Sim(t_i, t_{i'}) < \delta.$$

3.4.2.3   Advantages and Disadvantages.

In terms of the unifying framework for instance selection, leader sampling is a compound approach of clustering and prototyping. Leader sampling builds clusters (though not storing them) and leader tuples at the same time. Leader tuples correspond to prototypes, and each leader represents a single cluster. Thus, the set of all leaders defines a set of prototypes that represents the entire input.

The positive feature of leader sampling is that it only requires one sequential pass through the data. It only needs to retain leader tuples, not the clusters. On the other hand, several negative properties follow from immediate assignments of tuples to clusters as soon as the first leader exceeds similarity threshold δ.

The first negative property is that clusterings are not invariant with respect to reordering of tuples. For example, the first tuple is always the leader tuple of the first cluster. The second negative property is that the first clusters are always likely to contain more tuples than later clusters, since they get first chance at each tuple as it is allocated. Thirdly, representativeness of leaders mainly depends on similarity threshold δ. For small values of δ, leader clustering selects only a few leaders, whereas for large values of δ, the set of leaders is likely to contain many tuples. It is difficult to set δ to the most appropriate value in terms of representativeness of resulting leaders in advance.

As an extension to leader sampling that overcomes these drawbacks, we refer to *advanced leader sampling* (Reinartz, 1997). This novel approach provides two additional preparation steps, sorting and stratification (see section 3.5.1). Furthermore, we also point to *similarity-driven sampling* which supports automatic mechanisms for estimation and adaptation of similarity threshold δ during execution (Reinartz, 1998).

3.5. EVALUATION

In this section, we briefly present a condensed description of an experimental evaluation of different instance selection approaches. For this purpose, we utilize the *generic sampling* approach which results from providing arbitrary combinations of the unifying framework with enhancements in advanced leader sampling and similarity-driven sampling (Reinartz, 1999).

3.5.1. Generic Sampling

Algorithm GenSam which implements the generic sampling approach provides in comparison to the unifying framework for instance selection two additional preparation steps, sorting and stratification, plus an intelligent sampling step.

3.5.1.1 Sorting. *Sorting* uses the following order relation between tuples which considers attribute values in order of attribute relevance starting with the most important attribute.[2] If value t_{iN} for the most important attribute a_N in tuple t_i is larger than value $t_{i'N}$ in tuple $t_{i'}$, then sorting places t_i after $t_{i'}$ in the sorted table. If both tuples have the same value for a_N, sorting recursively proceeds with the next important attribute in the same way. Relative positions of completely identical tuples remain constant.

For local comparisons between attribute values, we distinguish between quantitative and qualitative attributes. For quantitative attributes, we use the regular numeric order, whereas for qualitative attributes, we employ the lexicographic order.

3.5.1.2   Stratification.   *Stratification* constructs a strata tree. It separates tuples into smaller subsets according to attribute values. Again, stratification considers attribute values in order of attribute relevance and also distinguishes between quantitative and qualitative attributes. For quantitative attributes, a stratum contains tuples with values in the same interval generated by discretization (e.g., (Dougherty et al., 1995)), whereas for qualitative attributes, a stratum includes tuples with the same value. Stratification treats missing or unknown values as special cases and builds extra strata for all tuples with special values for the same attribute.

The stratification procedure continues with the next important attribute at the next lower level in the hierarchy of strata as long as a pre-defined maximum number of tuples within a single stratum is exceeded or no more attributes for stratification are available.

3.5.1.3   Intelligent Sampling.   For the intelligent sampling step, generic sampling is able to supply any instantiation of the unifying framework for instance selection. In its implementation in GenSam, generic sampling provides random sampling, systematic sampling, leader sampling, and similarity-driven sampling.

All in all, generic sampling provides an enhanced framework which allows applications of many different instance selection approaches by combining the three steps, sorting, stratification, and sampling, in any way.

3.5.2.   Experimental Setting

For experimental studies, we selected the data mining goal classification, several data sets, representing varying data characteristics, from the UCI repository, and two different data mining algorithms for classification, C4.5 (Quinlan, 1993) and instance-based learning (Aha et al., 1991; Kohavi et al., 1996).

As instance selection approaches, we chose several different instantiations of generic sampling. In this chapter, we concentrate on simple random sampling (R), simple random sampling with stratification (RS), systematic sampling (S), systematic sampling with sorting (SS), leader sampling (L), and leader sampling with sorting and stratification (LS). For all approaches, we used various parameter settings, and selected the best results for each approach for evaluation.

For each data set, for each data mining algorithm, and for each instance selection approach, experimentation started with randomly splitting the data set into train and test data (about 80% and 20% of the data, respectively). Then, we applied each instance selection approach to the training set, run each data mining algorithm on the resulting output, and finally tested the generated classifiers on the test set to estimate predictive accuracy. Note, for non-deterministic instance selection approaches we conducted multiple runs and averaged results.

For evaluation and comparison of different instance selection approaches, we considered three different combined evaluation criteria which rank the results according to more than only a single aspect in filter and wrapper evaluation as well as in isolated and comparative modes. Below, we present the results for combined filter evaluation.

3.5.3. Focusing Advice

The overall goal of experimental evaluation is to generate focusing advice, i.e., to provide guidance in selecting the best suited instance selection approach given the context. Therefore, we defined *focusing scores* in the following way.

First, we considered situations where a specific instance selection approach behaves particularly good or particularly bad in terms of the evaluation criterion separately. We added a *plus* for good success and assigned a *minus* for poor performance. Thereafter, we analyzed comparisons between different approaches. For each pair of comparisons, we related results of the first approach to success of the second. The superior solution again got a plus, whereas the inferior solution deserved a minus. At the end, we summed up all plus and minus scores which resulted in final scores for each instance selection approach. Note, in individual cases, scores abstract from specific experimental results but keep most important relations among focusing solutions.

If we apply focusing solutions in KDD, we use the resulting score tables to decide which instance selection approach is preferable. Rows in each table allow comparisons among different approaches for specific data characteristics, whereas columns relate concrete approaches across varying data characteristics. Negative scores mean that this approach is generally less appropriate in this context, whereas positive scores recommend applications of this approach. The lower the score is, the worse is the solution, and the higher the value is, the better is the solution.

For example, table 3.1 shows focusing advice in terms of focusing scores for six instance selection approaches given different aspects of the focusing context according to combined filter evaluation. Here, we describe the context by the number of tuples, the number of attributes, the type of attributes, and the number of class labels.

Table 3.1   Focusing Advice for Filter Evaluation

Context	R	RS	S	SS	L	LS
few tuples	0	+1	-1	+4	+5	0
many tuples	-1	+1	+2	+5	+4	0
few attributes	-1	+1	+2	+5	+4	-2
many attributes	-3	-3	-1	0	+5	+6
no qual. attributes	-1	+1	+1	+5	+4	0
no quant. attributes	-1	+2	+3	+5	+2	0
more qual. attributes	-1	+1	+1	+5	+4	0
more quant. attributes	-1	+1	+1	+5	+4	0
many classes	+3	+3	+3	+5	-1	-4
Average	-0.6	+1.1	+1.4	+4.6	+3.4	-0.1

In summary, table 3.1 shows that on average systematic sampling with sorting (SS) is superior to other approaches, followed by leader sampling (L) without any preparation in advance. For full discussions of the experimental evaluation and more focusing advice, we refer to (Reinartz, 1999).

3.6. CONCLUSIONS

In summary, this chapter addressed different issues in the area of instance selection. First of all, we generally defined focusing tasks and showed that instance selection is one particular focusing task in the data preparation phase of the knowledge discovery and data mining process. We outlined evaluation strategies for instance selection and discussed an example criterion in detail.

Second, we presented a unifying framework for instance selection that covers existing attempts towards solutions. We discussed an example, demonstrated how it fits into the framework, and briefly analyzed its strengths and weaknesses.

Finally, we also outlined an enhanced framework, generic sampling, and summarized results of experimental studies with several instantiations of this approach in the context of classification tasks with varying data characteristics as focusing advice.

Although size and complexity of existing data and information sources grow at phenomenal rates and the importance of data reduction techniques became clear, it is interesting to see that there is relatively few research and technology in this area that directly contributes to solutions for these challenges. It is unclear whether the complexity of this problem is too high or whether the data mining community feels that statistical sampling approaches are sufficient. From our perspective, there is much more research necessary to overcome limitations of existing data mining algorithms that can not handle large scale databases properly. Data reduction is one particular important aspect to help in those situations where we are not able to scale up algorithms.

Below, we outline further research based on these considerations and raise more research challenges for the future.

3.6.1. More Research for Focusing Solutions

The overall work partly presented here also includes initial attempts for analytical studies to find out which instantiation of instance selection is best suited in given contexts as well as more detailed experimental results comparing different approaches in different domains. The analytical studies and experimental results imply first attempts to define heuristics for the selection of the most appropriate method for instance selection.

Initially, we started with theoretical attempts to prove the appropriateness of specific instance selection approaches in concrete data mining contexts. We selected random sampling, classification, and nearest neighbor classifiers as a starting point, and ended up in an average case analysis for the expected classification accuracy given specific sample sizes and nearest neighbor classification in even more constrained situations.

The average case analysis worked out well and comparisons between predicted accuracies and experimental results showed that this analysis is able to correctly foresee true results. Hence, this analysis either helps to predict which accuracy to expect for a given sample size, or to determine the sample size given an expected accuracy.

However, the assumptions on the data mining context which made this analysis feasible are too restricted and not applicable in general. Consequently, we went on the experimental way. We selected several alternative instance selection approaches, more than only a single data mining algorithm, and data sets with varying data characteristics. We run systematic experiments, compared the results using different evaluation strategies, and inferred heuristics for the relation between the characteristics of the focusing context and the appropriateness of specific approaches.

These heuristics, exemplified above, are now able to help in relatively many application scenarios to select the best suited focusing solution given the focusing context. However, these heuristics remain the result of an experimental study, and are consequently by no means theoretically proven.

3.6.2.   Open Issues and Research Challenges

In essence, we claim that the specification of focusing tasks defines the scope of potential research in the area of data reduction. The definition of various evaluation criteria allows systematic comparisons of approaches along different aspects of interest. The unifying framework of existing solutions helps the community to compare their developments in a methodological way and to discuss their strengths and weaknesses. The generic sampling approach outlined opens the perspective for future research to allow for flexible solutions that can be specialized according to specific contexts. Finally, the analytical studies and experimental results provide an initial attempt to help practitioners to find the best suited solution for instance selection.

The area of instance selection (or focusing in general) still needs more effort. There is no consistent and data mining oriented way for systematically evaluating the appropriateness of existing mechanisms or novel solutions. Current methods are still not able to really scale up data mining approaches to huge databases in terms of Giga-, Terra-, and Petabyte. Moreover (and possibly more important), there is also no theoretically proven guidance in selecting the right instance selection approach given a specific context.

Hence, we argue to work on the following issues for future research in the area of instance selection (and focusing in general):

- define standardized evaluation criteria to enable systematic comparison of existing and novel approaches;

- scale up intelligent focusing approaches by combining technologies from machine learning research and the database community;

- develop more intelligent focusing solutions that provide data reduction techniques beyond pure statistical sampling and make use of the specific characteristics of concrete contexts in data mining;

- extend attempts of analytical studies and experimental results to understand the relation between different instance selection techniques and to come up with reliable heuristics and guidelines for the selection of best suited focusing solutions given a specific data mining environment.

The work presented and outlined here is certainly a first step towards these directions but still does not provide an overall answer to all questions.

Notes

1. For simplicity, we do not explicitly compute average values in definition 3 but rather assume that t already covers the result of computations of average values for all outputs in F_{out}.

2. Attribute relevance is either defined by domain experts or computed automatically using existing state of the art approaches (e.g., (Wettschereck et al., 1995)).

References

Aha, D.W., Kibler, D., & Albert, M.K. (1991). Instance-Based Learning Algorithms. *Machine Learning*, 6, p. 37-66.

Barreis, E.R. (1989). *Exemplar-Based Knowledge Acquisition*. Boston, MA: Academic Press.

Cochran, W.G. (1977). *Sampling Techniques*. New York: John Wiley & Sons.

Chapman, P., Clinton, J., Khabaza, T., Reinartz, T., & Wirth, R. (1999). *The CRISP-DM Process Model*. www.crisp-dm.org/pub-paper.pdf.

Datta, P., & Kibler, D. (1995). Learning Prototypical Concept Descriptions. in: Prieditis, A., & Russell, S. (eds.). *Proceedings of the 12th International Conference on Machine Learning*. July, 9-12, Tahoe City, CA. San Mateo, CA: Morgan Kaufmann, pp. 158-166.

Dougherty, J., Kohavi, R., & Sahami, M. (1995). Supervised and Unsupervised Discretization of Continuous Features. in: Prieditis, A., & Russell, S. (eds.). *Proceedings of the 12th International Conference on Machine Learning*. July, 9-12, Tahoe City, CA. Menlo Park, CA: Morgan Kaufmann, pp. 194-202.

Fayyad, U., Piatetsky-Shapiro, G., & Smyth, P. (1996). Knowledge Discovery and Data Mining: Towards a Unifying Framework. in: Simoudis, E., Han, J., & Fayyad, U. (eds.). *Proceedings of the 2nd International Conference on Knowledge Discovery and Data Mining*. August, 2-4, Portland, Oregon. Menlo Park, CA: AAAI Press, pp. 82-88.

Genari, J.H. (1989). *A survey of clustering methods*. Technical Report 89-38, University of California, Irvine, CA.

Hartigan, J.A. (1975). *Clustering Algorithms*. New York, NY: John Wiley & Sons, Inc.

John, G.H., Kohavi, R., & Pfleger, K. (1994). Irrelevant Features and the Subset Selection Problem. in: Cohen, W.W., & Hirsh, H. (eds.). *Proceedings of the 11th International Conference on Machine Learning*. July, 10-13, Rutgers University, New Brunswick, N.J. San Mateo, CA: Morgan Kaufmann, pp. 121-129.

Kohavi, R., Sommerfield, D., & Dougherty, J. (1996). *Data Mining Using MLC++: A Machine Learning Library in C++*. http://robotics.stanford.edu/ ~ronnyk.

Linde, Y., Buzo, A., & Gray, R. (1980). An Algorithm for Vector Quantizer Design. *IEEE Transactions on Communications*, 28, pp. 85-95.

Matheus, C.J., Chan, P.K., & Piatetsky-Shapiro, G. (1993). Systems for Knowledge Discovery in Databases. *IEEE Transactions on Knowledge and Data Engineering*, Vol. 5, No. 6, pp. 903-913.

Quinlan, J.R. (1993). *C4.5: Programs for Machine Learning.* San Mateo, CA: Morgan Kaufmann.

Reinartz, T. (1997). Advanced Leader Sampling for Data Mining Algorithms. in: Kitsos, C.P. (ed.). *Proceedings of the ISI '97 Satellite Conference on Industrial Statistics: Aims and Computational Aspects.* August, 16-17, Athens, Greece. Athens, Greece: University of Economics and Business, Department of Statistics, pp. 137-139.

Reinartz, T. (1998). Similarity-Driven Sampling for Data Mining. in: Zytkow, J.M. & Quafafou, M. (eds.). *Principles of Data Mining and Knowledge Discovery: Second European Symposium, PKDD '98.* September, 23-26, Nantes, France. Heidelberg: Springer, pp. 423-431.

Reinartz, T. (1999). *Focusing Solutions for Data Mining: Analytical Studies and Experimental Results in Real-World Domains,* LNAI 1623, Heidelberg: Springer.

Scheaffer, R.L., Mendenhall, W., & Ott, R.L. (1996). *Elementary Survey Sampling,* 5th Edition. New York, NY: Duxbury Press.

Sen, S., & Knight, L. (1995). A Genetic Prototype Learner. in: Mellish, C.S. (ed.). *Proceedings of the 14th International Joint Conference on Artificial Intelligence.* August, 20-25, Montreal, Quebec, Canada. San Mateo, CA: Morgan Kaufmann, Vol. I, pp. 725-731.

Skalak, D.B. (1993). Using a Genetic Algorithm to Learn Prototypes for Case Retrieval and Classification. in: Leake, D. (ed.). *Proceedings of the AAAI '93 Case-based Reasoning Workshop.* July, 11-15, Washington, DC. Menlo Park, CA: American Association for Artificial Intelligence, Technical Report WS-93-01, pp. 64-69.

Skalak, D.B. (1994). Prototype and Feature Selection by Sampling and Random Mutation Hill Climbing Algorithms. in: Cohen, W.W., & Hirsh, H. (eds.). *Proceedings of the 11th International Conference on Machine Learning.* July, 10-13, Rutgers University, New Brunswick, N.J. San Mateo, CA: Morgan Kaufmann, pp. 293-301.

Smyth, B., & Keane, M.T. (1995). Remembering to Forget. in: Mellish, C.S. (ed.). *Proceedings of the 14th International Joint Conference on Artificial Intelligence.* August, 20-25, Montreal, Quebec, Canada. San Mateo, CA: Morgan Kaufmann, Vol. I, pp. 377-382.

Wettschereck, D., Aha, D., & Mohri, T. (1995). *A Review and Comparative Evaluation of Feature Weighting Methods for Lazy Learning Algorithms.* Technical Report AIC-95-012, Naval Research Laboratory, Navy Center for Applied Research in Artificial Intelligence, Washington, D.C.

Zhang, J. (1992). Selecting Typical Instances in Instance-Based Learning. in: Sleeman, D., & Edwards, P. (eds.). *Proceedings of the 9th International Conference on Machine Learning.* July, 1-3, Aberdeen, Scotland. San Mateo, CA: Morgan Kaufmann, pp. 470-479.

II

INSTANCE SELECTION METHODS

Chapter 4

COMPETENCE GUIDED INSTANCE SELECTION FOR CASE-BASED REASONING

Barry Smyth and Elizabeth McKenna

Department of Computer Science
University College Dublin
Belfield, Dublin 4, Ireland
{Barry.Smyth,Elizabeth.McKennna}@ucd.ie

Abstract   Case-based reasoning (CBR) solves problems by reusing the solutions to similar problems stored as cases in a case-base (Kolodner, 1993); (Leake, 1996); (Smyth and Keane, 1998). For reasons of efficiency it is often desirable to be able to reduce a large case-base to a much smaller edited subset without compromising the competence of the case-base. This is obviously related to instance selection tasks but existing algorithms have been developed mainly for classification problems with discrete solution classes. Many CBR applications address non-classification tasks such as prediction and estimation tasks, planning, or design so called synthesis tasks. In this paper we describe and evaluate a number of strategies, which are unique in that they are guided by an explicit model of competence for CBR systems.

Keywords: Case-based reasoning, instance selection, competence models.

4.1.   INTRODUCTION

Case-based reasoning (CBR) solves problems by reusing the solutions to similar problems stored as cases in a case-base (Kolodner, 1993); (Leake, 1996); (Smyth and Keane, 1998). A general performance goal for any CBR system is the construction and maintenance of a case-base that is optimal with respect to both competence and efficiency (Smyth, 1998); (Smyth and McKenna, 1999a); (McKenna and Smyth, 2000b).

To optimise competence and efficiency of a case-base techniques have been developed to reduce a large case-base to a much smaller subset without loss of competence.

In this paper we describe techniques for identifying what we call the *footprint* of a case-base, that is, a minimal set of cases which is representative of the entire case-base. Our footprinting strategies are novel in that they are guided by an explicit model of case competence thereby guaranteeing the selection of only those cases that actively contribute to the competence of a case-base.

The task of computing the case-base footprint is related to the instance selection problem from machine learning research. However, existing instance selection algorithms have been developed mainly for classification problems with discrete solution classes. Many CBR applications address non-classification tasks such as prediction and estimation tasks, planning, or design, so called synthesis tasks.

The remainder of this paper is organised as follows. The next section looks at related work on instance selection for classification problems. Section 4.3 describes an innovative model of competence for case-based reasoning systems and Section 4.4 explains how this model informs a variety of strategies for ordering and selecting cases according to their competence contributions. The results of an empirical study to evaluate the competence and efficiency characteristics of the case-bases produced by these competence-guided footprinting strategies are described in Section 4.5. Finally, Section 4.6 briefly describes our more recent research.

4.2.   RELATED WORK

The task of constructing a case-base footprint is closely related to the instance selection task which has been studied by the pattern recognition and machine learning community since the 1960's. Related research has mainly focussed on instance selection techniques for nearest-neighbour (NN) classification algorithms. NN classification algorithms assign a class label to a target instance based on the class labels assigned to the k closest training instances (this is the k-NN approach) (Aha et al., 1991); (Cover and Hart, 1967). While standard nearest-neighbour approaches have proved to be very effective classifiers they do suffer from significant storage and computational costs. The most common approach for reducing these costs is to use *instance selection* or *editing techniques* to reduce the training set to a much smaller edited set by selecting key instances (Hart, 1967); (Gates, 1972); (Tomek, 1976); (Aha et al., 1991), (Dasarathy, 1991); (Wilson, 1997); (Wilson and Martinez, 1997); (Wilson and Martinez, 1998); (Wilson, 1972). The resulting edited set should have the following three important properties: 1) **Size:** The set should contain as few instances as possible; 2) **Consistency:** The edited set should be capable of correctly classifying all of the instances in the original training set; 3) **Competency:** The edited set should be capable of correctly classifying unseen instances. In this section we will survey some of the more common and effective approaches for instance selection.

4.2.1. The CNN Family

The *condensed nearest-neighbour rule* (CNN) (Hart, 1967); (Dasarathy, 1991) is probably the simplest instance selection strategy. The basic approach is to build up a so-called *edited set* from scratch by adding instances that cannot be successfully solved by the edited set built so far; the algorithm makes multiple passes over the *training data* until no more additions can be made.

CNN tends to select training instances that fall near to class boundaries and it is guaranteed to produce an edited set that is consistent with the original training instances. However, it does not produce a minimal edited set because it tends to select redundant instances. The basic problem is that CNN is order-dependent and different initial orderings of instances lead to different edited sets. Specifically, redundant instances, which are examined early on in the editing process, tend to be added to the edited set.

To address this issue Gates described the *reduced nearest-neighbour* (RNN) method as an adaptation of CNN (Gates, 1972). RNN adds a post-processing step to CNN, which attempts to contract the final edited set by identifying and deleting redundant instances that were added too early on - cases that can be removed from the edited set without reducing its classification accuracy over the original training set.

Tomek (Tomek, 1976) describes another improvement to CNN by ordering instances prior to editing so that redundant instances tend to be examined late in the editing process when they are unlikely to be mistaken as useful. Tomek suggests the use of the distance to an instance's nearest neighbour in an opposing class (*nearest unlike neighbour*, or NUN, distance) as an ordering function. Instances with small NUN distances are examined first as they are likely to lie on class boundaries and therefore have a significant classification competence (Dasarathy, 1991). The CNN-NUN variation works well in that it significantly reduces edited case-base size, but still suffers from noise problems.

The edited sets produced by CNN and its variants do not always generalise well to unseen target instances. Specifically, these procedures are inherently sensitive to noisy data and will actively preserve noisy instances in the edited sets. This can have a significant impact on the classification accuracy (competence) of the edited sets on unseen data. For this reason, these procedures are not appropriate in noisy domains.

4.2.2.   Edited Nearest Neighbour

The *Edited Nearest Neighbour* (ENN) is a perfect counterpoint to CNN (Wilson, 1972). Its strategy is to filter out incorrectly classified instances in order to remove boundary instances (and noise) and preserve interior instances that are representative of the class being considered.

Unlike CNN the ENN approach begins with an edited set which contains all original training instances. Each instance is examined in turn and removed if its classification is not the same as the majority classification of its k nearest neighbours. This edits out noisy instances as well as boundary instances. However, ENN leaves redundancy in the edited set since all internal instances are retained. In theory, local groups of nearby instances with the same classifications could be reduced to a single representative instance. The repeated ENN (RENN) addresses this redundancy issue by repeatedly applying ENN until all instances have the majority classification of their neighbours. This has the effect of widening the gap between classes and smoothing the decision boundaries.

All-kNN is another extension of the basic ENN procedure (Tomek, 1976). All-kNN increases the value of k for each iteration of RENN so that more subtle examples of noise can be detected and removed. This again has the effect of removing boundary instances and preserving interior, representative instances for each class cluster.

ENN, RENN and All-kNN represent an advance on CNN type techniques as they produce edited sets with higher classification when noise is included in the training set.

4.2.3.   The IBL Family

The previous instance selection approaches are based on early research from the pattern classification community. More recently machine learning researchers have looked at the problem of instance selection with renewed interest because the development of recent inductive learning approaches such as instance-based learning. For example, Aha et al.(1991) describe a set of instance selection algorithms (IB1, IB2, and IB3) designed to significantly reduce the number of training instances without compromising classification accuracy. IB1 is the simplest algorithm and is almost identical to the nearest neighbour algorithm. IB2 takes its lead from CNN, selecting only those instances that could not otherwise be correctly classified. IB2 only makes one pass through the training instances and as such does not guarantee the construction of an edited set that is consistent with the original training instances. In general, IB2 suffers from the same problems as CNN - it preserves redundant instances and is sensitive to noisy data.

IB3 reduces the noise sensitivity of IB2 by only retaining *acceptable* misclassified instances (Aha et al., 1991). IB3 maintains a classification record for each selected instance which keeps track of the number of correct and incorrect classifications the instance makes with respect to subsequently presented training instances and suggests how this instance will perform in the future. IB3 employs a significance test to identify those instances that are likely to be good classifiers and those that are likely to be noisy (i.e. bad classifiers). The good classifiers are preserved in the edited set. In general, IB3 achieves greater storage reductions and higher classification accuracy than IB2 on unseen instances.

4.2.4. The DROP Family

Wilson and Martinez (1997a, 1997b, 1998) present a family of editing algorithms that are guided by two sets for each instances: the *k nearest neighbours* and the *associates* of the instance. The associates of an instance i are those cases which have i as one of their nearest neighbours. The algorithms begin with the entire training set and deletes instances to produce an edited set. An instance i is removed if at least as many of its associates can be correctly classified without i.

DROP1 is the simplest member of the family and applies the above deletion criterion to each case in turn, updating the nearest neighbour and associate sets for the remaining cases after each deletion. DROP1 tends to remove noise from the original case-base, as deleting a noisy case will usually result in an increase in the classification of its neighbours.

DROP2 is identical to the DROP1 algorithm except that cases are sorted in descending order of NUN distance (cf. CNN-NUN previously) prior to deletion. Therefore, cases in the interior of class regions are deleted before cases at the boundaries. DROP2 also differs from DROP1 in that there is no updating of the associates sets after deletion. Thus, a case can have associates that have already been deleted, but which can help to guide further deletions.

DROP3 procedure is a hybrid editing technique in that it combines an ENN pre-processing stage with DROP2 to remove noisy cases prior to DROP2 editing. DROP3 performs extremely well in terms of final edited case-base size and classification accuracy and Wilson and Martinez (1997) report that DROP3 is one of the best available instance-based classifiers over a wide variety of data sets and classification tasks.

4.2.5.   Instance Selection and CBR

In this paper we are concerned with the instance selection problem as it applies to case-based reasoning systems in general rather than nearest-neighbour classifiers in particular. CBR systems have been used to solve a wide range of problem types, not just classification, but also prediction, estimation, planning, and design. There are a number of implications for our study. Firstly, the requirement for an instance solution to be a simple atomic value (the class) is no longer valid. Case solutions can be continuous values, structured plans, or complex designs. Secondly, the simple classification-correctness criterion (simple class matching) is not appropriate for many case-based reasoning tasks especially when solutions are composite entities or where adaptation may be necessary after retrieval. Thirdly, the instance selection heuristics (nearest unlike neighbours and enemies, associates etc.) used above are tailored for classification problems and may not be appropriate in a more general CBR setting. Therefore, we argue the need for new instance selection strategies that are appropriate for CBR systems, and in the remainder of this paper we introduce and evaluate a number of such strategies based on an explicit model of competence for case-based reasoning.

4.3.   A COMPETENCE MODEL FOR CBR

The critical performance requirement for a selection algorithm is its ability to actively preserve competence in the face of significant data reduction. We argue that this requirement suggests the use of an explicit model of competence in order to guide the selection process. We have developed such a competence model (McKenna and Smyth, 2000b) and have previously shown the effectiveness of this model as an evaluation tool for CBR systems. In this paper we demonstrate how our competence model can be used to inform the case selection process. This section outlines the basic components of this model and the interested reader is referred to Smyth and Keane (1995); Smyth and McKenna (2000) .

4.3.1.   The Foundations of Competence

The local competence contributions of individual cases are charac-terised by two sets. The *coverage set* of a *case* is the set of all *target problems* that this case can be used to solve. The *reachability set* of a *target problem* is the set of all *cases* that can be used to solve it. It is not possible to enumerate all possible future target problems (T), but by using the case-base (C) itself as a representative of the target problem space we can efficiently estimate these sets as shown in definitions 4.1 and 4.2.

$$CoverageSet(c \in C) = \{c' \in C : Solves(c, c')\} \qquad (4.1)$$

$$ReachabilitySet(c \in C) = \{c' \in C : Solves(c', c)\} \qquad (4.2)$$

The assumption that the case-base is a representative sample of the target problem space (i.e. the *Representativeness Assumption*) may seem like a large step. However, we would argue that the representativeness assumption is one currently made, albeit implicitly, by CBR researchers; for if a case-base were not representative of the target problems to be solved then the system could not be forwarded as a valid solution to the task requirements. In short, if CBR system builders are not making these assumptions then they are constructing case-bases designed *not* to solve problems in the task domain.

4.3.2.   Competence Groups

Coverage and reachability sets provide a measure of local competence only. In order to estimate the true competence contributions of cases it is necessary to model the interactions between related cases, specifically in terms of how their coverage and reachability sets overlap.

$$RelatedSet(c) = CoverageSet(c) \cup ReachabilitySet(c) \qquad (4.3)$$

$$\begin{aligned} For \ c1, \ c2 &\in C, SharedCoverage(c1, c2) \ iff \\ & [RelatedSet(c1) \cap RelatedSet(c2)] \neq \{\} \end{aligned} \qquad (4.4)$$

$$\begin{aligned} For \ G = \{c1, .cn\} &\subseteq C, CompetenceGroup(G) \ iff \\ \forall ci \in G, \exists cj &\in G - \{ci\} : SharedCoverage(ci, cj) \\ \land \ \forall ck \in C - G, \neg \exists cl &\in G : SharedCoverage(ck, cl) \end{aligned} \qquad (4.5)$$

First we define the related set of a case to be the union of its coverage and reachability sets (4.3). When the related sets of two cases overlap we say that they exhibit shared coverage (4.4) and cases can be grouped together into so-called competence groups which are maximal sets of cases exhibiting shared coverage (4.5). In fact, every case-base can be

```
G, Competence Group
CNN-FP (G)
  G-Set ← set of cases in G
  FP ← {}
  For each c ∈ G-Set
    Add c to FP if c cannot be solved by a case in FP
  EndFor
Return FP
```

Figure 4.1 The CNN-FP Footprint Creation Algorithm

organised into a unique set of competence groups which, by definition, do not interact from a competence viewpoint - that is, while each case within a given competence group must share coverage with at least one other case in that group, no case from one group can share coverage with any case from another group.

The importance of the competence group concept is that each group makes a unique contribution to competence and, as a result, can be treated independently of all other groups in the case-base. This means that the competence of a case-base as a whole can be computed as the sum of the competence contributions of each competence group.

4.4.   COMPETENCE FOOTPRINTING

In this paper, we are interested in identifying algorithms for reducing a set of cases to a small, consistent and highly competent subset, which we call the competence footprint of a case-base. Since we now have a means of identifying groups of cases that make a unique and independent contribution to overall competence, the competence footprint of a case-base is simply the union of the competence footprints over all competence groups. The sections that follow describe a number of techniques for computing the competence footprint of a case-base by computing the competence footprint of each individual group.

4.4.1.   CNN Footprinting

In Section 4.2 we described the CNN rule whereby an instance is added to the edited set if it can not be solved by any other instance already present in this set (Hart, 1967); (Dasarathy, 1991). We have adapted this strategy as our first footprinting technique. In brief, we apply the basic CNN algorithm to each competence group in turn. The footprint set of cases is produced by combining the cases selected from

```
G, Competence Group
RC-FP (G)
  G-Set ← cases in G sorted in descending order of RC
  FP ← {}
  For each c ∈ G-Set
    Add c to FP if c cannot be solved by a case in FP
  EndFor
  Return FP
```

Figure 4.2 The RC-FP Footprint Creation Algorithm

each competence group. We call this the CNN footprint (CNN-FP) strategy and the algorithm is presented in figure 4.1.

As with the CNN strategy (Section 4.2.1) the CNN footprint is dependent on the presentation order of the competence group cases and it tends to preserve some redundant cases. These problems are addressed by subsequent footprinting techniques.

4.4.2.   RC Footprinting

In Section 4.2 we described a strategy that involved ordering instances (by their NUN distance) prior to the CNN selection process in order to reduce the redundancy of the final edited set. A similar approach can be used to improve the performance of CNN in a CBR setting. The ordering policy is based on the competence contribution of cases, thus allowing competence-rich cases to be examined (and selected) before competence-poor cases (which are likely to be redundant).

Coverage and reachability sets provide only local measures of case competence (Section 4.3). For a true picture of competence, a measure of the coverage of a case, relative to other nearby cases, is needed. We define a measure called *relative coverage* (RC), which estimates the unique competence contribution of an individual case, c, as a function of the size of the case's coverage set (see Def. 4.6).

$$RelativeCoverage(c) = \sum_{c' \in CoverageSet(c)} \frac{1}{|ReachabilitySet(c')|} \quad (4.6)$$

RC weights the contribution of each covered case by the degree to which these cases are themselves covered. It is based on the idea that if a case c' is covered by n other cases then each of the n cases will receive a contribution of 1/n from c' to their relative coverage measures.

The RC measure forms the basis for our ordering strategy - cases are arranged in descending order of their relative coverage contribu-

```
G, Competence Group
RFC-FP(G)
  G-Set ← G's cases in ascending order of reachability set size
  FP ← {}
  While G-Set is not empty
    ReachSet ← all c ∈ G-Set with minimum reachability set size
    c ← case in ReachSet with the largest coverage set
    FP ← FP ∪ c
    G-Set ← G-Set - CoverageSet(c)
    Update coverage and reachability sets of G-Set
  End While
Return FP
```

Figure 4.3   Reach for Cover Footprint Creation Algorithm

tions prior to the CNN-FP footprinting procedure (Smyth and McKenna, 1999a). This presents competent rich cases before less competent cases to maximise the rate at which competence increases during the case-base construction process. This modified footprint procedure is called RC-FP and it is outlined in figure 4.2.

4.4.3.   RFC Footprinting

While the reachability set size of a case does not provide an accurate measure of the competence contribution of this case in general, it can sometimes provide important competence information. Specifically, some cases will have reachability sets that only contain themselves. By definition these cases cannot be solved by any other case in the case-base and therefore they make a positive contribution to competence - these cases correspond to the pivotal cases described in (Smyth and Keane, 1995). Such cases must be present in a case-base footprint in order to preserve consistency.

The general idea behind the reach-for-cover (RFC-FP) algorithm is that cases with small reachability sets are interesting because they represent problems that are difficult to solve, and these cases should be considered for selection before cases that represent problems that are easy to solve (see figure 4.3). Thus, the RFC-FP method considers the remaining cases in a group in ascending order of their reachability set size. During each step the remaining case with the smallest reachability set and largest coverage set is added to the footprint. In addition, once a case is added to the footprint any cases that it covers are removed from the set of remaining group cases. Finally, the coverage and reachability sets of the remaining group cases are updated by removing all

```
G, Competence Group
COV-FP(G)
  G-Set ← cases in G in descending order of coverage set size
  FP ← {}
  While G-Set is not empty
    c ← case in G-Set with the largest coverage set size
    FP ← FP ∪ c
    G-Set ← G-Set - CoverageSet(c)
    Update coverage sets of cases in G-Set
  End While
Return FP
```

Figure 4.4   Coverage Footprint Creation Algorithm

references to elements of the coverage set of the newly added case. The RFC-FP technique tends to select those cases that lie on the boundary of the competence group and then work inward until the footprint set can solve all cases in the group.

4.4.4.   Coverage Footprinting

The COV-FP footprinting technique is related to the RFC-FP method except that instead of biasing cases by their reachability set size, it biases cases by their coverage set size. The cases in each competence group are examined in decreasing order of their coverage set size. Hence, cases with large coverage sets (that is, large local competence contributions) are added to the footprint and the cases that these cases cover are then removed from the original training set. The addition of cases continues in this way until the original set of group cases is empty (Figure 4.4).

The COV-FP strategy differs from the RFC-FP strategy in that it begins by adding internal cases from the competence groups and then working out to the boundary of the group until all cases in the group can be solved using the footprint.

4.5.   EXPERIMENTAL ANALYSIS

In this section we empirically test each of our footprinting techniques by analysing their ability to produce compact, consistent, and competent footprint sets. We use two publicly available case-bases. Firstly a 1400 case case-base from the Travel domain with each case describing a holiday using features such as destination, accommodation, price etc. (Lenz et al., 1996). Secondly a 500 case case-base from the Residential Property domain with each case describing a property using features such as location, number of room, price etc. (Blake et al., 1998). To facilitate experimentation with different levels of case redundancy we produced extra cases for each domain. For the Travel domain 400 duplicate cases and 400 near-miss (redundant) cases were added to generate a total of 2200 cases. For the Property domain 200 duplicates and 200 near-misses were added to give a total of 900 cases. Extra cases were produced to facilitate experimentation with different levels of case redundancy.

In the Travel domain we produced 9 case-base sizes ranging from 200 to 1800 cases with accompanying unseen target problem sets of 400 cases. For the Property domain we produced 7 case-base sizes ranging from 100 to 700 cases, and sets of 200 target problems. In each domain, for each case-base size we produced 30 different random case-bases and target problem sets to give 270 test case-bases for the Travel domain and 210 test case-bases from the Property domain. There was never any direct overlap between a case-base and its associated target problems.

4.5.1.   Footprint Correlation

Before directly investigating the size, consistency, and competency characteristics of our footprint techniques we begin by looking at the correlation between the sizes of the footprint sets created using the various footprinting strategies and true competence of the original training set on unseen target problems. True competence is defined to be the percentage of target problems that can be successfully solved by the training set where the solvability criterion is based on a similarity threshold. We argue that a close correlation between footprint set size and true competence is a necessary (but not sufficient) condition of an effective representative sample of cases. For example, a naive footprinting algorithm simply adds cases at random to the footprint set until consistency is achieved and is destined to produce footprints with varying degrees of redundancy. The sizes of these sets will not correlate well with their true competence as many of their selected cases are not representative.

Method: For each case-base of size n, four footprint sets were created using our four footprinting strategies. The correlation value for footprint size with competence was noted and the results were average over 30 runs for different initial case-bases.

Results: Table 1 reports the mean and standard deviations of these correlation values. There is clearly a very close relationship between footprint size and true competence for each of the footprinting techniques with correlations values of at least 0.913% in the Travel domain and at least 0.968% in the Property domain. For comparison we also show the correlation between actually training set size and true competence.

Table 4.1 Mean and standard deviation correlation values for Travel and Property domains

Correlation	Travel Domain		Property Domain	
	Mean	StdDev	Mean	StdDev
Training-Set	0.767	0.036	0.894	0.026
RC-FP	0.925	0.021	0.974	0.013
CNN-FP	0.927	0.026	0.973	0.015
RFC-FP	0.913	0.021	0.968	0.015
COV-FP	0.931	0.019	0.973	0.012

Discussion: We believe that the statistical correlation between the number of incidences in a footprint set and the true competence of the case-base is a measure of the representativeness of these incidences. If there is little or no redundancy in a set of cases then the set size should correlate well with competence. As each of the footprints produced using the various editing techniques have a high correlation value it would suggest that these subsets contain the cases which contribute most to competence. Conversely, the correlation value between training set size and competence is significantly smaller in both domains which suggests that there is some redundancy in the training set.

4.5.2. Footprint Size

One of the key objectives of an instance selection algorithm is the construction of a minimal subset of instances without compromising competence. In general, this is an intractable task and traditional instance selection techniques have employed heuristics and greedy search procedures to identify near-minimal subsets. Previously we confirmed a high correlation between footprint set size and true competence for each of our footprinting strategies. This implies that our footprint sets contain the right sort of cases. However, a high correlation on its own is not sufficient since a footprint may contain many redundant cases or may be missing important cases. In this experiment we address the first of these issues investigating the size of the footprint sets produced by our footprinting algorithms.

Method: As with the previous experiment for each case-base of size n, four footprint sets were created using our footprinting techniques.

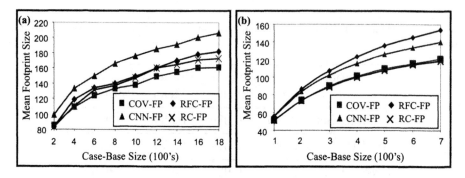

Figure 4.5 Mean Footprint Size vs. Case-Base Size for (a) the Travel, and (b) the Property domain.

The size of the footprint sets was noted for each case-base size and the results were averaged over the 30 runs.

Results: The results are displayed in Figure 4.5 (a) and (b) as graphs of mean footprint size versus case-base size for the Travel and Property domain respectively. Each graph shows the results for the four different footprint creation strategies. In both domains the RC-FP technique and the COV-FP technique perform well, consistently producing footprint sets with fewer cases than those produced by the other two techniques. For example, in the Travel domain the COV-FP technique produces footprint sets with a mean size of 165 cases from an original set of 1800 cases. In contrast, the CNN-FP technique produces footprint sets with a mean size of just over 202 cases, that is, 22.4% more cases than the COV-FP footprint.

Discussion: Interestingly, there is a switch in the ranking of the techniques between the Travel domain and the Property domain. For example, in Travel, RFC-FP out-performs CNN-FP and COV-FP out-performs RC-FP. However, in the Property domain CNN-FP out per-forms RFC-FP while RC-FP out-performs COV-FP (albeit marginally). The reason for this is that in the Travel domain there is a high degree of asymmetry between the coverage and reachability sets of cases, whereas in the Property domain there is a much lower degree of asymmetry. Thus, in the Property domain, if case c1 covers case c2 then there is a high probability that case c2 will cover c1. This is not true in the Travel domain. The effect of this decreased asymmetry in the Property domain is that cases with small reachability sets will also have small coverage sets and hence with the RFC-FP technique coverage will grow slowly and hence consistency will only be achieved with a large footprint set. A similar line of reasoning also explains the less pronounced RC-FP and COV-FP switch, since these algorithms are also sensitive to asymmetry between the reachability and coverage set sizes. Of course, any asymmetry between coverage and reachability sets is a direct affect of the asymmetry in the similarity assessment procedure.

4.5.3. Footprint Competence

In the last experiment we looked at the size of the footprint sets produced using the various footprinting strategies. However, when evaluating the quality of any editing or footprinting technique it is important to test the competence of the footprint with respect to unseen targets (that is, test the generalisation capability of the footprint set). In the final analysis a footprint set will only be useful as long as its competence over unseen problems remains at an acceptable level. Therefore, in this experiment we investigate the competence of the footprint sets produce by our footprinting techniques.

Method: For both domains, four footprint sets were created for each case-base of size n. Once again the size of the footprint sets created was noted. The competence of each footprint set was computed with respect to unseen target problems and the results were averaged over 30 runs.

Results: Figure 4.6 (a and b) show the results as mean competence versus mean footprint size for the Travel domain and Property domain, respectively. The important thing to notice is the rate at which competence accumulates during the footprint construction process. In particular, for both domains the RC-FP and COV-FP techniques accumulate competence more rapidly than the CNN-FP technique and, for the Property domain, more quickly than the RFC-FP method also. For example, in the Travel domain, for a footprint set size of 165 cases (the consistency limit reached by the COV-FP technique on 1800 cases) the CNN-FP, RC-FP, RFC-FP, and COV-FP methods reach 94%, 96%, 96.2%, and 96.73% competence, respectively.

Discussion: Again, as in the previous footprint size experiment, there is a similar switch in the ranking of the techniques between the Travel domain and the Property domain. Specifically, in Travel, RFC-FP out performs CNN-FP and COV-FP out performs RC-FP. However, in the Property domain CNN-FP out performs RFC-FP while RC-FP and COV-FP have similar competence characteristics. Again this can be attributed to an increased level of asymmetry between the coverage and reachability sets of Travel domain cases when compared to property domain cases. Reduced asymmetry in the Property domain degrades the competence building characteristics of the RFC-FP technique. The essential point in the experiment is not so much that any particular footprinting technique significantly outperforms another in terms of compe-

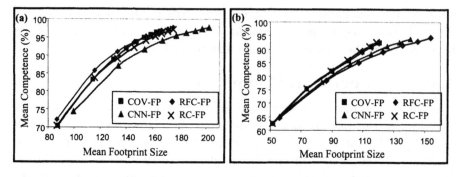

Figure 4.6 Mean Footprint Competence vs. Mean Footprint Size for (a) the Travel, and (b) the Property domains.

tence but rather that a target level of competence is achieved by some footprinting techniques with significantly fewer cases.

4.6.   CURRENT STATUS

In this paper we have looked at instance selection for synthesis tasks. However, to date most instance selection methods are primarily concerned with classification tasks. The question of whether our methods work well in classification tasks still remains. In our more recent research we have investigated this issue and the interested reader is referred to McKenna and Smyth 2000a for further details. Very briefly we have developed a family of hybrid instance selection techniques based on our competence model and we have shown that these techniques can out perform traditional state of the art methods under a range of experimental conditions and across many classification domains.

The instance selection techniques described in this paper rely on the availability of a comprehensive model of case competence and, of course, there is a cost associated with the construction and update of this model. While it is true that updating the competence model can involve a worst-case cost (of $O(n^2)$ in the size of the case-base), in recent work we have developed a new model building and update procedure that can significantly reduce this cost (see Smyth and McKenna (2000b)).

4.7.   CONCLUSIONS

The ability to edit training data prior to learning has been an important research goal for the machine learning community in order to improve the storage and run-time costs of nearest-neighbour algorithms. A broad range of effective techniques have been developed for classification tasks. Similar issues are now relevant to the CBR community as CBR systems are scaled-up to meet the requirements of real-world problems.

In this paper we have described a number of editing strategies designed specifically for case-based reasoning systems and synthesis tasks. These so-called footprinting strategies are innovative in their use of an explicit model of case competence, which provides valuable guidance for the footprinting algorithms. We have presented an empirical evaluation of these techniques to demonstrate their relative effectiveness in producing compact, competent footprint sets.

Current and future work involves investigating the role of competence models in instance selection for classification tasks and in ways of improving the efficiency of these models (McKenna and Smyth, 2000a); (Smyth and McKenna, 2000). In addition, we will also be looking at instance selection techniques for CBR systems which use multiple cases to solve a target problem.

References

Aha, D., Kibler, D., and Albert, M. (1991). Instance-Based Learning Algorithms. *Machine Learning*, 6:37–66.

Blake, C., Keogh, E., and Merz, C. (1998). *UCI Repository of Machine Learning Algorithms Databases*. Irvine, CA: University of California. Department of Information and Computer Science.

Cover, T. and Hart, P. (1967). Nearest Neighbor Pattern Classification. *IEEE Transactions on Information Theory*, IT-13:21–27.

Dasarathy, D., editor (1991). *Nearest Neighbor Norms: NN Pattern Classification Techniques*. IEEE Press.

Gates, G. (1972). The Reduced Nearest Neighbor Rule. *IEEE Transactions on Information Theory*, IT-18(3):431–433.

Hart, P. (1967). The Condensed Nearest Neighbor Rule. *IEEE Transactions on Information Theory*, IT-14:515–516.

Kolodner, J., editor (1993). *Case-Based Reasoning*. Morgan Kaufmann.

Leake, D., editor (1996). *Case-Based Reasoning: Experiences, Lessons, and Future Directions*. MIT Press.

Lenz, M., Burkhard, H., and Bruckner, S. (1996). Applying Case Retrieval Nets to Diagnostic Tasks in Technical Domains. In Smith, I. and Falthings, B., editors, *Advances in Case-Based Reasoning. Lecture Notes in Artificial Intelligence*, pages 219–233. Springer Verlag.

McKenna, E. and Smyth, B. (2000a). Competence-guided Editing Methods for Lazy Learning. In *Proceedings of the 14th European Conference on Artificial Intelligence*.

McKenna, E. and Smyth, B. (2000b). Visualising the Competence of Case-Based Reasoners. *Applied Intelligence: Special Issue on Interactive Case-Based Reasoning.*

Smyth, B. (1998). Case-Based Maintenance. In del Pobil, A. P., Mira, J., and Ali, M., editors, *Tasks and Methods in Applied Artificial Intelligence. Lecture Notes in Artificial Intelligence*, pages 507–516. Springer Verlag.

Smyth, B. and Keane, M. (1995). Remembering to Forget: A Competence Preserving Case Deletion Policy for CBR Systems. In Mellish, C., editor, *Proceedings of the 14th International Joint Conference on Artificial Intelligence*, pages 377–382. Morgan Kaufmann.

Smyth, B. and Keane, M. (1998). Adaptation-Guided Retrieval: Questioning the Similarity Assumption in Reasoning. *Artificial Intelligence*, 102:249–293.

Smyth, B. and McKenna, E. (1999a). Building Compact Competent Case-Bases. In Althoff, K. D., Bergmann, R., and Branting, L., editors, *Case-Based Reasoning Research and Development. Lecture Notes in Artificial Intelligence*, pages 329–342. Springer Verlag.

Smyth, B. and McKenna, E. (1999b). Footprint-Based Retrieval. In Althoff, K. D., Bergmann, R., and Branting, L., editors, *Case-Based Reasoning Research and Development. Lecture Notes in Artificial Intelligence*, pages 343–357. Springer Verlag.

Smyth, B. and McKenna, E. (2000). An Efficient and Effective Procedure for Updating a Competence Model for Case-Based Reasoners. In de Mantaras, R. L. and Plaza, E., editors, *Proceedings of the 11th European Conference on Machine Learning.* Springer Verlag.

Tomek, I. (1976). Two Modifications of CNN. *IEEE Transactions on Systems, Man, and Cybernetics*, 7(2):679–772.

Wilson, D. (1972). Asymptotic Properties of Nearest Neighbor Rules Using Edited Data. *IEEE Transactions on Systems, Man, and Cybernetics*, 2-3:408–421.

Wilson, D. (1997). *Advances in Instance-Based Learning Algorithms.* Phd thesis, Computer Science Department, Brigham Young University.

Wilson, D. and Martinez, T. (1997). Instance Pruning Techniques. In Fisher, D., editor, *Proceedings of the 14th International Conference on Machine Learning*, pages 403–411. Morgan Kaufmann.

Wilson, D. and Martinez, T. (1998). Reduction Techniques for Exemplar-Based Learning Algorithms. *Machine Learning*, 38(3):257–286.

Chapter 5

IDENTIFYING COMPETENCE-CRITICAL INSTANCES FOR INSTANCE-BASED LEARNERS

Henry Brighton and Chris Mellish
Department of Artificial Intelligence
The University of Edinburgh
Edinburgh, EH1 2QL UK.
{henryb,chrism}@ling.ed.ac.uk

Abstract   The basic nearest neighbour classifier suffers from the indiscriminate storage of all presented training instances. With a large database of instances classification response time can be slow. When noisy instances are present classification accuracy can suffer. Drawing on the large body of relevant work carried out in the past 30 years, we review the principle approaches to solving these problems. By deleting instances, both problems can be alleviated, but the criterion used is typically assumed to be all encompassing and effective over many domains. We argue against this position and introduce an algorithm that rivals the most successful existing algorithm. When evaluated on 30 different problems, neither algorithm consistently outperforms the other: consistency is very hard. To achieve the best results, we need to develop mechanisms that provide insights into the structure of class definitions. We discuss the possibility of these mechanisms and propose some initial measures that could be useful for the data miner.

Keywords: Instance-based learning, instance selection, forgetting, pruning, filtering, consistency.

5.1.   INTRODUCTION

The Nearest Neighbour Classifier is a simple supervised concept learning scheme which classifies unseen (i.e., unclassified) instances by finding the closest previously observed instance, taking note of its class, and predicting this class for the unseen instance (Cover and Hart, 1967).

Learners that employ this classification scheme are also termed Instance-Based Learners, Lazy Learners, Memory-Based Learners, and Case-Based Learners. They all suffer from the same problem: the instances used to train the classifier are stored indiscriminately. No process of selection is performed, and as result, harmful and superfluous instances are stored needlessly.

In this article we survey the chief efforts to alleviate this problem and review the criteria used to selectively store instances of the classification problem. We review this work using insights about the structure of classification problems in general. By viewing instances as feature vectors we can imagine an instance space where each instance is a point. We argue that the structure of the classes formed by the instances can be very different from problem to problem, which results in inconsistency when we apply one instance selection scheme over many problems. The thrust of this article is that the data miner needs to gain an insight into the structure of the classes within the instance space to effectively deploy an instance selection scheme. We shed light on possible class structures, and how they can be grouped. We aim to show that a knowledge of the class structures is an intrinsic part of designing and deploying instance selection algorithms.

The structure of this article is as follows. In Section 5.2 we characterise the problem by discussing exactly what an instance selection algorithm should achieve, and in what circumstances this is possible. Using these insights we review previous work in Section 5.3 We argue that three eras have occurred in the development of instance selection algorithms, with the most recent approaches being superior. Our contribution to the evaluation is a comparison of the ICF algorithm (Brighton and Mellish, 1999) with RT3 (Wilson and Martinez, 1997) over 30 domains. We argue that neither of these two algorithms is superior: both record the highest accuracy and space reduction on certain problems. Finally we discuss how our ICF algorithm offers insights into the structure of the instance space, and we discuss some future research directions.

5.2.   DEFINING THE PROBLEM

We want to isolate the smallest set of instances which enable us to predict the class of a query instance with the same (or higher) accuracy than the original set. Before reviewing the many methods one can employ to tackle this problem, we present two practical issues which are often neglected. First, we point out that instance selection is practically realised by instance removal in the context of nearest neighbour classification: we aim to retain only the critical instances. We argue that any scheme should aim to achieve what we term *unintrusive storage reduction*, which defines the position we should aim for in the trade-off between storage reduction and classification accuracy. Secondly, we argue that in the context of instance selection, we need to differentiate between certain types of classification problem: domains with *homogeneous* class definitions and those without. We argue that different removal criteria are required for the two opposing class structures. The second point reinforces the thrust of this article: consistency over many problems is hard when designing an instance filter. Instead of placing the whole solution on the algorithm, we argue it is largely placed on the data miner, as a knowledge of problem structure is required to select the best tool. In Section 5.5 we propose some measures to aid this process.

In general we define the problem of instance selection as the need to extract the most useful set of instances from a database which we know (or suspect) contains instances which are superfluous or harmful. In the context of instance-based learning, we seek to discard the cases which are superfluous or harmful to the classification process. Some instances of a class are just not telling us much, the job they do in informing classification decisions is done far better by other cases: they are superfluous. Similarly, some instances of a class might lead us to make false classification predictions if we rely on them: they are harmful.

In the context of instance-based learning, the problem of instance selection should be viewed more in terms of instance deletion as we remove superfluous and harmful instances and retain only the critical instances. By removing a set of instances from a database the response time for classification decisions will decrease, as fewer instances are examined when a query instance is presented. This objective is primary when we are working with large databases and have limited storage. The removal of instances can also lead to either an increase or decrease in classification competence. Therefore, when applying an instance selection scheme to a database of instances we must be clear about the degree to which we are willing to let the original classification accuracy depreciate. For example, if we have a fixed storage limit then the number of cases we are forced to remove might be too large, and unavoidably result in a degradation of classification accuracy (Markovitch and Scott, 1993; Smyth and Keane, 1995). Usually, the principle objective of an instance selection scheme is unintrusive storage reduction. Here, classification accuracy is primary: we desire the same (or higher) classification accuracy but we require it faster and taking up less space. Ideally, accuracy should not suffer at the expense of improved performance. Now, if our deletion decisions are not to harm the classification accuracy of the learner, we must be clear about the kind of deletion decisions that *introduce* erroneous classification decisions. Consider the following reasons why a k-nearest neighbour classifier might incorrectly classify a query instance:

1 When noise is present in locality of the query instance. The noisy instance(s) win the majority vote, resulting in the incorrect class being predicted.

2 When the query instance occupies a position close to an inter-class border where discrimination is harder due to the presence of multiple classes.

3 When the region defining the class, or fragment of the class, is so small that instances belonging to the class that surrounds the fragment win the majority vote. This situation depends on the value of k being large.

4 When the problem is unsolvable by an instance-based learner. This will be due to the nature of the underlying function, or due to the sparse data problem.

In the context of instance selection, we can address point (1) and try and improve classification accuracy by removing noise. We can do nothing about (4) as this situation is a given and defines the intrinsic difficulty of the problem. However, issues (2) and (3) should guide our removal decisions. Removing instances that are close to borders is not recommended as these instances are relevant to discrimination between classes. We should be aware of point (3), but as k is typically small, the occurrence of such a problem is likely to be rare.

5.2.1.   Instance Space Structure

Traditionally, the way in which critical instances are identified in an instance space is assumed to apply to all classification problems: we desire an algorithm which we can apply to any domain. We argue against this position and propose two broad categories of class structure which require dramatically different approaches. Fortunately, the vast majority of problems we encounter, especially in the field of data mining, fall into a single category. This category contains instance spaces whose classes are defined by homogeneous regions of instances. To illustrate such an instance space Figure 5.1(a) depicts the *2d-dataset* which we constructed to visualise instance selection decisions. The three classes (black, grey, and white) are each defined by regions of instances which share a class, i.e., each white instance is usually in the locality of other white instances. The second category is composed of those problems which have classes defined by non-homogeneous regions. For example, problems such as the *two-spirals* dataset depicted in Figure 5.1(c). Here, the classes are represented by a spiral structure which is not localised to one region of the space. In the past, the characterisation of a critical instance has not been problem dependent, partly due to rarity of non-homogeneous class structures amongst machine learning and data mining problems.

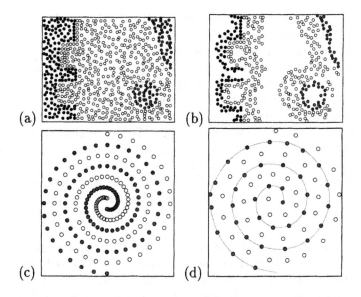

Figure 5.1 (a) The 2d-dataset, which is composed of homogeneous class definitions. (b) Removing instances from the interior of class definitions does not lead to a drop in classification accuracy as discrimination is still possible. (a) The two-spirals dataset, an example of a problem space not defined by homogeneous collections of cases. (b) Chang's prototype creation algorithm retains the class structure well

Given a class defined by a homogeneous collection of instances, which instances are critical to classifying instances of that class? A number of approaches have been proposed, and there is little consensus. For example, we might aim to identify instances that are *prototypes* (Chang, 1974; Zhang, 1992; Sebban et al., 1999) or instances with high utility (Markovitch and Scott, 1988; Markovitch and Scott, 1993; Smyth and Keane, 1995; Aha et al., 1991). We argue, as others have (Swonger, 1972; Wilson and Martinez, 1997), that instances which lie on borders between classes are almost always critical to the classification process. The instances located at the interior of class regions are superfluous as their removal does not lead to any loss in the ability of the nearest neighbour learner to discriminate between classes, which, for us, is the purpose of classification. To illustrate this point, the set of instances shown in Figure 5.1(b) will correctly predicts queries just as well as those instances in Figure 5.1(a). A prototype, an instance which in some way represents the essence, average, or typicality of a class, is useful in *characterising* a class, but not in characterising the differences between classes. Instances with high utility may well turn out to be critical border cases, but this is in no way guaranteed, as the manner in which we identify these instances is not *guided* by our knowledge that border cases are critical. The problem with utility-based methods is that we require a knowledge of the prior use of instances, i.e., classification feedback. When instances have been little used, they may have an inaccurate measure of utility. Indeed, border cases are less likely to be excel in this framework, as interior cases are by definition surrounded by cases of their own class and will therefore have a high probability of predicting queries correctly, as well as having a high probability of being used as a classifier. We can define a non-homogeneous class as one which is defined by a group of instances not sharing the same locality. We argue that in this kind of situation keeping only prototypical instances is the safest way to remove a number of instances. For example, we can dramatically reduce the number of instances in the two-spirals dataset by a employing prototype selection algorithm. Figure 5.1(d) shows the remaining prototypes after applying Chang's algorithm, discussed later. The class structure is still well defined.

To summarise, we have shown that the nature of a critical case depends on the structure of the class definitions. The majority of problems we find fall into the first category: the classes are defined by homogeneous regions of instances. However, we must be aware of other types of class structure. Given this skew towards homogeneous class structures we argue that prototypes might be good *classifiers* because they can classify many instances in the instance space. However, they are not good *discriminators*.

5.3. REVIEW

Selectively storing the set of presented instances has been an issue since the early work on nearest neighbour classification. The early schemes typically concentrate on either competence enhancement (noise removal) or competence preservation. We define these schemes as follows:

1 **Competence Enhancement:** By removing certain instances it is often possible to increase the classification accuracy of the learner. This is possible when noisy or corrupt instances are isolated and removed.

2 **Competence Preservation:** A superfluous instance is one which, when removed, will not lead to a decrease in classification accuracy. We can therefore remove it without any loss in classification competence.

In general, noise removal schemes will result in few cases being removed, and with little chance of competence depreciation and a high chance of competence enhancement. On the other hand, schemes which aim to preserve competence typically remove many cases, but are unlikely to result in competence enhancement. We will group previous studies on the basis of this distinction as well as introducing another distinction for those schemes that tackle both problems. We term the third group the hybrid approaches. Most modern instance selection algorithms are hybrid approaches. Few reviews have been compiled in this area, although a good collection of the early schemes, as well as a good overview is provided by Dasarathy (Dasarathy, 1991), but unfortunately, no experimental comparison is made between the methods.

5.3.1. Competence Enhancement

Noise can occur for a number of reasons, and takes many forms. However, we restrict ourselves here to the problem of removing stochastic noise. Wilson Editing (Wilson, 1972) attempts to remove noisy instances by making a pass through all the instances in the training set and removing those which satisfy an editing rule. The rule is simple: all instances which are incorrectly classified by their nearest neighbours are assumed to be noisy instances. The instances which satisfy such a rule will be those that have a different class to their neighbour(s). These instances will appear as exceptions within regions containing instances of the same class. Other candidates fulfilling this rule could be the odd instance lying on a border between two different classes. For this reason, Wilson Editing can be thought of as smoothing the instance-space at it removes instances that deviate from the coherent regions defined by instances sharing the same class. Wilson reported improved classification accuracy over a large class of problems when using the edited set rather than the original, unfiltered set. There are extensions to this approach, such as All k-NN and Repeated Wilson Editing (Tomek, 1976), but we have found little difference in performance.

An important point to note when removing harmful instances is that in certain domains we cannot differentiate between noise and genuine class exceptions. Recent work suggests that natural language domains, such as word pronunciation, are problematic in the context of instance deletion as the class definitions are not composed of large homogeneous regions but rather many small regions or *exceptions* (Daelemans et al., 1999). Deleting an instance in this kind of situation is a real problem, and reinforces the point we make in Section 5.2: we need a knowledge of the problem to effectively deploy a deletion scheme.

5.3.2. Competence Preservation

Hart's Condensed Nearest Neighbour rule (CNN) was an early attempt at finding, using Hart's terminology, a minimally consistent subset of the training set (Hart, 1968). A consistent subset of a training set T is some subset S of T that correctly classifies every case in T with the same accuracy as T itself. The deletion criteria used by the CNN is the opposite of that used in Wilson editing. Instead of looking to label cases which are misclassified by T as noise, we are looking for cases for which removal does not lead to additional miss-classifications. This criterion therefore results in superfluous cases being weeded out. The CNN algorithm seeks a minimal consistent subset but is not guaranteed to find one. Subsequent work addressed this problem, specifically the Reduced Nearest Neighbour (RNN) rule (Gates, 1972) and the Selective Nearest Neighbour Rule (SNN) (Ritter et al., 1975).

Chang's algorithm offers a novel approach to removing cases by repeatedly attempting to merge two existing cases into a new case (Chang, 1974). The process of merging cases results in a case-base containing cases which were not actually observed, but rather constructed. These cases are termed prototypes, which we can view as synthetic cases derived from the exemplars which a traditional nearest neighbour classification scheme would use. Chang's algorithm searches for candidates for merging: We seek two cases p and q which we can replace with a single case z. The merging process is permitted when p and q are of the same class, and after replacing them with z, the consistency of the case-base is not breached.

Another novel approach to competence preservation is the *Footprint Deletion* policy (Smyth and Keane, 1995) which is a filtering scheme designed for use within the paradigm of Case-Based Reasoning (CBR). We discuss this work here as Footprint Deletion provides a novel approach to the problem of case deletion which is relevant to our discussion. In previous work we have investigated how some of the concepts introduced by Smyth and Keane transfer to the simpler context of the nearest neighbour classification algorithm. CBR is an approach to solving reasoning and planning tasks on the basis of past solutions (Kolodner, 1993). The technicalities are much the same as instance-based learning, although the concept of case adaptation is usually used as a similarity metric. A CBR system aims to solve a new task by adapting a previously stored solution in such a way that it can be applied to the new problem. Much of Smyth and Keane's work relies on the notion of case adaptation. They use the property $Adaptable(c, c')$ to mean case c can be adapted to c'. Generally speaking, we can delete a case c when there are many other cases that can be adapted to solve c. In our previous work we introduced a nearest neighbour parallel to adaption which we term the *Local-Set* of a case c (Brighton, 1997). We define the *Local-set* of a case c as:

> The set of cases contained in the largest hypersphere centred on c such that only cases in the same class as c are contained in the hypersphere.

The originality of Smyth and Keane's work stems from their proposed taxonomy of case groups. By defining four case categories, which reflect the contribution to overall competence the case provides, we gain an insight into the effect of removing a case. We define these categories in terms of two properties: *Reachability* and *Coverage*. These properties are important, as the relationship between them has been used in crucial work which we discuss later. For a case-base $\mathcal{CB} = \{c_1, c_2, \ldots, c_n\}$ we define *Coverage* and *Reachability* as follows:

$$Coverage(c) = \{c' \in \mathcal{CB} : \text{Adaptable}(c, c')\}$$

$$Reachable(c) = \{c' \in \mathcal{CB} : \text{Adaptable}(c', c)\}$$

Using these two properties we can define the four groups in the taxonomy using set theory. For example, a case in the *pivotal* group is defined as a case with an empty reachable set. For a more thorough definition we refer the reader to the original article. Our investigation into the instance-based learning parallel of Footprint Deletion differs only in the replacement of *Adaptable* with the *Local-set* property. Whether a case c can be adapted to a case c' relies on whether c is relevant to the solution of c'. In the context of instance-based learning, this means that c is a member of nearest neighbours of c'. However, we cannot assume that a case of a differing class is relevant to the solution (correct prediction) of c'. We therefore bound the neighbourhood of c' by the first case of a differing class. Armed with this parallel we found that Footprint deletion performed well. Perhaps more interestingly, we found that a simpler method which uses only the local-set property, and not the case taxonomies, performs just as well. With local-set deletion, we choose to delete cases with large local-sets, as these are cases located at the interior of class regions. The issue of deciding how many cases to delete is the problem. Local-set deletion has subsequently been employed in the context of natural language processing (Daelemans et al., 1999).

5.3.3.   Hybrid Approaches

Aha et al. introduced the incremental lazy learning algorithms IB1, IB2, IB3, and IB4 (Aha et al., 1991). We concentrate on IB2 and IB3 as their primary function is to filter training cases. With IB2, if a new case to be added can already be classified correctly on the basis of the current case-base, then the case is discarded and not stored at all. Only those cases which the learner can not classify correctly are stored. This is a measure employed to weed out superfluous cases, and is a good one as the cases never need to be stored, unlike the other algorithms reviewed here which operate on a batch of cases.

IB3 is IB2 augmented with a "wait and see" policy for removing noisy cases. IB3 does this by keeping a record of how well the stored cases are classifying. Noisy cases are likely to be bad classifiers, so we can try and spot them after their inclusion in the case-base. Stored cases that miss-classify to a statistically significant degree are removed. Note that these cases could also be useful exceptions to the class definitions. A number of workers have augmented the IBn algorithms (Cameron-Jones, 1992; Zhang, 1992; Brodley, 1993). To summarise, Aha's algorithms offer an incremental approach to filtering, and for this reason offer improved efficiency, but suffer from the order of case presentation. Crucial cases could be rejected early on when the class definitions are poorly defined.

Wilson and Martinez present three algorithms for reducing the size of case-bases: RT1, RT2 and RT3 (Wilson and Martinez, 1997). RT1 is the basic removal scheme. The algorithm proceeds by computing, for each case, the set of k nearest neighbours (where k is small and odd). Then, another set of cases is computed for each case p, termed the *associates* of the case p. The associates of case p are the set of stored cases which have p as one of their nearest neighbours. The set of nearest neighbours is always of size k, whereas the size of the set of associates can be larger. RT1 removes a case p if at least as many of its associates, after the removal of p, would be classified correctly without it, i.e., we look to see if removing a case p has a detrimental effect on those cases which have p as a nearest neighbour.

The removal of noise is implicit in this scheme. Noise will typically not lead to an increase in misclassification of its surrounding neighbours. It will therefore often be deleted by RT1. RT2 is identical to RT1, only the cases in the training set are sorted by the distance from their nearest enemy (a case of another class). Cases furthest from a case of another class are therefore deleted first. This means that cases furthest from boundary positions will be removed before cases in border areas. RT2 also differs from RT1 in that deletion decisions still rely on the original set of associates. A case can therefore have associates which have already been deleted, but are still used to guide case deletion as we continue to test the ability to classify them. RT3 differs from RT2 though the introduction of a noise filtering pass that is executed before the RT2 procedure is carried out. The noise filtering procedure is similar to that of Wilson's (Wilson, 1972): remove those cases which are misclassified by their k nearest neighbours. The RT algorithms are driven by the relationship between the nearest neighbours and the associates of each case. The relationship is analogous to that introduced by Smyth and Keane, where Coverage and Reachability are defined in terms of the *Adaptable* property. The properties used in the RT algorithm, those of bounded neighbourhood and associate sets, are similar to the relationships we used in implementing Smyth and Keane's work in the context of instance-based learning. The algorithms differ in how they use these relationships, however. Wilson and Martinez have shown RT3 to consistently be the best case filter in a comparison with IB3.

5.3.4. An Iterative Case Filtering Algorithm

We now introduce an algorithm which we term the *Iterative Case Filtering Algorithm* (ICF). The ICF algorithm uses the instance-based learning parallels of case coverage and reachability we developed when transferring the CBR footprint deletion policy, discussed above. Rather like the Repeated Wilson Algorithm investigated by Tomek, we apply a rule which identifies cases that should be deleted. These cases are then removed, and the rule is applied again, iteratively, until no more cases fulfil the pre-conditions of the rule.

The ICF algorithm uses the reachable and coverage sets described above, which we can liken to the *neighbourhood* and *associate* sets used by Wilson and Martinez. An important difference is that the reachable set is not fixed in size but rather bounded by the nearest case of different class. This difference is crucial as our algorithm relies on the relative sizes of these sets. Our deletion rule is simple: we remove cases which have a reachable set size greater than the coverage set size. A more intuitive reading of this rule is that a case c is removed when more cases can solve c than c can solve itself. These cases will be those furthest from the class borders as their reachable sets will be large. After removing these cases, the case-space will typically contain thick bands of cases either side of class borders.

```
ICF(T)
1    > Perform Wilson Editing
2    for all x ∈ T do
3         if x classified incorrectly by k nearest neighbours then
4              flag x for removal
5    for all x ∈ T do
6         if x flagged for removal then T = T − {x}
7    > Iterate until no cases flagged for removal:
8    repeat
9         for all x ∈ T do
10             compute reachable(x)
11             compute coverage(x)
12        progress = false
13        for all x ∈ T do
14             if |reachable(x)| > |coverage(x)| then
15                  flag x for removal
16                  progress = true
17        for all x ∈ T do
18             if x flagged for removal then T = T − {x}
19   until not progress
20   return T
```

Figure 5.2 The Iterative Case Filtering Algorithm.

This is the deletion criterion the algorithm uses; the algorithm proceeds by repeatedly computing these properties after filtering has occurred. Usually, additional cases will begin to fulfil the criteria as thinning proceeds and the bands surrounding the class boundaries narrow.

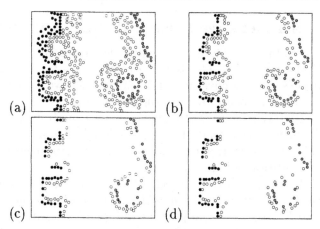

Figure 5.3 (a) The cases remaining from the 2d-dataset after 1 iteration of the ICF algorithm. (b) after 2 iterations, (c) after 3 iterations, and (d) after 4 iterations.

After a few iterations of removing cases and re-computing the sets, the criterion no longer holds. This point turns out to be a very good point to stop removing cases as removing more cases tends to breach our objective of unintrusive storage reduction. Figure 5.3(a)-(d) illustrates how the algorithm progresses. The algorithm is depicted in Figure 5.2. As with the majority of algorithms that concentrate on removing superfluous cases, ours is likely to protect noisy cases. A noisy case will have a singleton reachable set and a singleton coverage set. This property protects the case from removal. For this reason we employ the noise filtering scheme based on Wilson Editing and adopted by Wilson and Martinez. Lines 2-6 of the algorithm perform this task. The remainder of the algorithm concentrates on removing superfluous cases in the manner described above. A check is carried out to make sure progress is being made after each iteration. The algorithm is decremental in nature, like the RT algorithms, but it differs in that more than one pass is required to thin the dataset.

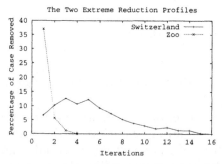

Figure 5.4 The two most extreme reduction profiles found after applying the ICF algorithm to 30 domains.

We evaluated the ICF algorithm on 30 datasets taken from the UCI repository of machine learning databases (Blake and Merz, 1998). The maximum number of iterations performed, of the 30 datasets, was 17. This number of iterations was required for the switzerland database, where the algorithm removed an average of 98% of cases. However, a number of the datasets consistently require as little as 3 iterations. Examining each iteration of the algorithm, specifically the percentage of cases removed after each iteration, provides us with an important insight into how the algorithm is working. We call this the *reduction profile* and is a characteristic of the dataset. Of the 30 datasets used, we isolated the two extreme reduction profiles which can be seen in Figure 5.4. These were found for the switzerland database and the zoo database. The switzerland database exhibits a slow path to convergence. On average, a maximum of 17 iterations are required, each one removing at most 13% of the case-base and at minimum 2% of the case-base. The zoo database, on the other hand, exhibits fast convergence. An average of two iterations are required, with an average of 37% of cases being removed on the first pass.

By examining how many cases are removed after each iteration, we can imagine the possible nature of the class structures. For example, with the switzerland database many iterations are required, with a small number of cases being removed each time. This observation would indicate that a high proportion of inter-related regions exist, as in order for one region to be thinned, a series of others must be thinned first. The length of the series reflects the complexity, rather than the size of the regions being filtered. Profiles exhibiting a short series of iterations, each one removing a large number of cases, would indicate a simple class structure containing little inter-dependency between regions. The most problematic of class structures would be characterised by a long series of iterations which results in few cases being removed.

5.4.   COMPARATIVE EVALUATION

Throughout the long development of instance pruning schemes one problem persists: little comparative evaluation between methods has been carried out, and those that have are not experimentally consistent with each other. To a degree, this problem still persists. In this section we aim to provide a comparison of the methods, and in doing so, we draw heavily of recent work (Wilson and Martinez, 1997; Brighton and Mellish, 1999). For the purposes of comparison it is useful to group the approaches into three chronological groups: (1) Early approaches: CNN, RNN, SNN, Chang, Wilson Editing, Repeated Wilson Editing, and All k-NN; (2) Recent additions: IB2, IB3, TIBLE, IBL-MDL (Cameron-Jones, 1992); (3) State of the Art: RT3, ICF. Roughly speaking, these three groups also encapsulate three classes of performance. Wilson and

Martinez compared many of the early approaches with the recent additions, as well as their own RT3. They found RT3 to be consistently superior over 30 different domains. Brighton and Mellish carried out an similar study comparing RT3 with the ICF algorithm. In earlier work we also compared the ICF and RT3 algorithms with some of those algorithms drawn from the early approaches (Brighton, 1997). Our results agreed with those of Wilson and Martinez. Given this evidence it is apparent that progress has been made despite the lack of comparison: performance has got progressively better, we are closer to achieving our goal of unintrusive storage reduction. In this section we concentrate on a comparison between RT3 and ICF. The reader is referred to the article by Wilson and Martinez for their comparison between the early methods, the recent additions, and RT3.

Table 5.1 The classification accuracy and storage requirements for each dataset. The benchmark competence, which is the accuracy achieved without any filtering, is compared with Wilson Editing, RT3, and ICF.

Dataset	Benchmark* Acc.	Stor.	Wilson Editing* Acc.	Stor.	RT3* Acc.	Stor.	ICF* Acc.	Stor.
abalone	19.53	100	22.01	19.64	22.11	40.95	22.74	15.11
anneal	95.28	100	93.24	95.46	91.82	20.72	91.35	22.59
balance-scale	77.36	100	86.04	77.48	83.40	18.23	81.47	14.67
breast-cancer-w	95.76	100	96.33	95.56	95.26	3.13	95.14	4.27
breast-cancer-l	62.46	100	68.42	64.69	74.42	19.94	72.81	23.51
bupa	59.71	100	61.81	60.49	61.23	35.07	60.75	24.79
cleveland	77.67	100	78.67	77.39	78.89	20.92	72.08	15.60
credit	82.32	100	84.46	81.12	83.15	19.9	82.28	16.89
ecoli	81.94	100	86.27	81.77	82.84	15.76	81.34	14.06
fleas	100.00	100	99.64	100.00	98.21	19.64	98.21	30.28
glass	71.43	100	69.05	70.17	69.05	23.26	69.64	31.40
hepatitis	85.16	100	82.10	84.48	83.33	19.15	82.26	16.33
hungarian	76.55	100	79.91	77.03	80.17	9.81	78.30	12.15
iris	95.00	100	95.33	96.21	93.61	16.04	92.56	42.08
led	63.77	100	68.27	66.11	69.62	18.04	71.74	41.81
led-17	42.82	100	43.00	43.09	41.48	46.78	42.33	27.50
lymphography	77.59	100	76.38	79.41	72.70	26.73	77.59	25.63
mushrooms	99.92	100	99.24	99.64	98.89	5.50	98.64	12.80
pima-indians	69.54	100	71.27	69.20	71.08	22.38	69.17	17.22
post-operative	57.78	100	66.94	54.65	69.44	6.45	65.28	7.18
primary-tumor	36.57	100	36.57	35.81	39.43	30.76	37.06	18.32
switzerland	92.08	100	93.54	90.45	91.67	2.15	92.28	2.02
thyroid	90.93	100	89.30	91.48	77.91	16.23	86.63	21.85
voting	92.99	100	93.28	92.76	93.77	7.43	91.19	8.88
waveform	75.36	100	76.62	76.37	76.14	22.79	73.93	18.98
wine	84.57	100	86.43	85.17	86.43	15.37	83.81	12.00
wisconsin-bc-di	93.01	100	93.85	92.94	92.92	6.95	92.99	6.38
wisconsin-bc-pr	67.18	100	75.90	72.64	76.28	15.43	75.64	18.24
yeast	52.70	100	55.39	52.97	55.32	27.03	52.25	16.62
zoo	95.50	100	96.25	95.31	87.08	26.13	92.42	52.78
Average	75.75	100	77.52	75.98	76.59	19.29	76.13	19.73

*These results represent the average values after 20 independent trials. Each trial, 20% of the instances were drawn at random to form the testing set.

Comparing the ICF algorithm with RT3, the most successful of Wilson and Martinez's algorithms, we found that the average case behaviours over the 30 datasets were very similar (See Table 5.1). Neither algorithm consistently outperformed the other. Both algorithms narrowly achieved an average case generalisation accuracy greater than that of the basic nearest-neighbour classifier. Both algorithms achieved approximately 80% reduction over the 30 domains. More interestingly, the behaviour of the two algorithms differ considerably on some problems. We find that no one deletion criterion consistently wins out. If we refer back to the theoretical limits discussed in Subsection 5.2, we notice that this is exactly what our average case results should look like. In our experiments, we retain 20% of the instances for testing, which means that (theoretically) only 20% of the training set is required to achieve competence preservation, and this is what we achieve in the average case. We also found that the domains which suffer a competence degradation as a result of filtering using ICF and RT3 are exactly those for which competence degrades as a result of noise removal. This would indicate that noise removal is sometimes harmful, and both ICF and RT3 suffer as a consequence. This result supports the conclusions of Daelemans et al. who argue that filtering natural language problems is unwise due to the number of class exceptions (Daelemans et al., 1999). Class exceptions in the domains we consider would appear as noise to the filters that we employ, and would therefore be removed. However, Daelemans et al. do not use a filtering criterion that sufficiently ensures the retention of border cases, so the only real conclusion we can draw is that noise removal is unwise when datasets contain many class exceptions, rather than filtering in general. This does not bode well when we consider our objective of finding a consistent case filtering criteria: the characterisation of noise in some domains will be analogous to the characterisation of class exceptions in other domains. If the exceptions are single case exceptions, then it is impossible to differentiate.

To summarise, we have presented an algorithm which iteratively filters a case-base using an instance-based learning parallel of the two case properties used in the CBR Footprint Deletion policy. Due to the iterative nature of the algorithm, we have gained an insight into how the deletion of regions depend on each other. The point at which our deletion criterion ceases to hold (quite elegantly) results in unintrusive storage reduction. Our algorithm rivals the most successful scheme of those devised by Wilson and Martinez. Our results indicate that in some problems, noise cannot be differentiated from class exceptions.

5.5. CONCLUSIONS

We began by outlining some practical issues. We argued that different domains can sometimes have drastically different class structures and classified these domains into those with either homogeneous or non-homogeneous class structures. This is important as the notion of an instance critical to the classification process depends on this distinction.

In the field of data mining homogeneous class definitions are the norm, and this article concentrates on those schemes that perform instance selection on these problems. After reviewing the principle approaches we grouped them into three classes: early schemes, recent additions, and the state of the art. The degree to which each class of algorithm achieves unintrusive storage reduction approximately mirrors this chronological order. We found that our ICF algorithm and Wilson and Martinez' RT3 algorithm achieve the highest degree of instance set reduction as well as the retention of classification accuracy: they are close to achieving unintrusive storage reduction. The degree to which these algorithms perform is quite impressive: an average of 80% of cases are removed and classification accuracy does not drop significantly. The comparison we provide is important as, considering the number of approaches, few consistent comparisons have been made. In our review we also direct the reader to the work located on the fringes of this area. The chief point we wish to address is that, traditionally, reduction schemes have been seen as general solutions to the problem of instance selection. Our observations on how these schemes work, and how well they work in different problems, suggest that the success of a scheme is highly dependent on the structure of the instance-space. We argue that one selection criterion is not enough for high performance across the board. Our results reinforce this point, especially when we consider that the problem coverage in our experiments is minimal in comparison to the variety of databases we might encounter. If we look at larger and more complex datasets, the point is likely to reinforced still. Similarly, in the context of noise removal, problem specific dependency is also a problem. We do not have a full understanding of the problem dependency, but the reduction profile provided by our ICF algorithm is a first step in achieving more perspicuous insights into problem structure. For example, some domains may contain *both* homogeneous and non-homogeneous class structures, in which case we have a problem because certain parts of the instance space are best served by different reduction criteria: both prototypical and border cases are required for the most effective solution. The local-set construction we introduced in Section 5.3.2 could also be used as a measure of how homogeneous the class structures in instance space are. By computing the average local-set size, we would have a measure of how local instances of the same class are to each other. For example, an instance space with a low average local-set size might either contain lots of noise, or plenty of class exceptions.

To summarise, with the majority of classification problems, border cases are critical to discrimination between classes. However, this is not always the case, and it is a knowledge of the class structure should guide the deployment of an instance selection policy.

Acknowledgments

Richard Nock, David Willshaw, and Pete Whitelock provided invaluable advice during the preparation of this work. The three anonymous reviewers also helped to clarify the work, as well as pointing out areas for improvement. We would also like to thank both the donors and maintainers of the UCI Repository for Machine Learning Databases.

References

Aha, D. W., Kibler, D., and Albert, M. K. (1991). Instance based learning algorithms. *Machine Learning*, 6(1):37–66.

Blake, C. and Merz, C. (1998). UCI repository of machine learning databases.

Brighton, H. (1997). Information filtering for lazy learning algorithms. Masters Thesis, Centre for Cognitive Science, University of Edinburgh, Scotland.

Brighton, H. and Mellish, C. (1999). On the consistency of information filters for lazy learning algorithms. In Zytkow, J. M. and Rauch, J., editors, *Principles of Data Mining and Knowledge Discovery: 3rd European Conference*, LNAI 1704, pages 283 – 288, Prague, Czech Republic. Springer.

Brodley, C. (1993). Addressing the selective superiority problem: Automatic algorithm/mode class selection. In *Proceedings of the Tenth International Machine Learning Conference*, pages 17 – 24, Amherst, MA.

Cameron-Jones, R. M. (1992). Minimum description length instance-based learning. In *Proceedings of the Fifth Australian Joint Conference on Artificial Intelligence*, pages 368–373, Hobart, Australia. World Scientific.

Chang, C.-L. (1974). Finding prototypes for nearest neighbor classifiers. In *IEEE Transactions on Computers*, volume C-23, pages 1179 – 1184. IEEE.

Cover, T. M. and Hart, P. E. (1967). Nearest neighbor pattern classification. *Institute of Electrical and Electronics Engineers Transactions on Information Theory*, IT-13:21 – 27.

Daelemans, W., van den Bosch, A., and Zavrel, J. (1999). Forgetting exceptions is harmful in language learning. *Machine Learning*, 34(1/3):11–41.

Dasarathy, B. (1991). *Nearest Neighbor (NN) norms: NN Pattern Classification Techniques*. IEEE Computer Society Press, Los Alimos, CA.

Gates, G. W. (1972). The reduced nearest neighbor rule. *Institute of Electrical and Electronics Engineers Transactions on Information Theory*, 18(3):431–433.

Hart, P. E. (1968). The condensed nearest neighbor rule. *Institute of Electrical and Electronics Engineers Transactions on Information Theory*, 14(3):515–516.

King, R. D., Feng, C., and Sutherland, A. (1995). Statlog: Comparison of classification algorithms on large real-world problems. *Applied Artificial Intelligence*, 9(3):289–333.

Kolodner, J. L. (1993). *Case-based reasoning*. Morgan Kaufmann Publishers, San Mateo, Calif.

Markovitch, S. and Scott, P. D. (1988). The role of forgetting in learning. In *Proceedings of the Fifth International Conference on Machine Learning*, pages 459 – 465, Ann Arbor, MI. Morgan Kaufmann Publishers.

Markovitch, S. and Scott, P. D. (1993). Information filtering: Selection mechanisms in learning systems. *Machine Learning*, 10(2):113–151.

Ritter, G. L., Woodruff, H. B., Lowry, S. R., and Isenhour, T. L. (1975). An algorithm for the selective nearest neighbour decision rule. *Institute of Electrical and Electronics Engineers Transactions on Information Theory*, 21(6):665 – 669.

Sebban, M., Zighed, D. A., and Di Palma, S. (1999). Selection and statistical validation of features and prototypes. In Zytkow, J. M. and Rauch, J., editors, *Principles of Data Mining and Knowledge Discovery: 3rd European Conference*, LNAI 1704, pages 184 – 192, Prague, Czech Republic. Springer.

Smyth, B. and Keane, M. T. (1995). Remembering to forget. In Mellish, C. S., editor, *IJCAI-95: Proceedings of the Fourteenth International Conference on Artificial Intelligence*, volume 1, pages 377 – 382. Morgan Kaufmann Publishers.

Swonger, C. W. (1972). Sample set condensation for a condensed nearest neighbour decision rule for pattern recognition. In Watanabe, S., editor, *Frontiers of Pattern Recognition*, pages 511 – 519. Academic Press, Orlando, Fla.

Tomek, I. (1976). An experiment with the edited nearest-neighbor rule. *IEEE Transactions on Systems, Man, and Cybernetics*, SMC-6(6):448 – 452.

Wilson, D. L. (1972). Asymptotic properties of nearest neighbor rules using edited data. *IEEE Transactions on Systems, Man, and Cybernetics*, SMC-2(3):408 – 421.

Wilson, D. R. and Martinez, A. R. (1997). Instance pruning techniques. In Fisher, D., editor, *Machine Learning: Proceedings of the Fourteenth International Conference*, San Francisco, CA. Morgan Kaufmann.

Zhang, J. (1992). Selecting typical instances in instance-based learning. In *Proceedings of the Ninth International Machine Learning Conference*, pages 470–479, Aberdeen, Scotland. Morgan Kaufmann.

Chapter 6

GENETIC-ALGORITHM-BASED INSTANCE AND FEATURE SELECTION

Hisao Ishibuchi, Tomoharu Nakashima, and Manabu Nii
Department of Industrial Engineering, Osaka Prefecture University,
Gakuen-cho 1-1, Sakai, Osaka 599-8531, Japan
{hisaoi,nakashi,manabu}@ie.osakafu-u.ac.jp

Abstract   This chapter discusses a genetic-algorithm-based approach for selecting a small number of instances from a given data set in a pattern classification problem. Our genetic algorithm also selects a small number of features. The selected instances and features are used as a reference set in a nearest neighbor classifier. Our goal is to improve the classification ability of our nearest neighbor classifier by searching for an appropriate reference set. We first describe the implementation of our genetic algorithm for the instance and feature selection. Next we discuss the definition of a fitness function in our genetic algorithm. Then we examine the classification ability of nearest neighbor classifiers designed by our approach through computer simulations on some data sets. We also examine the effect of the instance and feature selection on the learning of neural networks. It is shown that the instance and feature selection prevents the overfitting of neural networks.

Keywords: Genetic algorithm, pattern classification, nearest neighbor classifier, instance selection, feature selection.

6.1.   INTRODUCTION

Genetic algorithms (Holland, 1975) have been successfully applied to various problems (Goldberg, 1989). Genetic algorithms can be viewed as a general-purpose optimization technique in discrete search spaces. They are suitable for complex problems with multi-modal objective functions.

Their application to instance selection was proposed by (Kuncheva, 1995) for designing nearest neighbor classifiers. In her approach, the classification performance of selected instances was maximized. A penalty term with respect to the number of selected instances was added to the fitness function of her genetic algorithm in (Kuncheva, 1997). In the design of nearest neighbor classifiers, genetic algorithms were also used for selecting features in (Siedlecki and Sklansky, 1989) and finding an appropriate weight of each feature in (Kelly, Jr. and Davis, 1991, and Punch et al., 1993). The effectiveness of genetic algorithms for feature selection was examined in (Kudo and Sklansky, 2000). Recently genetic algorithms were used for simultaneously selecting instances and features in (Kuncheva and Jain, 1999). Such simultaneous selection was also discussed in (Skalak, 1994) where random mutation hill climbing algorithms (i.e., a kind of local search) were proposed.

In our former work (Ishibuchi and Nakashima, 1999, 2000), we proposed a genetic-algorithm-based approach to the design of compact reference sets by simultaneous instance and feature selection. Our approach incorporated several ideas such as instance selection (Kuncheva, 1997), feature selection (Siedlecki and Sklansky, 1989), and biased mutation probabilities (Ishibuchi et al., 1997) into a single algorithm. Our approach is the same as the instance and feature selection method in (Kuncheva and Jain, 1999) except for the definition of a fitness function and the use of biased mutation probabilities. Our fitness function is defined by a weighted sum of the following three criteria: (i) the classification performance of a selected reference set, (ii) the number of selected instances, and (iii) the number of selected features. In this chapter, we examine two definitions of the classification performance. One definition, which was used in our former work, is based on the classification results of the given instances by a reference set. This definition is to find compact reference sets that can correctly classify almost all the given instances. In the other definition, the classification of each instance is performed by a reference set excluding that instance. This definition of the fitness function is to find compact reference sets with high generalization ability. The same idea as the second definition has been used in some instance selection methods. We demonstrate that the instance and feature selection with the second definition can improve the generalization ability of nearest neighbor classifiers.

6.2.   GENETIC ALGORITHMS

6.2.1.   Coding

Let us assume that m labeled instances $\mathbf{x}_p = (x_{p1}, \ldots, x_{pn})$, $p = 1, 2, \ldots, m$ are given from c classes in an n-dimensional pattern space where x_{pi} is the value of the i-th feature in the p-th instance. Our task is to select a small number of instances together with a few features for designing a compact nearest neighbor classifier. Let P_{ALL} be the set of the given m instances: $P_{ALL} = \{\mathbf{x}_1, \mathbf{x}_2, \ldots, \mathbf{x}_m\}$. We also denote the set of the given n features as $F_{ALL} = \{f_1, f_2, \ldots, f_n\}$ where f_i is the label of the i-th feature. Let F and P be the set of selected features and the set of selected instances, respectively, where $F \subseteq F_{ALL}$ and $P \subseteq P_{ALL}$. We denote the reference set as $S = (F, P)$. In the standard formulation of nearest neighbor classification, the reference set S is specified as $S = (F_{ALL}, P_{ALL})$ because all the given features and instances are used for classifying new instances. In our genetic algorithm, every reference set $S = (F, P)$ is coded by a binary string of the length $(n + m)$ as

$$S = a_1 a_2 \cdots a_n s_1 s_2 \cdots s_m, \tag{6.1}$$

where a_i denotes the inclusion $(a_i = 1)$ or the exclusion $(a_i = 0)$ of the i-th feature f_i, and s_p denotes the inclusion $(s_p = 1)$ or the exclusion $(s_p = 0)$ of the p-th instance $\mathbf{x}_p$.

6.2.2.   Fitness Function

In our nearest neighbor classification with the reference set $S = (F, P)$, the nearest neighbor $\mathbf{x}_{p*}$ of a new instance $\mathbf{x}$ is found from the instance set P as

$$d_F(\mathbf{x}_{p*}, \mathbf{x}) = \min\{d_F(\mathbf{x}_p, \mathbf{x}) \mid \mathbf{x}_p \in P\}, \tag{6.2}$$

where $d_F(\mathbf{x}_p, \mathbf{x})$ is the distance between $\mathbf{x}_p$ and $\mathbf{x}$, which is defined as

$$d_F(\mathbf{x}_p, \mathbf{x}) = \sqrt{\sum_{i \in F} (x_{pi} - x_i)^2}. \tag{6.3}$$

The new instance $\mathbf{x}$ is classified by its nearest neighbor $\mathbf{x}_{p*}$.

In our instance and feature selection problem, the number of selected instances and the number of selected features are to be minimized, and the classification performance of the reference set $S = (F, P)$ is to be maximized. Thus our problem is written as

$$\text{Minimize } |F|, \text{ minimize } |P|, \text{ and maximize } g(S), \tag{6.4}$$

where $|F|$ is the number of features in F, $|P|$ is the number of instances in P, and $g(S)$ is a performance measure of the reference set $S = (F, P)$. The performance measure is defined based on the classification results of the given m instances by the reference set $S = (F, P)$.

In our former work (Ishibuchi and Nakashima, 1999, 2000), we defined the performance measure $g(S)$ by the number of correctly classified instances by $S = (F, P)$. Each instance $\mathbf{x}_q$ ($q = 1, 2, \ldots, m$) was classified by its nearest neighbor $\mathbf{x}_{p*}$ defined as

$$d_F(\mathbf{x}_{p*}, \mathbf{x}_q) = \min\{d_F(\mathbf{x}_p, \mathbf{x}_q) \mid \mathbf{x}_p \in P\}. \tag{6.5}$$

We denote this performance measure as $g_A(S)$. That is, $g_A(S)$ is the number of correctly classified instances by $S = (F, P)$ when the nearest neighbor is defined by (6.5). The following formulation corresponds to the instance selection problem for finding the minimum consistent set that can correctly classify all the given instances:

$$\text{Minimize } |P| \text{ subject to } g_A(S) = m, \tag{6.6}$$

where $S = (F_{\text{ALL}}, P)$ and $P \subseteq P_{\text{ALL}}$.

In the definition of the performance measure in (Kuncheva, 1995), when an instance $\mathbf{x}_q$ was included in the reference set, $\mathbf{x}_q$ was not selected as its own nearest neighbor. In the context of our instance and feature selection, this means that the nearest neighbor $\mathbf{x}_{p*}$ of $\mathbf{x}_q$ is selected as

$$d_F(\mathbf{x}_{p*}, \mathbf{x}_q) = \begin{cases} \min\{d_F(\mathbf{x}_p, \mathbf{x}_q) \mid \mathbf{x}_p \in P\}, \text{if } \mathbf{x}_q \notin P \\ \min\{d_F(\mathbf{x}_p, \mathbf{x}_q) \mid \mathbf{x}_p \in P - \{\mathbf{x}_q\}\}, \text{if } \mathbf{x}_q \in P \end{cases} \tag{6.7}$$

We denote the performance measure defined in this manner as $g_B(S)$. That is, $g_B(S)$ is the number of correctly classified instances when the nearest neighbor is defined by (6.7).

The fitness value of the reference set $S = (F, P)$ is defined by the three objectives of our instance and feature selection problem in (6.4) as

$$fitness(S) = W_g \cdot g(S) - W_F \cdot |F| - W_P \cdot |P|, \tag{6.8}$$

where W_g, W_F, and W_P are user definable non-negative weights.

6.2.3.   Basic Algorithm

We use a genetic algorithm to maximize the fitness function in (6.8). Our genetic algorithm is similar to (Kuncheva and Jain, 1999). In our genetic algorithm, first a number of binary strings (say, N_{pop} strings) of the length $(n+m)$ are randomly generated to form an initial population.

Of course, we can incorporate any heuristic procedures in our genetic algorithm for generating good initial strings. Next a pair of strings are randomly selected from the current population to generate two strings by crossover and mutation. The selection, crossover, and mutation are iterated to generate N_{pop} strings. The newly generated N_{pop} strings are added to the current population to form an enlarged population of the size $2 \cdot N_{pop}$. The next population is constructed by selecting the best N_{pop} strings from the enlarged population. The population update is iterated until a pre-specified stopping condition is satisfied. Our genetic algorithm is written as follows:

Step 1 (Initialization): Generate N_{pop} strings of the length $(n + m)$.

Step 2 (Genetic Operations): Iterate the following procedures $N_{pop}/2$ times to generate N_{pop} strings.

> 1 Randomly select a pair of strings from the current population.
>
> 2 Apply a crossover operation to the selected pair of strings to generate two offspring. In our computer simulations, we use the uniform crossover to avoid the dependency of the performance on the order of n features and m instances in the string.
>
> 3 Apply a mutation operation to each bit value of the two stings generated by the crossover operation.

Step 3 (Generation Update): Add the newly generated N_{pop} strings in Step 2 to the current population of the N_{pop} strings to form an enlarged population of the size $2 \cdot N_{pop}$. Select the N_{pop} best strings from the enlarged population to form the next population.

Step 4 (Termination Test): If a pre-specified stopping condition is not satisfied, return to Step 2. In our computer simulations, we use the total number of generations (i.e., iterations of Step 2 and Step 3) as the stopping condition of our genetic algorithm.

Our genetic algorithm is different from the standard implementation (Goldberg, 1989) in the selection and generation update procedures. In our algorithm, the selection of parent strings for the crossover is performed randomly. The selection of good strings is actually performed in the generation update procedure. We adopted this implementation according to the first attempt of the application of genetic algorithms to the instance selection in (Kuncheva, 1995). We also examined a more standard implementation based on the roulette wheel selection with the linear scaling and a single elite strategy. Simulation results of these two implementations were almost the same. So we only report simulation results by the above implementation.

6.2.4.   Numerical Example

Let us illustrate our approach to the instance and feature selection by a simple numerical example in Fig. 6.1 where 30 instances from each class are given. In Fig. 6.1, the classification boundary is drawn by the nearest neighbor classification using all the given instances. We applied our genetic algorithm with the following parameter specifications to this example.

String length: 62 (2 features and 60 instances),
Population size: $N_{pop} = 50$,
Crossover probability: 1.0,
Mutation probability: 0.01,
Stopping condition: 1000 generations,
Weight values: $W_g = 10$; $W_F = 1$; $W_P = 1$,
Performance measure: $g_A(S)$.

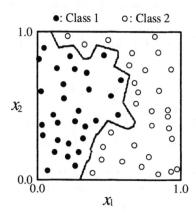

Figure 6.1   A numerical example with 30 instances from each class.

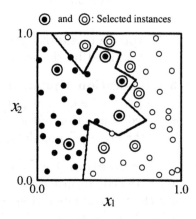

Figure 6.2   Selected instances using the first definition.

Our genetic algorithm selected 11 instances and the two features (i.e., no feature was removed). The selected instances are shown in Fig. 6.2 together with the classification boundary generated by them. From Fig. 6.2, we can see that all the given instances are correctly classified. Since we used the large weight value (i.e., $W_g = 10$) for the performance measure $g_A(S)$, we had a 100% classification rate on the given instances by the selected reference set. This is not always the case especially when we use a small weight value W_g and/or the second definition of the performance measure (i.e., $g_B(S)$). In Fig. 6.3, we show a simulation result by the performance measure $g_B(S)$. The other parameters were specified in the same manner as in Fig. 6.2. In Fig. 6.3, a single instance is misclassified by the selected nine instances. The classification boundary in Fig. 6.3 was drawn by the selected instances. In the case of the second definition of the performance measure, each instance in the reference set is not classified by itself when the fitness value is evaluated. Thus the inclusion of misclassified instances in the reference set does not always improve the performance measure.

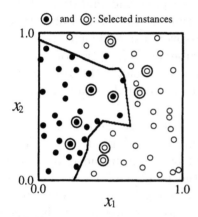

Figure 6.3 Selected instances using the second definition.

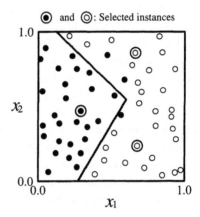

Figure 6.4 Selected instances using a small weight value $W_g = 0.5$.

Our genetic algorithm with different weight values generates different reference sets. In Fig. 6.4, we show selected instances by our genetic algorithm with a small weight value (i.e., $W_g = 0.5$) for the performance measure $g_A(S)$. We used the same specifications as in Fig. 6.2 for the other parameters. In Fig. 6.4, three instances are selected, and three instances are misclassified by the selected instances. Since the weight value W_g for $g_A(S)$ is small, a reference set with a 100% classification rate on the given instances does not always have the maximum fitness value. As a result, our genetic algorithm selected the reference set in Fig. 6.4, which can not correctly classify all the given instances. In such computer simulations with the small weight value for the performance measure, the second feature f_2 (i.e., x_2-axis in Fig. 6.4) was often removed. Actually, it was not selected by our genetic algorithm with $W_g = 0.5$ in 25 out of 30 independent trials.

6.2.5.   Biased Mutation

As we can see from the coding mechanism, our instance and feature selection method is computationally intensive. The string length is $(n + m)$ where n is the number of features and m is the number of instances. Thus the size of the search space is 2^{n+m}, which is terribly large especially in the case of pattern classification problems with many instances. Since the number of features is usually much smaller than the number of instances in many real-world pattern classification problems, we concentrate on how to effectively decrease the number of selected instances by our genetic algorithm in this subsection.

Our genetic algorithm has two genetic operations for generating new strings by modifying existing strings: crossover and mutation. Let us examine the effect of these genetic operations on the number of instances included in each string. Since the crossover just exchanges bit values between two parent strings, the total number of 1's (i.e., the total number of selected instances) in the parent strings is exactly the same as that in their offspring. This means that the crossover does not change the number of instances on the average.

On the contrary, the mutation tends to increase the average number of selected instances. We illustrate this fact using simple numerical calculations. Let m_1 be the number of instances included in a string before the mutation. That is, m_1 is the number of 1's in the string. We also denote the number of 0's in the string by m_0. Among the m_1 instances included in the string, the mutation removes $p_m \cdot m_1$ instances from the string on the average where p_m is the mutation probability. At the same time, the mutation adds some instances to the string by changing some 0's to 1's. The expected value of the number of added instances is $p_m \cdot m_0$. Thus the expected value of the number of 1's after the mutation is calculated as

$$\hat{m}_1 = m_1 - p_m \cdot m_1 + p_m \cdot m_0. \tag{6.9}$$

Since a small number of instances are to be selected from a large number of given instances in our instance and feature selection, m_1 should be much smaller than m_0. For example, let us assume that we have a binary string with 10 instances out of 1000 instances (i.e., $m_1 = 10$ and $m_0 = 990$). In this case, the expected value $\hat{m}_1$ of the number of 1's after the mutation is calculated as

$$\hat{m}_1 = 108 \quad \text{when} \quad p_m = 0.1, \tag{6.10}$$

$$\hat{m}_1 = 19.8 \quad \text{when} \quad p_m = 0.01, \tag{6.11}$$

$$\hat{m}_1 = 10.98 \quad \text{when} \quad p_m = 0.001. \tag{6.12}$$

These calculations show that large mutation probabilities prevent our genetic algorithm from decreasing the number of selected instances.

For demonstrating the effect of the mutation on the number of selected instances, we applied our genetic algorithm to a numerical example of a two-class pattern classification problem. In this numerical example, we generated 500 instances from each class using the normal distribution $N(\mu_k, \Sigma_k)$ where μ_k is a mean vector, Σ_k is a covariance matrix, and k is the class label ($k = 1, 2$). We specified μ_k and Σ_k as follows:

$$\mu_1 = (0,\ 1), \quad \mu_2 = (1,\ 0), \quad \Sigma_1 = \Sigma_2 = \begin{pmatrix} 0.3^2 & 0 \\ 0 & 0.3^2 \end{pmatrix}. \qquad (6.13)$$

In our genetic algorithm, we examined three specifications of the mutation probability, $p_m = 0.1, 0.01, 0.001$, for instance selection. The other parameters were specified in the same manner as in Fig. 6.2. Using each mutation probability, we applied our genetic algorithm to 1000 instances generated by the above normal distributions. Simulation results are shown in Table 6.1 where the CPU time was measured by a PC with a Pentium II 400MHz processor. From this table, we can see that the large mutation probability prevented our genetic algorithm from finding compact reference sets. We can also see that the larger the size of reference sets is, the longer the computation time is.

Table 6.1 Average simulation results over ten trials.

Mutation	$g_A(S)$	$|P|$	CPU time
0.1	9649.4	348.6	316.9 (min.)
0.01	9965.7	32.3	110.4 (min.)
0.001	9973.0	18.0	85.5 (min.)

Our trick for effectively decreasing the number of selected instances is to bias the mutation probability (Ishibuchi and Nakashima, 1999, 2000). In the biased mutation, a much larger probability is assigned to the mutation from "$s_p = 1$" to "$s_p = 0$" than the mutation from "$s_p = 0$" to "$s_p = 1$". That is, we use two different mutation probabilities $p_m(1 \to 0)$ and $p_m(0 \to 1)$ for the instance selection (i.e., for the last m bits of each binary string: see Eq.(6.1)). Since the number of features is usually much smaller than the number of instances, we use the standard unbiased mutation for the feature selection.

In the same manner as in the above computer simulations, we applied our genetic algorithm with the biased mutation to the pattern classification problem with 1000 instances. The three mutation probabilities were specified as $p_m(1 \to 0) = 0.1$, $p_m(0 \to 1) = 0.001$, and $p_m = 0.1$. The following average results were obtained from ten independent trials (compare the following results with those in Table 6.1).

$$g_A(S) = 9958.0, \quad |P| = 7.0, \quad \text{CPU time: 50.6 min.}$$

6.3.   PERFORMANCE EVALUATION

6.3.1.   Data Sets

We use six data sets: two artificially generated data sets and four real-world data sets in the literature. In our computer simulations, we applied our genetic algorithm to each data set after normalizing given attribute values to real numbers in the unit interval [0, 1].

Data Set I from Normal Distributions with Small Overlap: We generated a two-class pattern classification problem in the unit square $[0, 1] \times [0, 1]$. For each class, we generated 50 instances using the normal distribution $N(\mu_k, \Sigma_k)$ where μ_k and Σ_k were specified as

$$\mu_1 = (0, 1), \quad \mu_2 = (1, 0), \quad \Sigma_1 = \Sigma_2 = \begin{pmatrix} 0.4^2 & 0 \\ 0 & 0.4^2 \end{pmatrix}. \tag{6.14}$$

Data Set II from Normal Distributions with Large Overlap: This data set has much larger overlap between the two classes than the above data set. We specified the normal distribution of each class as

$$\mu_1 = (0, 1), \quad \mu_2 = (1, 0), \quad \Sigma_1 = \Sigma_2 = \begin{pmatrix} 0.6^2 & 0 \\ 0 & 0.6^2 \end{pmatrix}. \tag{6.15}$$

Iris Data: This data set consists of 150 instances with four features from three classes (50 instances from each class). A small overlap exists between the second and third classes. In (Weiss and Kulikowski, 1991), the performance of ten classification methods such as statistical techniques, neural networks, and machine learning techniques was examined by the leaving-one-out procedure on the iris data.

Appendicitis Data: The appendicitis data set consists of 106 instances with eight features from two classes. Since one feature has some missing values, we used seven features as in (Weiss and Kulikowski, 1991).

Cancer Data: The cancer data set consists of 286 instances with nine features from two classes. This data set was also used in (Weiss and Kulikowski, 1991). It should be noted that this data set is not the well-known Wisconsin breast cancer data.

Wine Data: The wine data set consists of 178 instances with 13 features from three classes, which is available from the machine learning database in the University of California, Irvine.

6.3.2. Performance on Training Data

We applied our genetic algorithm to the six data sets using the following parameter specifications:

Population size: $N_{pop} = 50$,
Crossover probability: 1.0,
Mutation probabilities: $p_m = 0.01$ for feature selection,
$\qquad\qquad\qquad\quad p_m(1 \to 0) = 0.1$ for instance selection,
$\qquad\qquad\qquad\quad p_m(0 \to 1) = 0.01$ for instance selection,
Stopping condition: 500 generations,
Weight values: $W_g = 5$; $W_F = 1$; $W_P = 1$,
Performance measure: $g_A(S)$ or $g_B(S)$.

All the given instances in each data set were used as training data in this subsection. Our genetic algorithm was applied to each data set 30 times. Average simulation results over 30 trials are summarized in Table 6.2 and Table 6.3 where each figure in parentheses denotes the number of given features or given instances in each data set. From Table 6.2 and Table 6.3, we can see that a small number of instances were selected by our genetic algorithm. From the comparison between the simulation results for the two artificial data sets (i.e., Data Set I with small overlap and Data Set II with large overlap), we can see that the smaller the class overlap is, the smaller the number of selected instances is. This is also observed from the comparison between the wine data with no overlap and the cancer data with large overlap. From the projection of the iris data into the x_3-x_4 plane, we can see that these two features are important for the classification purpose of the iris data. These two features were selected by our genetic algorithm in 29 out of the 30 trials in Table 6.2. In Table 6.3, $\{f_3, f_4\}$ were selected in 16 trials, and $\{f_2, f_3, f_4\}$ were selected in the other 14 trials.

In Table 6.2 and Table 6.3, we used the first definition $g_A(S)$ and the second definition $g_B(S)$, respectively. The first definition directly evaluates the classification ability on training data. As a result, we obtained higher classification rates on training data in Table 6.2 than Table 6.3. Such higher classification rates were realized by selecting much more instances for constructing reference sets in the case of data sets with large overlaps such as Data Set II, the appendicitis data, and the cancer data. On the other hand, in Table 6.3, the generalization ability of each reference set on unseen data was estimated in our genetic algorithm by the second definition. Thus the classification rates on training data in Table 6.3 are inferior to those in Table 6.2. In the case of data sets with small overlaps (e.g., the wine data), simulation results with the second definition in Table 6.3 are similar to those with the first definition in Table 6.2. On the contrary, these two definitions lead to different results in the case of data sets with large overlaps (e.g., Data Set II and the cancer data).

Table 6.2 Simulation results on training data using the first performance measure.

Data set	Features	Instances	Classification
Data Set I	1.9 (2)	14.5 (100)	96.7 %
Data Set II	1.8 (2)	31.0 (100)	94.4 %
Iris	2.0 (4)	6.1 (150)	99.4 %
Appendicitis	3.3 (7)	16.0 (106)	97.5 %
Cancer	5.1 (9)	54.3 (286)	89.2 %
Wine	6.3 (13)	5.9 (178)	100.0 %

Table 6.3 Simulation results on training data using the second performance measure.

Data set	Features	Instances	Classification
Data Set I	2.0 (2)	6.2 (100)	92.3 %
Data Set II	1.8 (2)	12.3 (100)	80.9 %
Iris	2.6 (4)	7.6 (150)	94.2 %
Appendicitis	3.2 (7)	4.4 (106)	91.8 %
Cancer	2.9 (9)	27.2 (286)	81.3 %
Wine	6.3 (13)	7.3 (178)	99.9 %

6.3.3.　Performance on Test Data

While we examined classification rates on training data in the previous subsection, the performance of classification systems should be evaluated by classification rates on test data. In this subsection, we examine the generalization ability of selected reference sets.

Since the first two data sets were artificially generated from the given normal distributions, we can generate unseen instances from the same normal distributions. In our computer simulations in this subsection, a reference set generated from 100 training instances was examined by 1000 test instances at each trial. This procedure was iterated 50 times for Data Set I and Data Set II. For the iris data and the appendicitis data, we used the leaving-one-out (LV1) procedure as in (Weiss and Kulikowski, 1991). The LV1 procedure was performed ten times for each data set. For the cancer data and the wine data, we used the

10-fold cross-validation (10CV) procedure. The 10CV procedure was performed ten times.

We used the same parameter values of our genetic algorithm as in the previous subsection. We examined the two definitions of the performance measure. Simulation results are summarized in Table 6.4. For comparison, we also examined the generalization ability of the original data sets before the instance and feature selection. From Table 6.4, we can see that the generalization ability was improved by the instance and feature selection with $g_B(S)$ in many data sets. The improvement of the generalization ability is clear in the appendicitis data and the cancer data with large overlaps between different classes. On the contrary, we can not observe such clear improvement in the iris data and the wine data. Those data sets, for which we obtained high classification rates on test data by the original nearest neighbor classifiers, do not have large overlaps between different classes in the pattern spaces. From the comparison between the two definitions of the performance measure, we can see that higher classification rates on test data were obtained by the second definition $g_B(S)$ than the first definition $g_A(S)$.

Table 6.4 Average classification rates on test data.

Data set	Original data	$g_A(S)$	$g_B(S)$
Data Set I	81.3 %	80.2 %	84.7 %
Data Set II	60.6 %	60.2 %	64.8 %
Iris	95.3 %	96.9 %	94.2 %
Appendicitis	80.2 %	77.0 %	85.7 %
Cancer	65.3 %	68.3 %	73.6 %
Wine	95.3 %	94.8 %	96.5 %

6.3.4.   Effect of Feature Selection

In Table 6.4, we demonstrated that the selected reference sets had higher generalization ability than the original data sets. Such improvement in the generalization ability was realized by the simultaneous feature and instance selection. In this subsection, we examine the necessity of such simultaneous selection by computer simulations without feature selection. Our genetic algorithm was used only for instance selection. In such computer simulations, we used the second performance measure because we obtained higher generalization ability in Table 6.4 from $g_B(S)$ than $g_A(S)$. Simulation results are summarized in Table 6.5. Since the two artificial data sets (i.e., Data Sets I and II) include no redundant features, the effect of removing the feature selection mechanism was slight. The effect was also slight in computer simulations on the iris data set and the wine data set for which the generalization ability was not improved by the simultaneous feature and instance selection. The effect of removing the feature selection mechanism is clear in computer simulations on the other data sets (i.e., appendicitis and cancer).

Table 6.5   Average classification rates on test data.

Data set	Original data	Instance selection	Simultaneous selection
Data Set I	81.3 %	84.5 %	84.7 %
Data Set II	60.6 %	64.7 %	64.8 %
Iris	95.3 %	95.0 %	94.2 %
Appendicitis	80.2 %	83.2 %	85.7 %
Cancer	65.3 %	69.6 %	73.6 %
Wine	95.3 %	95.0 %	96.5 %

6.4.   EFFECT ON NEURAL NETWORKS

We have already demonstrated that the instance and feature selection can improve the generalization ability of nearest neighbor classifiers. In this section, we use the selected reference sets as training data for the learning of neural networks by the back-propagation algorithm (Rumelhart et al., 1986). We examine the effect of the instance and feature selection on the generalization ability of trained neural networks.

In the same manner as in Section 6.3, we examined the generalization ability of trained neural networks by the LV1 procedure and the 10CV procedure. The generalization ability of neural networks depends on various factors such as the number of hidden units, the choice of a stopping condition of the learning, and the selection of training data. Thus it is very difficult to discuss the effect of the instance and feature selection on the generalization ability of trained neural networks. Here we just suggest that our instance and feature selection method can improve the generalization ability of trained neural networks in some data sets. In Fig. 6.5, we summarize simulation results on the cancer data. We used a three-layer feedforward neural network with 15 hidden units and two output units. The number of input units was the same as the number of selected features by our genetic algorithm. Fig. 6.5 shows average classification rates on test data evaluated by ten independent trials of the 10CV procedure. From Fig. 6.5, we can see that the instance and feature selection using the second definition prevented the overfitting of the neural networks. The instance and feature selection using the first definition, however, deteriorated the performance of the trained neural networks.

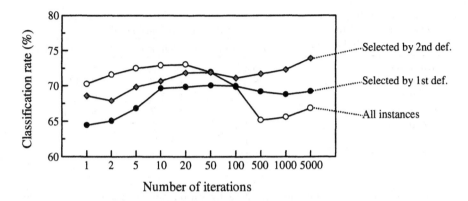

Figure 6.5 Classification rates on test data during the learning for the cancer data.

In Table 6.6, we summarize simulation results for all the six data sets. Table 6.6 shows the best average classification rate on test data for each data set through 5000 iterations of the learning algorithm. That is, we monitored the classification rate on test data for each data set at each iteration of the learning algorithm, and calculated the average classification rate at each iteration over multiple trials. Among those average classification rates, we chose the highest value. That is, Table 6.6 can be viewed as classification rates on test data by neural networks with an ideal stopping condition. From Table 6.6, we can see that the instance and feature selection deteriorated the generalization ability of neural networks in the case of data sets without large overlaps (e.g., the iris data and the wine data). From the simulation results in this section, we can see that selected instances with high generalization ability as nearest neighbor classifiers are likely to lead to high generalization ability of trained neural networks.

6.5. SOME VARIANTS

In our genetic algorithm, we have to pre-specify the weight values for the three objectives. From the definition of the fitness function in (6.8), it is obvious that we should assign a large weight value to the most important objective. If our main aim is to find a few representative instances, we will assign a large weight value to the number of selected instances. On the other hand, we will assign a large weight value to the classification performance if we do not want to degrade the classification ability. In our computer simulations, we used a large weight value for the

Table 6.6   Best classification rates on test data through 5000 iterations of the learning algorithm.

Data set	Number of hidden units	Instance and feature selection		
		No selection	1st definition	2nd definition
Data Set I	5	86.3 %	86.3 %	87.5 %
Data Set II	5	68.6 %	67.4 %	69.6 %
Iris	5	97.3 %	96.3 %	94.9 %
Appendicitis	10	88.7 %	83.2 %	88.8 %
Cancer	15	73.1 %	70.5 %	74.1 %
Wine	5	98.1 %	95.4 %	96.5 %

Table 6.7   Average results on the wine data by various specifications of the weight values.

Weight vector	Classification	Features	Instances
(1, 1, 1)	99.4 %	4.7	3.7
(1, 1, 10)	99.0 %	5.4	3.0
(1, 10, 1)	96.2 %	2.0	4.5
(10, 1, 1)	100.0 %	7.0	7.0
(1, 10, 10)	92.8 %	2.0	3.0
(10, 1, 10)	99.7 %	6.6	3.5
(10, 10, 1)	99.9 %	4.5	7.6

classification performance in order to demonstrate the improvement of the generalization ability by the instance and feature selection. In Table 6.7, we show average simulation results on the wine data by various specifications of the weight values. The weight vector denotes (W_g, W_F, W_P). Table 6.7 shows the average results on training data over 30 trials. The other parameter values were specified in the same manner as in Section 6.3. From Table 6.7, we can see that our genetic algorithm found reference sets according to the specifications of the weight values. For example, when only the weight for the number of instances was large, only three instances were selected in all the 30 trials (see the third row with the weight vector (1, 1, 10)).

When desired levels are given to some of our three objectives, we can modify the definition of the fitness function. The desired levels can be used as constraint conditions. When user's preference is not explicitly given, it is difficult to specify a desired level of any objective. It is also difficult to assign appropriate weight values to the three objectives in this case. Thus the instance and feature selection requires some trial-and-error steps using various weight values. Then we show several alternatives from such multiple trials to the user. The user then chooses the most preferable reference set. The generation of alternative reference sets can be also performed by a single run of a multi-objective genetic algorithm (Ishibuchi et al., 1997). In general, multi-objective genetic algorithms are designed for finding a number of non-dominated solutions of optimization problems with multiple objectives.

6.6. CONCLUDING REMARKS

In this chapter, we discussed the instance and feature selection for the design of compact nearest neighbor classifiers. Through computer simulations, we demonstrated that a small number of instances were selected together with only significant features by our genetic algorithm. That is, our genetic algorithm simultaneously performs instance and feature selection. We also demonstrated that the generalization ability of nearest neighbor classifiers was improved by the instance and feature selection in some data sets. This improvement was clear for data sets with large overlaps between different classes. For defining the fitness value of each reference set in our genetic algorithm, we examined two performance measures. One performance measure was defined by the classification performance of the reference set on given instances. That is, the classification performance was evaluated by classifying all the given instances by the reference set. This performance measure is suitable for finding the minimum reference set that can correctly classify all the given instances. The other performance measure was also based on the classification results of all the given instances by the reference set. The point is that every instance included in the reference set is not used as its own nearest neighbor. In this manner, the generalization ability of each reference set is evaluated in the execution of our genetic algorithm. As shown by our computer simulations, this performance measure is suitable for selecting compact reference sets with high generalization ability from data sets with large overlaps between different classes. We also suggested that the instance and feature selection can improve the generalization ability of trained neural networks when selected reference sets are used as training data.

References

Corcoran, A. L. and Sen, S. (1994). "Using real-valued genetic algorithms to evolve rule sets for classification", *Proc. of 1st IEEE International Conference on Evolutionary Computation*, 120–124.

Goldberg, D. E. (1989). *Genetic Algorithms in Search, Optimization, and Machine Learning.* Reading, MA: Addison-Wesley.

Holland, J. H. (1975). *Adaptation in Natural and Artificial Systems.* Ann Arbor, MI: University of Michigan Press.

Ishibuchi, H., Murata, T., and Turksen, I. B. (1997). "Single-objective and two-objective genetic algorithms for selecting linguistic rules for pattern classification problems", *Fuzzy Sets and Systems,* 89: 135–150.

Ishibuchi, H. and Nakashima, T. (1999). "Evolution of reference sets in nearest neighbor classification", in B. McKay et al.(eds.) *Lecture Notes in Artificial Intelligence 1585: Simulated Evolution and Learning* (2nd Asian-Pacific Conference on Simulated Evolution and Learning, Canberra, 1998, Selected Papers), 82–89.

Ishibuchi, H. and Nakashima, T. (2000). "Pattern and feature selection by genetic algorithms in nearest neighbor classification", *International Journal of Advanced Computational Intelligence* to appear.

Kelly, Jr., J. D. and Davis, L. (1991). "Hybridizing the genetic algorithm and the k nearest neighbors classification algorithm", *Proc. of 4th International Conference on Genetic Algorithms,* 377–383.

Kudo, M. and Sklansky, J. (2000). "Comparison of algorithms that select features for pattern classifiers", *Pattern Recognition,* 33: 25–41.

Kuncheva, L. I. (1995). "Editing for the k-nearest neighbors rule by a genetic algorithm", *Pattern Recognition Letters,* 16: 809–814.

Kuncheva, L. I. and Jain, L. C. (1999). "Nearest neighbor classifier: Simultaneous editing and feature selection", *Pattern Recognition Letters,* 20: 1149–1156.

Punch, W. F., Goodman, E. D., Pei, M., Chia-Shun, L., Hovland, P., and Enbody, R. (1993). "Further research on feature selection and classification using genetic algorithms", *Proc. of 5th International Conference on Genetic Algorithms,* 557–564.

Rumelhart, D. E., McClelland, J. L., and the PDP Research Group. (1986). *Parallel Distributed Processing.* Cambridge, MA: MIT Press.

Siedlecki, W. and Sklansky, J. (1989). "A note on genetic algorithms for large-scale feature selection", *Pattern Recognition Letters,* 10: 335–347.

Skalak, D. B. (1994). "Prototype and feature selection by sampling and random mutation hill climbing algorithms", *Proc. of Eleventh International Conference on Machine Learning,* 293–301.

Weiss, S. M. and Kulikowski, C. A. (1991). *Computer Systems That Learn.* San Mateo, CA: Morgan Kaufmann.

Chapter 7

THE LANDMARK MODEL: AN INSTANCE SELECTION METHOD FOR TIME SERIES DATA

Chang-Shing Perng

IBM T. J. Watson Research Center, 30 Sawmill River Road, Hawthorne, New York

perng@us.ibm.com

Sylvia R. Zhang

Candle Corporation, 701 Westchester Avenue, White Plains, New York

sylvia_zhang@candle.com

D. Stott Parker

Computer Science Department, University of California, Los Angeles

stott@cs.ucla.edu

Abstract

We present the Landmark Model, an instance selection method for time series data that has applications in querying, indexing, and mining time series. The Landmark Model gives a new foundation for time series data reduction, similarity measurement, smoothing and pattern representation. A by-product of the model is Landmark Similarity, a general model of similarity that is often consistent with human intuition and episodic memory. People often appear to find time series similar if they differ only by certain kinds of transformations (e.g., amplitude scaling). By tracking different specific subsets of features of landmarks, we can efficiently compute different similarity measures that are invariant under corresponding subsets of six transformations. We discuss a generalized approach for removing noise from raw time series without smoothing out the peaks and bottoms, and present a pattern representation based on the Landmark Model.

Keywords: Landmark, time series, instance selection, similarity, patterns, invariance, episodic memory, noise removal, data smoothing.

7.1.   INTRODUCTION

Time series data is ubiquitous in science, engineering and business. Recently there has been a surge of interest in managing this kind of data, and in mining temporal association rules in time series databases (Fayyad et al., 1996). However, difficult problems face development of systems for extracting knowledge from time series data. In the following subsections, we review some important problems and our contribution to solving them.

7.1.1.   Complexity

Even more than other types of information, time series data has grown in abundance. Voluminous streams of data are now available for everything ranging from tick-by-tick quotes for financial instruments to sensor network snapshots of meteorological data collected by the NOAA (the National Oceanic and Atmospheric Administration).

However, problems of complexity in processing time series data are more severe than for other types of data, because time series require processing at different granularities. Individual time series records usually do not carry enough information to be basic units of association rules. Real-world events often either have time spans longer than the sampling interval, or are defined as 'derivatives' of successive records. Thus, the reflection of events in time series is better represented by patterns — continuous time series segments with particular features. This is especially obvious in financial data, where changes in prices are often of more concern than the prices themselves.

The complexity of processing patterns in time series is challenging. Adopting a 'events = patterns' paradigm, so events correspond to time intervals, the number of all possible segments for a time series of length N is $N(N+1)/2$. A simple inspection of each of these segments therefore takes $O(N^3)$ time. Since N is often enormous, this complexity renders many data mining and data processing procedures useless.

Good instance selection algorithms are especially helpful here, since they can greatly reduce complexity by reducing the volume of data.

7.1.2.   Similarity Model

Perhaps the simplest way for someone to analyze patterns in a time series is to specify an exemplary pattern and then review the set of time series segments that are found to be similar to the pattern. The similarity measure plays a key role in pattern semantics because it defines what is a 'match'.

Many similarity models have been proposed. For example, the pioneering work of Agrawal (Agrawal et al., 1993) and Faloutsos (Faloutsos et al., 1994) uses Euclidean distance as the basic similarity measure. However, Euclidean distance is a very restrictive measure, and often does not match human intuition. The two segments $\langle 1,2,3,4,3 \rangle$ and $\langle 3,4,5,6,5 \rangle$ are identical in shape and length, and are visually recognized as similar, but the Euclidean distance between them is quite large because their 'amplitudes' are scaled differently. So this naive similarity measure is inadequate in many applications.

It is interesting to study human perception of similarity in time series. Figure 7.1 shows several instances of the double bottom pattern in daily stock charts. Humans can spot the resemblance between these charts almost immediately, which means these charts are similar to some degree although they are noisy and have different levels, scales, and time spans.

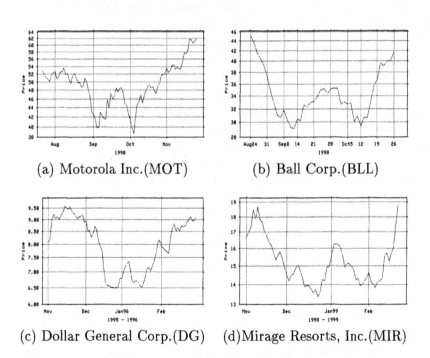

(a) Motorola Inc.(MOT)   (b) Ball Corp.(BLL)

(c) Dollar General Corp.(DG)   (d)Mirage Resorts, Inc.(MIR)

Figure 7.1 Instances of *Double Bottom* pattern.

Since 1993, mainstream research in the time series data management area has concentrated great deal of effort on extending similarity models: taking time warping into account (Berndt and Clifford, 1996; Shatkay and Zdonik, 1996; Rafiei and Mendelzon, 1997; Yi et al.,

1998; Park et al., 2000); allowing amplitude shifting (Goldin and Kanellakis, 1995; Shatkay and Zdonik, 1996; Chu and Wong, 1999); allowing time series segments of different amplitude scales to be similar (Goldin and Kanellakis, 1995; Agrawal et al., 1995; Das et al., 1997; Chu and Wong, 1999). Some work also takes smoothing or noise removal into account. Rafiei and Mendelzon (Rafiei and Mendelzon, 1997) proposed a similarity measure based on moving averages. Agrawal et al. (Agrawal et al., 1995) suggested eliminating gaps before time series segments are compared.

Although similarity measures between two time series can be computed directly in theory, doing so is usually too expensive in practice. A popular strategy for avoiding this expense is to extract sets of features with good properties from the raw data. Each feature set is used to represent a portion of the original time series. Then feature sets are indexed and stored based on multi-dimensional indexing structures. For example, Agrawal et al. (Agrawal et al., 1993) and Faloutsos et al. (Faloutsos et al., 1994) use the coefficients of the moving-window Discrete Fourier Transform (DFT) as the data representation.

Today it is common for even the simplest similarity measures to be too expensive to apply on raw data. Assuming the total length of the time series in a database is N, the search space is $O(N)$ for fixed-length pattern querying and $O(N^2)$ for variable-length pattern querying. With a linear time comparison algorithm, the overall time complexity can be $O(N^2)$ and $O(N^3)$, respectively. The situation is still worse if the similarity model is to be invariant under certain transformations. For example, Berndt and Clifford (Berndt and Clifford, 1996) use an algorithm with $O(N^3)$ time complexity to handle time warping. Real time series databases are not queryable without a sub-linear time algorithm. So various feature extraction methods have been proposed in order to provide an 'indexable' search space. The majority of these (Faloutsos et al., 1994; Agrawal et al., 1993; Goldin and Kanellakis, 1995; Rafiei and Mendelzon, 1997; Chu and Wong, 1999) use a few DFT coefficients for each time window. Wavelet coefficients are used in (Chan and Fu, 1999). Shatkay and Zdonik (Shatkay and Zdonik, 1996) suggested breaking sequences into meaningful subsequences and representing them using real-valued functions.

To our best knowledge, none of these proposed techniques supports a similarity model that can both capture the similarity of the charts in Figure 7.1 and support efficient pattern querying of time series. Furthermore, although previous work has generalized the similarity model in different directions, there has been no apparent way to unify this work under a generalized similarity model.

7.1.3. Pattern Representation

There has been little discussion about association rule mining in time series data, although association rules have been extensively studied in other contexts. An immediate issue is to define the formats of time series association rules.

Cause-effect relations capture valuable knowledge about events. We propose two formats for temporal association rules to verify the cause-effect relation: forward association and backward association. Assume $C_1, \ldots, C_{n,}, E_1, \ldots, E_m$ are events. A forward association rule is of the form $C_1, \cdots, C_n \Rightarrow E_1, \cdots, E_m$, which means that when $C_1, \cdots, C_n$ are observed appearing sequentially, $E_1, \cdots, E_m$ can also be observed appearing right after them. A backward association rule is of the form $C_1, \cdots, C_n \Leftarrow E_1, \cdots, E_m$, which means that when $E_1, \cdots, E_m$ are observed appearing sequentially, $C_1, \cdots, C_n$ can also be observed appearing right before them.

Association rules can be either formulated as hypotheses and verified with data, or be discovered by data mining process. We have discussed why time series segments offer a good representation for events. But it is still not clear what kind of segments can represent events. Taking a simple case as an example, a single univariate time series of length N (stock quotes, say), which segments among all $N(N+1)/2$ of them are suitable for representing events? In other words, what is the basic vocabulary for spelling association rules? What makes a meaningful event boundary in a time series?

7.1.4. Noise Removal and Data Smoothing

Noise accompanies almost every real measurement, and time series data are usually noisy. Often noise removal or some kind of smoothing is applied before analyzing this data. In some situations, smoothing is used in order to transform data to a particular granularity. For example, for a middle- or long-term stock trader, even though tick-by-tick stock quotations are completely accurate, moving averages give longer-term trends.

The presence of noise significantly affects the meaning of similarity. Humans usually perceive similarity of patterns with an implicit smoothing procedure. Most chart readers have long known that every pattern is recognizable only on certain time scales. In charts with long time scales, small fluctuations are treated as noise. The parameters used in current smoothing techniques often lack clear meaning. Smoothing is an essential issue to resolve in defining patterns.

Commonly-used smoothing techniques, such as moving averages, often lag or miss the most significant peaks and bottoms.[1] These peaks and bottoms can be very meaningful, and smoothing or removing them can lose a great deal of information.

Despite the importance of noise removal and data smoothing, little previous work takes smoothing as an integral part of the process of pattern definition, index construction, and query processing. Instead, it is common to apply smoothing techniques (usually done by application programs) first, and then build an index on the result. This not only increases the difficulty of using the time series data, it also creates ambiguity in the semantics of queries. Is there a better approach?

7.2.   THE LANDMARK DATA MODEL AND SIMILARITY MODEL

Most previous work separates similarity models from data models. It is important to establish a connection between the two. For example, Parseval's theorem is used to relate Euclidean similarity with the Fourier transform model.

This separation also makes completeness (no false dismissals) and soundness (no false alarms) two serious issues in pattern querying. Soundness can be guaranteed by checking the original data. Completeness is usually more important in time series, because when a search through indices fails, there is no way to avoid scanning the whole database. A common strategy is to relax the error tolerance and allow more false alarms in order to reduce or eliminate false dismissals. Eventually, both completeness and soundness grow into performance problems.

The separation is not necessary with a data model for which there is a natural corresponding similarity model. In this section, we introduce the concept of the Landmark Model, which is both a similarity model and a data model. It addresses the questions raised above with both a general model of events (Landmarks), and a generic method for noise removal and data smoothing (MDPP).

7.2.1.   The Landmark Concept

Researchers in Psychology and Cognitive Science have amassed considerable evidence that human and animals depend on **landmarks** in organizing their spatial memory (Cheng and Spetch, 1998). Research into *episodic memory* has also produced results for organizing memory around 'landmark events' (Humphreys et al., 1994; Glenberg, 1997). This all conforms to our daily experience. If one is asked to look at Figure 7.1(a) for a short period and then duplicate the chart, a relatively successful strategy is to memorize the positions of the turning points and reconnect them. These turning points serve as the landmarks in their charts. The success of this strategy also suggests that humans, to some extent, consider two charts similar if their turning points are similar and the rest of the charts are smooth curves that connect the turning points.

Extreme points also are significant to chart readers. Taking stock prices as an example, every trader would wish he/she had bought (covered) at every local minimum, sold (shorted) at every local maximum, and otherwise did little. The curves between the extreme points are indifferent to the maximal potential profit or the optimal trading strategy.

Based on this observation, we define Landmarks in time series to be those points (times, events) of greatest importance. The gist of the Landmark Model is to use landmarks instead of the raw data for processing. Different landmarks arise in different application domains, and their definition can range from simple predicates (for example, local maxima, local minima, inflection points, etc.) to more sophisticated constructs. Since most important points possess some mathematical properties, a more generic way is to categorize them mathematically. We call a point an *n*-th order landmark of a curve if the *n*-th order derivative is 0 on the point. So local maxima and minima are first-order landmarks, and inflection points are second-order landmarks.

The decision as to which kinds of points can be landmarks amounts to a tradeoff between two extremes. The more different types of landmarks in use, the more accurately a time series will be represented. However, using fewer landmarks will result in storage savings and smaller index trees. The decision about where to balance this tradeoff should be based on the nature of the data.

In our empirical study in stock market data, this decision was resolved easily. As shown in Table 7.1, even for IBM stock (which is supposed to be comparably more stable than other stocks), 4553 points out of 9950 — almost half of the records — are either local minima or maxima. Also, the normalized error[2] is reasonably small when the curve is reconstructed from the landmarks. So, we restrict discussion to only 'first-order landmarks' (although in other applications different landmarks might be more useful).

A somewhat surprising fact about landmarks is that the more volatile the time series, the less significant the higher-order landmarks. Only slowly changing time series, in which the distances between extrema are long, require higher-order landmarks for accurate reconstruction.

Given a sequence of landmarks, the curve can be reconstructed by segments of real-valued functions. Figure 7.2 comparse the time series reconstructed from landmarks and DFT coefficients. Note that only 4

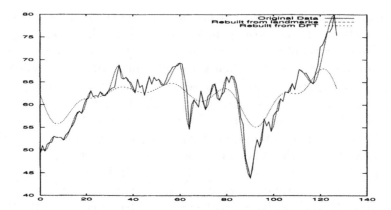

Figure 7.2 Cisco (CSCO) stock prices from June 1,1998 to November 30,1998. Shown are the original time series and two time series reconstructed from first-order landmarks and from 4 DFT coefficients.

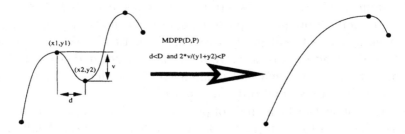

Figure 7.3 Minimal Distance/Percentage Principle

DFT coefficients are used to represent a window of length 128. For a time series of length n, there are roughly $4n$ coefficients (or n 4-dimensional index points) because the DFT has to be performed on every trailing window. Our study of stocks in the S&P500 index showed that the average number of landmarks was less than $n/2$, regardless of the time span.

The Landmark Model has another desirable property that all the peaks (local maxima) and bottoms (local minima) are preserved, while they are typically filtered out by both the DFT and wavelet transforms (being captured in coefficients of higher frequencies), as shown in Figure 7.2.

7.2.2.   Smoothing

Real world data are usually noisy. Even for a typical pattern like Figure 7.1, one cannot expect smooth transitions from each major landmark to the next. Low-pass filters like the DFT and moving averages are often introduced to eliminate noise in these transitions. Moving averages, like the DFT, tend to smooth out peaks and bottoms along with noise. Moving averages are also known to be *lagging indicators*, which have a phase delay comparing to the original data.

While there are infinitely many possible ways to classify landmarks, we introduce the **Minimal Distance/Percentage Principle (MDPP)**. MDPP is a smoothing process that can be implemented as a linear time algorithm. It is defined as following: Given a sequence of landmarks $(x_1, x_2), \cdots, (x_n, y_n)$, a minimal distance D and a minimal percentage P, remove landmarks (x_i, y_i) and (x_{i+1}, y_{i+1}) if

$$x_{i+1} - x_i < D \text{ and } \frac{|(y_{i+1} - y_i)|}{(|y_i| + |y_{i+1}|)/2} < P.$$

We use **MDPP**(D,P) to represent this process.

Figure 7.3 illustrates how MDPP works. Figure 7.4 shows the effect of MDPP while using different distances and percentages. Table 7.1 shows how the parameters affect the number of remaining landmarks and the normalized error. The real power of the Landmark Model and MDPP can be illustrated by the last cell in Table 7.1. We can use 1.5% of the original points to represent the whole time series with only 6% normalized error! This is not a special case. Our studies on financial data shows almost every stock and index with sufficiently long history gives similar results.

The parameters of MDPP have intuitive meanings. For example, if a stock trader trades once a week (5 business days) and regards a 5% gain or loss as significant, then he/she simply use MDPP(5, 5%) to smooth the data. This approach ensures that no price movement larger than 5% is smoothed out.

In contrast, the DFT approach does not scale as well as the MDPP. Figure 7.5 shows the error generated from DFT and MDPP. Again, this is a fair comparison because DFT must be performed on every trailing window (assuming the DFT is performed on all elements in a sliding fixed-size window).

A difficult decision to make with the DFT approach is the choice of window size. In contrast, MDPP is almost insensitive to the window size. In fact, neither raw landmarks nor MDPP is based on moving windows, so the length of the time series has little effect on the quality of the Landmark Model.

The MDPP preserves the offsets of each landmark. It is possible to design different smoothing methods that remove 'noisy' segments and support a similarity model similar to the one introduced by Agrawal et al. (Agrawal et al., 1995).

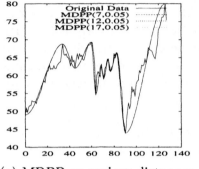
(a) MDPP on various distances

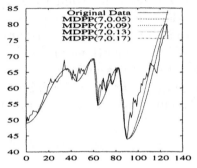
(b) MDPP on various percentages

Figure 7.4   The effect of the Minimal Distance and Percentage Principle.

Table 7.1   The percentage of remaining landmarks and the normalized error generated by MDPP with different minimal distances(D) and minimal percentages(P). The original data contains 9550 closing prices of IBM(1/2/62-12/9/99). The number of raw landmarks is 4553.

D/P	2%	6%	10%	14%	20%
2	20.7%/1.7%	18.0%/1.9%	18%/2%	18%/2%	18%/2%
4	13.4%/2.0%	7.2%/3.0%	6.8%/3.2%	6.8%/3.3%	6.7%/3.3%
6	13.0%/2.1%	5.1%/4.1%	4.5%/4.4%	4.5%/4.4%	4.4%/4.4%
8	13.0%/2.1%	4.2%/4.5%	3.4%/4.6%	3.3%/5.9%	3.3%/5.9%
10	13.0%/2.1%	3.8%/4.6%	2.9%/4.8%	2.7%/5.6%	2.7%/5.6%
12	13.0%/2.1%	3.7%/4.7%	2.6%/4.9%	2.4%/5.7%	2.4%/5.7%
14	13.0%/2.1%	3.5%/4.8%	2.4%/5.0%	2.1%/5.8%	2.1%/5.7%
16	13.0%/2.1%	3.3%/4.9%	2.2%/5.2%	1.9%/6.0%	1.8%/5.9%
18	13.0%/2.1%	3.3%/4.9%	2%/5.2%	1.7%/6.0%	1.7%/5.7%
20	13.0%/2.1%	3.2%/4.9%	1.9%/5.3%	1.5%/6.0%	1.5%/6.0%

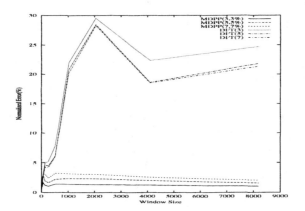

Figure 7.5 Normalized error generated by the MDPP and DFT. DFT(k) is the time series reconstructed from k coefficients. The data used here reflects different time series window lengths for the Dow Jones Industrial Average ending on April 23,1999.

7.2.3. Transformations

As previously mentioned, the more transformations included in a similarity model, the more powerful the similarity model. Most related work has considered two to three transformations. We consider six kinds of transformations. Given an univariate time series s, assume $f(t)$ is a continuous function generated by interpolating s. The transformations are each defined by a family of functionals:

1 **Shifting (SH)**
   $SH_k(f)$ such that $SH_k(f(t)) = f(t) + k$ where k is a constant.

2 **Uniform Amplitude Scaling (UAS)**
   $UAS_k(f)$ such that $UAS_k(f(t)) = k\,f(t)$ where k is a constant.

3 **Uniform Time Scaling (UTS)**
   $UTS_k(f)$ such that $UTS_k(f(t)) = f(k\,t)$ where k is a positive constant.

4 **Uniform Bi-scaling (UBS)**
   $UBS_k(f)$ such that $UBS_k(f(t)) = k\,f(t/k)$ where k is a positive constant.

5 **Time Warping (TW)**(or **Non-uniform Time Scaling**)
   $TW_g(f)$ such that $TW_g(f(t)) = f(g(t))$ where g is positive and monotonically increasing.

6 **Non-uniform Amplitude Scaling (NAS)**
   $NAS_g(f)$ such that $NAS_g(f(t)) = g(t)$ where for every t, $g'(t) = 0$ if and only if $f'(t) = 0$.

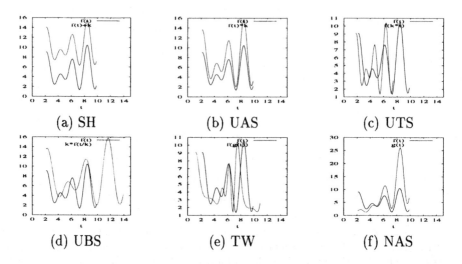

Figure 7.6 The six transformations in the Landmark Model: (a) Shifting (b) Uniform Amplitude Scaling (c) Uniform Time Scaling (d) Uniform Bi-scaling (e) Time Warping (f) Non-uniform Amplitude Scaling.

These transformations can be composed to form new transformations (although not every composition is meaningful, as we show later). The composition order is flexible, in the sense that for any two transformations F_u and G_v, there exist alternative u' and v' such that $F_u \circ G_v = G_{u'} \circ F_{v'}$. The composition is also idempotent, in the sense that for any transformation F and parameters u and v, there exists a parameter w such that $F_w = F_u \circ F_v$. With these two properties, we can use basic transformations to represent a composite transformation.

These transformations are not actually performed, but instead are *ignored*: two time series are defined to be similar if they differ only by a transform. For example, time series segments $f_1(t)$ and $f_2(t)$ are similar (actually: identical) modulo *Shifting* if there exist a constant k such that for all t in the domain, $f_1(t) = f_2(t) + k$. Putting it another way, the set of functions that are similar modulo Shifting is a set that is *invariant* under Shifting transformations. There is no need to find a specific value for a constant k or function g in the definitions above.

7.2.4.   Landmark Similarity

The error tolerance in most similarity models is a single value ϵ that is computed from pointwise differences in amplitude. This simple error measurement is no longer adequate when transformations like Uniform Time Scaling and Uniform Bi-scaling are taken into account. In the Landmark Model, drift on the time axis is also important. Furthermore, the scales on the amplitude-axis and time-axis are incomparable, which means the 2-dimensional Euclidean distance is meaningless. Hence we must generalize the dissimilarity measure.

Definition 1. Given two sequences of landmarks $L = \langle L_1, \cdots, L_n \rangle$ and $L' = \langle L'_1, \cdots, L'_n \rangle$ where $L_i = (x_i, y_i)$ and $L'_i = (x'_i, y'_i)$, the distance between the k-th landmarks is defined by $\Delta_k(L, L') = (\delta_k^{time}(L, L'), \delta_k^{amp}(L, L'))$ where

$$\delta_k^{time}(L, L') = \begin{cases} \frac{|(x_k - x_{k-1}) - (x'_k - x'_{k-1})|}{(|x_k - x_{k-1}| + |x'_k - x'_{k-1}|)/2} & \text{if } 1 < k \leq n \\ 0 & \text{otherwise} \end{cases}$$

$$\delta_k^{amp}(L, L') = \begin{cases} 0 & \text{if } y_k = y'_k \\ \frac{|y_k - y'_k|}{(|y_k| + |y'_k|)/2} & \text{otherwise} \end{cases}$$

The distance between the two sequences is

$$\Delta(L, L') = (\| \delta^{time}(L, L') \|, \| \delta^{amp}(L, L') \|) = (\delta^{time}, \delta^{amp})$$

where $\| \cdot \|$ is a vector norm, viewing both $\delta^{time}(L, L')$ and $\delta^{amp}(L, L')$ as n-vectors. The max norm $\| \delta \|_\infty = \max_k \delta_k$ often works well on financial time series.

Abusing language, we use $\delta = (\delta^{time}, \delta^{amp})$ to denote the distance between two time series segments if the rest of the parameters are clear from context. We define $(\delta^{time}, \delta^{amp}) \leq (\delta'^{time}, \delta'^{amp})$ if $\delta^{time} \leq \delta'^{time}$ and $\delta^{amp} \leq \delta'^{amp}$.

With this dissimilarity measure, we can define similarity in the Landmark Model.

Definition 2. A *landmark similarity measure* is a binary relation on time series segments defined by a 5-tuple $LMS = \langle D, P, T, \epsilon^{time}, \epsilon^{amp} \rangle$ where D and P are MDPP parameters, T is a set of basic transformations, ϵ^{time} is an error tolerance on the time-axis and ϵ^{amp} is an error tolerance on the amplitude-axis. Given two time series segments s_1 and s_2, let L_1 and L_2 be the landmark sequences after MDPP(D, P) smoothing. Then $(s_1, s_2) \in LMS$ if and only if $|L_1| = |L_2|$ and there exist two parameterized transformations T_1 and T_2 of T whose dissimilarity satisfies $\delta^{time}(T_1(L_1), T_2(L_2)) < \epsilon^{time}$ and $\delta^{amp}(T_1(L_1), T_2(L_2)) < \epsilon^{amp}$.

Figure 7.7 illustrates the operational structure of landmark similarity.

7.3.   DATA REPRESENTATION

Up to this point, we have used only simple coordinates of landmarks in modeling time series. But a sequence of landmarks denoted by co-ordinates represents only a single, specific time series segment. The

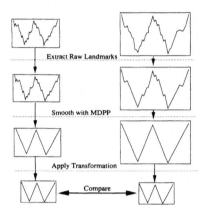

Figure 7.7 The operational structure of Landmark similarity. In comparing two time series segments s_1 and s_2, we first extract landmarks and apply MDPP on the raw landmarks. The dissimilarity of the two time series segments is the minimal distance error between the landmark sequences under the given set of transformations.

similarity we seek treats a *family* of time series segments as equivalent under the six transformations we introduced. One way to realize this similarity is to replace naive landmark coordinates with various *features* of landmarks that are invariant under these transformations.

Given a sequence of landmarks $L_1, \cdots, L_n$ where $L_i = (x_i, y_i)$, we can define as many features as we like. For demonstration purposes, we use the small feature set $F = \{y, h, v, hr, vr, vhr, pv\}$ defined by:[3]

$$h_i = x_i - x_{i-1} \qquad v_i = y_i - y_{i-1}$$

$$hr_i = h_{i+1} \,/\, h_i \qquad vr_i = v_{i+1} \,/\, v_i$$

$$vhr_i = v_i \,/\, h_i \qquad pv_i = v_i \,/\, y_i.$$

All these features are generated from the coordinates of landmarks, but each has different characteristics. In particular, every feature is invariant under some time series transformations. Table 7.2 indicates which features are invariant under each transformation:

The invariant feature set of a composite transformation is the intersection of the invariant feature sets of its components.

By observing the invariant sets, it is easy to see that not every composition of these transformations is meaningful. Time series might become *over-transformed*, and the similarity relation become a complete relation (in which each segment is similar to all others) if the time series segments are long enough. This happens when the transformation has an empty invariant feature set. For example, under Time Warping and Non-uniform Amplitude Scaling of a time series, segments can be trans-

Table 7.2 Invariant features under transformations

	y	h	v	hr	vr	vhr	pv
Shifting (SH)		•	•	•	•	•	
Uniform Amplitude Scaling (UAS)		•		•	•		•
Uniform Time Scaling (UTS)	•		•	•	•		•
Uniform Bi-scaling (UBS)				•	•	•	•
Time Warping (TW)	•		•		•		•
Non-uniform Amplitude Scaling (NAS)		•		•			

formed to any shape if they are sufficiently long that the intersection of their invariant set is empty.

On the other hand, sometimes one basic transformation can be subsumed by another. For example, Uniform Time Scaling is subsumed by Time Warping. The composition of *UHS* and *TW* is identical to *TW*.

A family of time series can be reconstructed from the values of features. Assume $F = \{F^1, F^2, \cdots, F^n\}$ is a feature set. Given a multivariate sequence $L = \ell_1, \cdots, \ell_m$ where $\ell_i = \{F_i^1, \cdots, F_i^n\}$, we define the quotient function Θ such that

$$\Theta(L) = \{\text{time series segment } s \,|\, \text{the landmarks of } s \text{ have the same feature values as } L\}$$

Abusing language slightly, we let $\Theta(F)$ denote the family of time series segments defined by values of feature set F of a sequence of landmarks L where L is clear from context. By observing the dependency relation, we have the following lemma.

Lemma 1. If F is a set of features, and '$\cup$' denotes disjoint union:

1 $\Theta(F \cup \{y, v\}) = \Theta(F \cup \{y\})$

2 $\Theta(F \cup \{h, hr\}) = \Theta(F \cup \{h\})$

3 $\Theta(F \cup \{v, vr\}) = \Theta(F \cup \{v\})$

4 $\Theta(F \cup \{vhr, y, h\}) = \Theta(F \cup \{y, h\})$

5 $\Theta(F \cup \{vhr, h, v\}) = \Theta(F \cup \{h, v\})$

6 $\Theta(F \cup \{pv, y\}) = \Theta(F \cup \{y\})$

The above lemma should be interpreted as a set of rewrite rules that reduces the number of features. Having fewer features to extract and manipulate leads to more efficient execution.

Example 1. A user chooses to construct a landmark set under Shifting, Uniform Time Scaling and Time Warping. The feature set is $\{h, v, hr,$ $vr, vhr\} \cap \{y, v, hr, vr, pv\} \cap \{y, v, vr, pv\} = \{v, vr\}$. By Lemma 1, we can use only $\{v\}$ as the feature set.

Based on what we have proposed in this section, a landmark representation of patterns is obvious. The values of a landmark's invariant set are features describing the segment between this and the previous landmark. For example, suppose the minimal invariant set is $\{h, vr\}$, then a landmark with invariant set value $\{10, 1.2\}$ represent a segment of length 10 and the height of the ending point is 1.2 times the height of the starting point. A pattern containing more turning points can be represented using concatenated landmarks. This is our answer to the questions asked in Section 7.1.3.

Unlike set-oriented data representations, landmarks are sequential. Based on this fact, landmark sequences are more like strings than multi-dimensional objects. So, a pattern in the Landmark Model is defined by a set of transformations, a sequence of values of invariant set, a pair of MDPP parameters (D, P), and an error tolerance $(\epsilon^{time}, \epsilon^{amp})$. A time series segment matches the pattern if it is landmark-similar to the pattern under the MDPP alternative to smoothing. This gives an answer to the questions about noise and smoothing in Section 7.1.4.

7.4.　CONCLUSION

We have presented the Landmark Model, an instance selection system for time series. The Landmark Model is also new model for similarity measurement in time series. Landmark Model model integrates similarity measures, data representation and smoothing techniques in a single framework. The Landmark Model concept is based on the fact that people recognize similar time series charts by comparing their important points. The idea of using landmarks also turns out to have good mathematical properties. It can use less information to represent time series more accurately and reconstruct any segments from the representation. In comparison, DFT-based techniques have to store low-frequency coefficients for every sliding window in order to track local fluctuations, which typically results in much greater index size.

We have also introduced the Minimal Distance/Percentage Principle (MDPP) as the smoothing method for the Landmark Model. The selection of the MDPP parameters is intuitive. We have shown that the MDPP is very scalable and linear-time computable.

The Landmark Model supports a generalized similarity model which can ignore differences corresponding to six basic transformations. This is accomplished by comparing features that are invariant under the transformations. We have also identified a connection between error tolerance and invariant features.

The Landmark Model is not only intuitive to humans. It is designed in a way that every parameter, error tolerance and transformation has an intuitive meaning. Furthermore it is possible to incorporate into a time series database without knowledge about Euclidean distance, the discrete Fourier Transform, or database indexing.

Notes

1. The smoothness of a curve is measured by the frequency of direction changes. So removing major peaks and bottoms is not necessary when smoothing a curve.

2. Given an original sequence of length n, $X = \langle x_i, \cdots, x_n \rangle$, the normalized error of the reconstructed sequence $Y = \langle y_i, \cdots, y_n \rangle$ is defined as: $\sqrt{(\sum_{i=1}^{n}(x_i - y_i)^2)/(\sum_{i=1}^{n}(x_i)^2)}$. An alternative definition is $\sqrt{(1/n)\sum_{i=1}^{n}(x_i - y_i)/x_i}$.

3. x is used only when a user requires a pattern to appear at a certain offset. We found this happened only rarely, so x is not included in the feature list.

References

Agrawal, R., Faloutsos, C., and Swami, A. N. (1993). Efficient similarity search in sequence databases. In *FODO*.

Agrawal, R., Lin, K.-I., Sawhney, H. S., and Shim, K. (1995). Fast similarity search in the presence of noise, scaling, and translation in time-series databases. In *VLDB*.

Berndt, D. J. and Clifford, J. (1996). Finding patterns in time series: A dynamic programming approach. In *Advances in Knowledge Discovery and Data Mining*, pages 229–248. MIT Press.

Chan, K.-P. and Fu, A.-C. (1999). Efficient time series matching by wavelets. In *ICDE*.

Cheng, K. and Spetch, M. (1998). Mechanisms of landmark use in mammals and birds. In Healy, S., editor, *Spatial Representation in Animals*. Oxford University Press.

Chu, K. K. W. and Wong, M. H. (1999). Fast time-series searching with scaling and shifting. In *PODS*.

Das, G., Gunopulos, D., and Mannila, H. (1997). Finding similar time series. In *PKDD*.

Faloutsos, C., Ranganathan, M., and Manolopoulos, Y. (1994). Fast subsequence matching in time-series databases. In *SIGMOD*.

Fayyad, U. M., Piatetsky-Shapiro, G., Smyth, P., and Uthurusamy, R., editors (1996). *Advances in Knowledge Discovery and Data Mining*. MIT Press.

Glenberg, A. (1997). What memory is for. *Behavioral and Brain Sciences*, 20(1).

Goldin, D. and Kanellakis, P. (1995). On similarity queries for time-series data: Constraint specification and implementation. In *International Conference on the Principles and Practice of Constraint Programming*.

Humphreys, M., Wiles, J., and Dennis, S. (1994). Toward a theory of human memory: Data structures and access processes. *Behavioral and Brain Sciences*, 17(4).

Park, S., Chu, W. W., Yoon, J., and Hsu, C. (2000). Efficient searches for similar subsequences of different lengths in sequence databases. In *IEEE Data Engineering*.

Rafiei, D. and Mendelzon, A. O. (1997). Similarity-based queries for time series data. In *SIGMOD*.

Shatkay, H. and Zdonik, S. B. (1996). Approximate queries and representations for large data sequences. In *ICDE*.

Yi, B.-K., Jagadish, H., and Faloutsos, C. (1998). Efficient retrieval of similar time sequences under time warping. In *ICDE*.

III

USE OF SAMPLING METHODS

Chapter 8

ADAPTIVE SAMPLING METHODS FOR SCALING UP KNOWLEDGE DISCOVERY ALGORITHMS

Carlos Domingo
Dept. of Math. and Comp. Science, Tokyo Institute of Technology, Tokyo, Japan
carlos@is.titech.ac.jp

Ricard Gavaldà
Dept. of LSI, Universitat Politècnica de Catalunya, Barcelona, Spain
gavalda@lsi.upc.es

Osamu Watanabe
Dept. of Math. and Comp. Science, Tokyo Institute of Technology, Tokyo, Japan
watanabe@is.titech.ac.jp

Abstract   Scalability is a key requirement for any KDD and data mining algorithm, and one of the biggest research challenges is to develop methods that allow to use large amounts of data. One possible approach for dealing with huge amounts of data is to take a random sample and do data mining on it, since for many data mining applications approximate answers are acceptable. However, as argued by several researchers, random sampling is difficult to use due to the difficulty of determining an appropriate sample size. In this paper, we take a sequential sampling approach for solving this difficulty, and propose an adaptive sampling method that solves a general problem covering many actual problems arising in applications of discovery science, prove theorems concerning its properties and show its efficiency by experimental evaluation.

Keywords: Data mining, knowledge discovery, scalability, adaptive sampling, concentration bounds.

8.1.　INTRODUCTION

Scalability is a key requirement for any knowledge discovery and data mining algorithm. It has been previously observed that many well known machine learning algorithms do not scale well. Therefore, one of the biggest research challenges is to develop new methods that allow to use machine learning techniques with large amount of data.

Once we are facing with the problem of having a huge input data set, there are typically two possible ways to address it. One way could be to redesign known algorithms so that, while almost maintaining its performance, can be run efficiently with much larger input data sets. The second possible approach is reducing the size of the data in such a way that the reduced set produced a result almost exactly the same one would obtain using all the data. One such a data reduction method is random sampling since for many data mining applications approximately optimal answers are acceptable. Thus, we could take a random sample and do data mining on it. However, as argued by several researchers this approach is less recommendable due to the difficulty of determining appropriate sample size needed. In this paper, we advocate for this second approach of reducing the dimensionality of the data through random sampling. For this, we propose a general problem that covers many situations arising in data mining algorithms, and a general sampling algorithm for solving it.

A typical task of knowledge discovery and data mining is to find out some "rule" or "law" explaining a huge set of examples well. It is often the case that the size of possible candidates for such rules is still manageable. Then the task is simply to select a rule among all candidates that has certain "utility" on the dataset. This is the problem we discuss in this paper, and we call it *General Rule Selection*. More specifically, we are given an input data set X of examples, a set H of *rules*, and a *utility function U* that measures "usefulness" of each rule on X. The problem is to find a nearly the best rule h, more precisely, a rule h satisfying $U(h) \geq (1 - \epsilon)U(h_\star)$, where $h_\star$ is the best rule and ϵ is a given accuracy parameter. Though simple, this problem covers several key problems occurring in data mining.

We would like to solve the General Rule Selection problem by random sampling. From a statistical point of view, this problem can be solved by taking *first* a random sample S from the domain X and *then* selecting $h \in H$ with the largest $U(h)$ on S. If we choose enough number of examples from X randomly, then we can guarantee that the selected h is nearly best within a certain confidence level. We will refer to this simple method as a *batch sampling* method.

One of the most important issues when doing random sampling is choosing proper sample size, i.e., the number of examples. Any sampling method must take into account problem parameters, an accuracy parameter, and a confidence parameter to determine appropriate sample size needed to solve the desired problem.

Widely used and theoretically sound tools to determine appropriate sample size for given accuracy and confidence parameters are the so called concentration bounds or large deviation bounds like the Chernoff or the Hoeffding bounds. They are commonly used in most of the theoretical learning research as well as in many other branches of computer science. For some examples of sample size calculated with concentration bounds for data mining problems, see, e.g., (Kivinen and Mannila, 1994) or (Toivonen, 1996). While these bounds usually allow us to calculate sample size needed in many situations, it is usually the case that resulting sample size is immense to obtain a reasonably good accuracy and confidence. Moreover, in most of the situations, to apply these bounds, we need to assume the knowledge of certain problem parameters that are unknown in practical applications.

It is important to notice that, in the batch sampling method, the sample size is calculated a priori and thus, it must be big enough so that it will work well in all the situations we might encounter. In other words, the sample size provided by the above bounds for the batch sampling should be the worst case sample size and thus, it is overestimated for most of the situations. This is one of the main reasons why researchers have found that, in practice, these bounds are overestimating necessary sample size for many non worst-case situations; see, e.g., discussion by Toivonen (Toivonen, 1996) for sampling for association rule discovery.

For overcoming this problem, we propose in this paper to do sampling in an on-line sequential fashion instead of batch. That is, an algorithm obtains examples sequentially one by one (or block by block), and it determines from those obtained examples whether it has already received enough examples for issuing the currently best rule as nearly the best with high confidence. Thus, we do not fix sample size a priori. Instead, sample size depends *adaptively* in the situation at hand. Due to this adaptiveness, if we are not in a worst case situation as fortunately happens in most of the practical cases, we may be able to use significantly fewer examples than in the worst case. Following this approach, we propose a general algorithm — AdaSelect — for solving the General Rule Selection problem, which provides us with an efficient tool for many knowledge discovery applications. To show the applicability of our approach, we show an instantiation of the approach to a particular problem, boosting decision stumps and present some experiments to verify the correctness and advantage of our approach.

The idea of adaptive sampling is quite natural, and various methods for implementing this idea have been proposed in the literature. In statistics, in particular, these methods have been studied in depth under the name of "sequential test" or "sequential analysis" (see, for instance, the pioneer book of Wald (Wald, 1947)). Their main goal has been, however, to test statistical hypotheses. Thus, even though some of their methods are applicable for some instances of the General Rule Selection problem, as far as the authors know, there has been no method that is as reliable and efficient as AdaSelect for the General Rule Selection problem. More recent work on adaptive sampling comes from the database community (Lipton and Naughton, 1995; Lipton et al., 1995). The problem they address is that of estimating the size of a database query by using adaptive sampling. While their problem and algorithms are very similar in spirit to ours, they do not deal with selecting competing hypotheses that may be arbitrarily close in value, and this makes their algorithms simpler but not applicable to our problem. From the data mining community, the work of John and Langley (John and Langley, 1996) and Provost et al. (Provost et al., 1999) is related to ours although their technique for stopping sampling is based on fitting the learning curve of some assumed learning algorithm, while our stopping condition of sampling is from purely statistical bounds. Other crucial difference is that their methods are used on top of the assumed learning algorithm, whereas our method is for a simpler problem and it is used to speed-up some part, not a whole, of a data mining algorithm.

8.2. GENERAL RULE SELECTION PROBLEM

We begin introducing some notation. Let $X = \{x_1, x_2, \ldots, x_k\}$ be a (large) set of examples and let $H = \{h_1, \ldots, h_n\}$ be a (finite, not too large) set of n functions such that $h_i : X \mapsto \mathbb{R}$. That is, $h \in H$ can be thought as a function that can be evaluated on an example x producing a real value $y_{h,x}$ as a result. Intuitively, each $h \in H$ corresponds to a "rule" or "law" explaining examples, which we call below a *rule*, and $y_{h,x}$ measures the "goodness" of the rule on x. For example, if the task is to predict a particular Boolean feature of example x in terms of its other features, then we could set $y_{h,x} = 1$ if the feature is predicted correctly by h, and $y_{h,x} = 0$ if it is predicted incorrectly. We also assume that there is some fixed real-valued and nonnegative *utility function* $U(h)$, measuring some global "goodness" of the rule (corresponding to) h on the set X. More specifically, $U(h)$ is defined by

$$U(h) = F(\text{the average value of } y_{h,x} \text{ in } X),$$

where F is some function $\mathbb{R} \mapsto \mathbb{R}$. That is, the "goodness" of h measured by the utility function U is not just the average of $y_{h,x}$, but it could be something else that is computed by F from the average of $y_{h,x}$. For the "average" of $y_{h,x}$, we simply use here the arithmetic average $\sum_{x \in X} y_{h,x}/\|X\|$, which we denote $\text{avg}(y_{h,x} : x \in X)$.

Now we are ready to state our problem.

<u>General Rule Selection</u>
Given: X, H, and ϵ, $0 < \epsilon < 1$.
Goal: Find $h \in H$ such that $U(h) \geq (1 - \epsilon) \cdot U(h_*)$,
where $h_* \in H$ be the rule with maximum value of $U(h_*)$.

For any $S \subseteq X$, we define $U(h, S) = F(\text{avg}(y_{h,x} : x \in S))$. Then $U(h)$ is simply $U(h, X)$. Thus, a trivial way to solve our problem is evaluating all h in H over all examples x in X, hence computing $U(h, X)$ for all h, and then finding the h that maximizes this value. Obviously, if X is large, this method might be extremely inefficient. We want to solve this task much more efficiently by random sampling. That is, we want to look only at a fairly small, randomly drawn subset $S \subseteq X$, find the h that maximizes $U(h, S)$, and still be sure with *high* probability that h is *close enough* to the best one. (This is similar to the PAC learning, whose goal is to obtain a *probably approximately* correct hypothesis.)

Remark 1: (Accuracy Parameter ϵ) Intuitively, our task is to find some $h \in H$ whose utility is reasonably high compared with the maximum $U(h_*)$, where the *accuracy* of $U(h)$ to $U(h_*)$ is specified by the parameter ϵ. Certainly, the closer $U(h)$ is to $U(h_*)$ the better. However, depending on the choice of U, the accuracy is not essential in some cases, and we may be able to use a large ϵ. The advantage of our algorithm becomes clear in such cases. (See discussion at the end of the next subsection and the application in the next section.)

Remark 2: (Confidence Parameter δ) We want to achieve the goal above by "random sampling", i.e., by using examples randomly selected from X. Then there must be some chance of selecting bad examples that make our algorithm to yield unsatisfactory $h \in H$. Thus, we introduce one more parameter $\delta > 0$ for specifying *confidence* and require that the probability of such error is bounded by δ.

Remark 3: (Distribution on X) In order to simplify our discussion, we assume the uniform distribution over X, in which case $U(h)$ is just the same as $U(h, X)$. But our method works as well on any distribution D over X so long as we can get each example independently following the distribution D. In fact, this is the case for the example we will consider in Section 8.4. There is no difference at all in such cases except that $U(h)$

is now defined as $U(h) = F(\mathrm{E}[y_{h,x}])$, where $\mathrm{E}[y_{h,x}]$ is the expectation of $y_{h,x}$ under the distribution that we assume over X.

Remark 4: (Condition on H) In order to simplify our discussion, we assume in the following that the value of $y_{h,x}$ is in $[0, d]$ for some constant $d > 0$. (From now on, d will denote this constant.)

Remark 5: (Condition on U) Our goal does not make sense if $U(h_*)$ is negative. Thus, we assume that $U(h_*)$ is positive. Also in order for (any sort of) random sampling to work, it cannot happen that a single example changes drastically the value of U; otherwise, we would be forced to look at all examples of X to even approximate the value of $U(h)$. Thus, we require that the function F that defines U is smooth. Formally, F need to be c-Lipschitz for some constant $c \geq 0$, as defined below. (From now on, c will denote the Lipschitz constant of F.)

Definition. 1. Function $F : \mathbb{R} \mapsto \mathbb{R}$ is *c-Lipschitz* if for all x, y it holds $|F(x) - F(y)| \leq c \cdot |x - y|$. *The Lipschitz constant of F* is the minimum $c \geq 0$ such that F is c-Lipschitz (if there is any).

All Lipschitz functions are continuous, and all differentiable functions with a bounded derivative are Lipschitz. In fact, if F is differentiable, then by Mean Value Theorem, the Lipschitz constant of F is $\max_x |F'(x)|$. Also, from the above conditions, we have $0 \leq U(h) \leq cd$ for any $h \in H$.

8.3.   ADAPTIVE SAMPLING ALGORITHM

One can easily think of the following simple batch sampling approach. Obtain a random sample S from X of a priori fixed size m and output the function from H that has the highest utility in S. There are several statistical bounds to calculate an appropriate number m of examples. While this batch sampling solves the problem, its efficiency is not satisfactory because, as discussed in the previous section, it has to choose sample size for the worst case. For overcoming this inefficiency, we take a sequential sampling approach. Instead of statically deciding the sample size, our algorithm obtains examples sequentially one by one and it stops according to some condition based on the number of examples seen and the values of the functions on the examples seen so far. That is, the algorithm *adapts* to the situation at hand, and thus if we are not in the worst case, it would be able to realize of that and stop earlier. Figure 8.1 shows a pseudo-code of the algorithm we propose, called AdaSelect, for solving the General Rule Selection problem.

Statistical bounds used to determine sample size for the batch sampling still plays a key role for designing our algorithm. Here we choose the Hoeffding bound. One can use any reasonable bound, but the reason

```
AdaSelect(X, H, ε, δ)
    t ← 0; St ← ∅; n ← ||H||
    repeat
        x ← randomly drawn example from X;
        t ← t + 1;
        αt   =   cd√(ln(nt(t + 1)/δ)/(2t));
        St := St−1 ∪ {x};
    until   ∃h ∈ H [U(h, St) ≥ αt · (2/ε − 1)];
    output h ∈ H with the largest U(h, St);
```

Figure 8.1 Pseudo-code of our on-line sampling algorithm AdaSelect.

that we choose the Hoeffding bound is that basically no assumption is necessary[1] for using this bound to estimate the error probability and calculate the sample size. Other bounds, like the bound from the Central Limit Theorem might be also appropriate for some practical situations since it behaves better. We refer the reader to (Domingo et al., 2000) where this issue is discussed.

Now we provide two theorems discussing the reliability and the complexity of the algorithm AdaSelect. For our analysis, we derive the following bounds from the Hoeffding bound.

Lemma. 2. Let $S \subseteq X$ be a set of size t obtained by independently drawing t elements of X at random. For any $h \in H$ and $\alpha \geq 0$, we have

$$\Pr\{ U(h, S) \geq U(h) + \alpha \} < \exp\left(-2\frac{\alpha^2}{(cd)^2}t\right).$$

The same bound holds for $\Pr\{ U(h, S) \leq U(h) - \alpha \}$.

Proof. We prove the first inequality. The second one is proved symmetrically. Let g be the random variable $\text{avg}(y_{h,x} : x \in S)$, i.e., the arithmetic average of t independent random variables. Observe that $U(h) = F(\text{E}[g])$, and $U(h, S) = F(g)$, where $\text{E}[g]$ is g's expectation. Then, using the fact that F is c-Lipschitz, we have $\Pr\{ U(h, S) \geq U(h) + \alpha \} \leq \Pr\{ g \geq \text{E}(g) + \alpha/c \}$. Note that g is the average of t independent random variables with range bounded by d. Hence, by the Hoeffding bound, the above probability is less than $\exp\left(-2\frac{\alpha^2}{(cd)^2}t\right)$, as claimed.

We first prove the reliability of Algorithm AdaSelect.

Theorem 1. With probability $1 - \delta$, AdaSelect(X, H, ϵ, δ) outputs a function $h \in H$ such that $U(h) \geq (1 - \epsilon)U(h_\star)$.

Proof. For a fixed $\epsilon > 0$, we will show that, with probability less than δ, the function output by AdaSelect(X, H, ϵ, δ) is in H_{bad}, defined as $H_{\text{bad}} = \{ h \in H \mid U(h) < (1 - \epsilon)U(h_\star) \}$. That is, we want to bound the following error probability P_{error} by δ.

$$P_{\text{error}} = \text{Pr}\{ \text{AdaSelect yields some } h \in H_{\text{bad}} \}.$$

Here we regard one repeat-loop iteration as a basic step of the algorithm and measure the algorithm's running time in terms of the number of repeat-loop iterations. Note that the variable t keeps the number of executed repeat-loop iterations. Let t_0 be the integer such that the following inequalities hold. (We assume that $\alpha_0 = \infty$.)

$$\alpha_{t_0 - 1} > U(h_\star)\frac{\epsilon}{2} \quad \text{and} \quad \alpha_{t_0} \leq U(h_\star)\frac{\epsilon}{2}.$$

Note that α_t is strictly decreasing as a function of t; hence, t_0 is uniquely determined. As we see below, the algorithm terminates by the t_0th step (i.e., within t_0th repeat-loop iterations) with high probability.

For deriving the bound, we consider the following two cases: (Case 1) Some $h \in H_{\text{bad}}$ satisfies the stopping condition of the repeat-loop before the t_0th step, and (Case 2) $h_\star$ does not satisfy the stopping condition during the first t_0 steps. Clearly, whenever the algorithm makes an error, one of the above cases certainly occurs. Thus, by bounding the probability that either (Case 1) or (Case 2) occurs, we can bound the error probability P_{error} of the algorithm.

First we bound the probability of (Case 1). Let h_{bad} be a rule in H_{bad} with the largest utility, and let $EV(h, \tilde{t})$ be the event that h satisfies the stopping condition of AdaSelect at $\tilde{t}$th step. Then we have

$$\text{Pr}\{ \text{(Case 1) holds} \} \leq \sum_{1 \leq \tilde{t} < t_0} \sum_{h \in H_{\text{bad}}} \text{Pr}\{ EV(h, \tilde{t}) \} \leq \sum_{1 \leq \tilde{t} < t_0} n \cdot \text{Pr}\{ EV(h_{\text{bad}}, \tilde{t}) \}.$$

Note that $\text{Pr}\{EV(h_{\text{bad}}, \tilde{t})\}$ is bounded by $P_1(\tilde{t}) = \text{Pr}\{ U(h_{\text{bad}}, S_{\tilde{t}}) \geq \alpha_{\tilde{t}} \cdot (2/\epsilon - 1) \}$. Now we would like to bound this $P_1(\tilde{t})$ by using Lemma 2. Note that the Hoeffding bound and hence Lemma 2 is applicable only for the case that the size of set $S \subseteq X$ is fixed in advance. On the other hand, in our algorithm t itself is a random variable. Thus, precisely speaking, we cannot use Lemma 2 directly, and we argue as follows. First fix any $\tilde{t}$, $1 \leq \tilde{t} < t_0$. We modify our algorithm by replacing the stopping condition of the repeat-loop with "$t \geq \tilde{t}$?"; that is, modify the algorithm so that it always sees $\tilde{t}$ examples. Then it is easy to show that if $U(h_{bad}, S_{\tilde{t}}) \geq \alpha_{\tilde{t}} \cdot (2/\epsilon - 1)$ in the original algorithm, then with the same sampling sequence we have $U(h_{bad}, S_{\tilde{t}}) \geq \alpha_{\tilde{t}} \cdot (2/\epsilon - 1)$ in the modified algorithm. Thus, $P_1(\tilde{t})$ is at most $P_1'(\tilde{t}) = \Pr\{ U(h_{bad}, S_{\tilde{t}}) \geq \alpha_{\tilde{t}} \cdot (2/\epsilon - 1)$ in the modified algorithm$\}$. Note that $\alpha_{\tilde{t}} > U(h_*)(\epsilon/2)$ for any $\tilde{t} < t_0$ and that $U(h_*)(1 - \epsilon) > U(h_{bad})$. From these we have

$$P_1'(\tilde{t}) \leq \Pr\{ U(h_{bad}, S_{\tilde{t}}) > U(h_{bad}) + \alpha_{\tilde{t}} \}.$$

On the other hand, since the sample size is fixed in the modified algorithm, we can now use Lemma 2 bound to the righthand side. Also by our choice of α_t, we have

$$P_1'(\tilde{t}) < \exp\left(-2\frac{\alpha_{\tilde{t}}^2}{(cd)^2}\tilde{t}\right) \leq \frac{\delta}{n\tilde{t}(\tilde{t} + 1)}$$

In fact, α_t is defined so that the last inequality holds.

Now by using the above bound, we can bound $\Pr\{(\text{Case 1}) \text{ holds}\}$ by $\sum_{1 \leq \tilde{t} < t_0} \frac{\delta}{\tilde{t}(\tilde{t}+1)} \leq \delta(1 - 1/t_0)$.

Next consider (Case 2). Clearly, (Case 2) implies $U(h_*, S_{t_0}) < \alpha_{t_0} \cdot (2/\epsilon - 1)$, which implies, as in (Case 1), that $U(h_*, S_{t_0}) < \alpha_{t_0} \cdot (2/\epsilon - 1)$ in a modified algorithm that always sees t_0 examples. From this observation, an argument as in the (Case 1) shows that the probability of (Case 2) is at most $\exp\left(-2\frac{\alpha_{t_0}^2}{(cd)^2}t_0\right) = \frac{\delta}{nt_0(t_0+1)} \leq \delta/t_0$.

Therefore, the probability that either (Case 1) or (Case 2) holds is bounded by δ.

Next we estimate the running time. As above we regard one repeat-loop iteration as a basic step of the algorithm and measure the algorithm's running time in terms of the number of repeat-loop iterations that is exactly the number of required examples. In the above proof, we have already showed that the probability that the algorithm does not terminate within t_0 steps, that is, (Case 2) occurs, is at most δ. Thus, the following theorem is immediate from the above proof.

Theorem 2. With probability $1 - \delta$, AdaSelect(X, H, ϵ, δ) halts within t_0 steps (in other words, AdaSelect(X, H, ϵ, δ) needs at most t_0 examples), where t_0 is the smallest integer such that $\alpha_{t_0} \leq U(h_*)(\epsilon/2)$.

From the fact that t_0 is the smallest integer satisfying $\alpha_t \leq U(h_*)(\epsilon/2)$, we may assume that $\alpha_{t_0} \approx U(h_*)(\epsilon/2)$. Then by using some approximate unfolding of the logarithm factor, we can estimate t_0 as follows.

$$t_0 \approx 2 \left(\frac{cd}{\epsilon U(h_\star)} \right)^2 \cdot \left(\ln \frac{4n}{\delta} + 4 \ln \frac{cd}{\epsilon U(h_\star)} \right).$$

Let us discuss the meaning of this formula. Since both n and δ are within the log function, their influence to the complexity is small. In other words, we can handle the case with a relatively large number of rules and/or the case requiring a very high confidence without increasing too much the sample size needed. Or we may roughly consider that the sample size is proportional to $(1/\epsilon)^2$ and $((cd)/U(h_\star))^2$. Depending on the choice of U, in some cases we may assume that $(cd)/U(h_\star)$ is not so large; or, in some other cases, we may not need small ϵ, and thus, $1/\epsilon$ is not so large. AdaSelect performs particularly well in the latter case.

It should be noted here that in the case that $y_{h,x}$ is either 0 or 1 and $U(h, S)$ is defined as $U(h, S) = \text{avg}(y_{h,x} : x \in S)$, we had better use the algorithm of Lipton et al. (Lipton et al., 1995). Of course, we could also use AdaSelect even for this case, but the sample size is roughly $\mathcal{O}(1/(\epsilon U(h_\star))^2)$, whereas their sample size is $\mathcal{O}(1/\epsilon^2 U(h_\star))$.

We should also mention that a similar result could be obtained for the case where $\|H\|$ is infinite using the VC-dimension of H instead of its cardinality. However, for this case there are certain computational problems that have to be addressed since it does not exists a general way to compute which rule is the best at each iteration of the algorithm.

In summary, AdaSelect shows its advantage most when U is chosen so that (1) relatively large ϵ is sufficient, and (2) though $(cd)/U(h_\star)$ is not bounded in general, it is not so large in lucky cases, which happen more often than the bad cases. We will see such an example in Section 8.4.

8.4. AN APPLICATION OF ADASELECT

In this section we describe in detail how our method can be instantiated into a particular application and provide an experimental evaluation of the algorithms proposed.

Before describing our application, let us make some remarks about the kind of problems where our algorithms can be applied. First, recall that our algorithms are designed for the General Rule Selection problem described in Section 8.3. This problem is simple; it may be too simple to capture any sort of actual data mining problem. We do not propose to use our sampling algorithms for solving some data mining problems, but instead, we do propose to use them as a tool. As explained below, some subtasks of data mining algorithms can be casted as instances of the General Rule Selection problem and then, those parts can be scaled by using our sampling algorithms instead of all the data. Also as remarked in the previous sections, our method is not the one that we should always use. It shows its advantages most when an appropriate utility function U can be used, and there are some cases, though only very particular cases, that the other method performs better.

The example we choose is a boosting based classification algorithm that uses a simple decision stump learner as a base learner, and our sampling algorithms are used to scale-up the base learner. *Boosting* (Freund and Schapire, 1997) is a technique to make a sufficiently "strong" learning algorithm based on a "weak" learning algorithm. Since a weak base learning algorithm (a *base learner*, in short) does not need to obtain a highly accurate classification rule, we can use for a base learner a learning ·algorithm that produces a not-so-accurate but simple classification rule (which we call a *weak hypothesis*). Then since a set H of such simple weak hypotheses is not so large, we may be able to select the best one just by searching through H exhaustively, which is indeed one instance of our General Rule Selection problem. That is, any rule selection algorithm can be used for a base learner, and we do this by random sampling. Here we use a set H_{DS} of decision stumps for our weak hypothesis class, where a *decision stump* is a single-split decision tree based on a single attribute with only two terminal nodes. We consider the case that attributes used for a split node are discrete attributes and that the number of all possible decision stumps is not so large. On the other hand, a set X of examples is huge and it is just infeasible to see all examples in X. In such a situation, the best possible approach is the selection via random sampling, i.e., selecting a weak hypothesis from H_{DS} that performs the best classification on randomly selected examples from X. We use our sampling algorithms for this selection.

We should also clarify the type of boosting technique we use here. For the boosting part, the obvious choice would be the AdaBoost algorithm of Freund and Schapire (Freund and Schapire, 1997) since this algorithm has been repeatedly reported to outperform any other voting method. AdaBoost is designed for "boosting by subsampling" or "boosting by re-weighting" where its base learner is required to produce a hypothesis that tries to minimize error with respect to a weighted training set. However, we are aiming to use our sampling method to speed up the base learner by changing the size of a training set adaptively. Thus, instead of fixing a training set, we have to draw random examples, filter them according to their current weights, and pass some of them to the base learner, what is usually called "boosting by filtering" or "boosting by re-sampling". As discussed in (Domingo and Watanabe, 2000) AdaBoost is not appropiate or the later framework since the weights become very extreme and it might take a long time to generate a new sample at each step. Thus, we will use the modification of AdaBoost (denoted by

MadaBoost) proposed in (Domingo and Watanabe, 2000) that is suitable for boosting by re-weighting and re-sampling while keeping the desired boosting property. MadaBoost uses basically the same weighting scheme as AdaBoost but using the initial weight as a saturation bound so the weights cannot grow uncontrolled. The combined function is a weighted majority of the base functions defined in the same way as AdaBoost. (For more details on MadaBoost we refer the reader to (Domingo and Watanabe, 2000).)

8.4.1.   The Problem and The Algorithm

As we have already explained, we use our sampling algorithm for a base learner used by the boosting algorithm MadaBoost. Our task is to select, from a weak hypothesis class, a "good" hypothesis that performs nearly the best classification on randomly selected examples from X. Here note that the boosting algorithm modifies a distribution over X at each boosting step, and the goodness of hypotheses has to be measured based on the current distribution. But on the other hand, we can assume some example generator **EX**, provided by the boosting algorithm, that generates x from X according to the current distribution.

Now we define each item of the General Rule Selection problem specifically. We assume some set of discrete attributes. Each example is regarded as a vector of the values of each attribute on the instance plus its class. Thus, set X contains all such labeled examples. The class H of rules is H_{DS}, the set of all possible decision stumps over the set of discrete attributes, and a decision stump is used as a classification rule. For each $h \in H_{DS}$ and each $x = (a, b) \in X$ (a being the example and b its class), we define $y_{h,x} = 1$ if x is correctly classified by h (that is, $h(a) = b$) and $y_{h,x} = 0$ otherwise. That is, $y_{h,x}$ indicates "goodness" of h on x. Boosting techniques require a base learner to provide a weak hypothesis that performs better than random guess. Since they were originally designed for binary classification the worst classifier is assumed to have accuracy equal to 50%. A weak hypothesis with error less than 50% would force the boosting process to stop. Thus, in order to make sure we get a weak hypothesis with prediction error less than $1/2$, we should measure the goodness of a hypothesis by its "advantage" over the random guess, i.e., its correct probability $- 1/2$. Here it is quite easy to adjust our goodness measure. What we need is to define the utility function U by $U(h) = \mathrm{E}[y_{h,x}] - 1/2$; that is, we use $F(y) = y - 1/2$ to calculate our total goodness of h from the average goodness of h. Accordingly, $U(h, S)$ is defined by

$$U(h, S) = \mathrm{avg}(y_{h,x} : x \in S) - \frac{1}{2} = \frac{\sum_{x \in S} y_{h,x}}{\|S\|} - \frac{1}{2}.$$

Decision Stump Selector
% We set $\epsilon = 0.5$ and $\delta = 0.1$.
   $t \leftarrow 0$; $S \leftarrow \emptyset$; $n \leftarrow ||H_{DS}||$
  repeat
    use **EX** to generate one example and add it to S;
    $t \leftarrow t + 1$;
    $\alpha_t = \sqrt{(2 \ln \tau - \ln \ln \tau + 1)/t}$,   where $\tau = nt(t+1)/(2\delta\sqrt{\pi})$;
    $U(h,S) \leftarrow ||\{x \in S : h \text{ classifies } x \text{ correctly}\}||/t - 1/2$;
  until $(\exists h \in H[\ \ U(h,S) \geq \alpha_t(2/\epsilon - 1)\ \ ])$
  output $h_0 \in H$ with largest $U(h,S)$ and $U(h_0,S) + 1/2$ as an
estimation of h_0's success prob.;

Figure 8.2   Decision Stump Selector.

This is the specification of the problem that we want to solve. One important point here is that we do not have to worry so much about the accuracy parameter ϵ, which is due to our choice of the utility function. For example, we can fix $\epsilon = 1/2$. Then our sampling algorithm may choose some h whose utility $U(h)$ is $U(h_\star)/2$, just the half of the best one. But we can still guarantee that the misclassification probability of h is smaller than $1/2$, in fact, it is $1/2 - U(h_\star)/2$, whereas that of the best one is $1/2 - U(h_\star)$. On the other hand, if we measured the goodness of each hypothesis by its correct classification probability, then we would have to choose ϵ small enough (depending on the best performance) to ensure that the selected hypothesis is better than the random guess. Introducing the notion of "utility", we could discuss general enough selection problems that allow us to attack problems like this one easily.

Next we define some other parameters and instantiate our algorithm. Note that the function F used here to define U from $\text{avg}(y_{h,x} : x \in S)$ is just $F(z) = z - 1/2$; hence, it is 1-Lipschitz and the parameter c is set to 1. Also since $y_{h,x}$ is either 0 or 1, the parameter d is set to 1. With all necessary parameters being fixed, we can now state our algorithm as in Figure 8.2.

One additional remark. We modified the algorithm to output not only the hypothesis h_0 but also an estimation of the correct classification probability of h_0, because it is necessary in the boosting algorithm.

8.4.2.   Experimental Results

We have conducted our experiments on a collection of datasets from the repository at University of California at Irvine. Two points need to be clarified on the way we used these datasets.

Firstly, some attributes in these datasets are continuous, while our learning algorithm is designed for discrete attributes. Hence, we needed to discretize some of the data. For a discretization algorithm, we have used equal-width interval binning discretization with 5 intervals. Although this method has been shown to be inferior to more sophisticated methods like entropy discretization of Fayad and Irani (Fayad and Irani, 1993), it is very easy to implement, very fast, and the performance difference is small (Dougherty et al., 1995). Missing attributes are handled by treating "missing" as a legitimate value.

Secondly, we had to inflate the datasets since we need large datasets but most of the UCI datasets are quite small. That is, following John and Langley (John and Langley, 1996), we have artificially inflated the training set (the test set has been left unchanged to avoid making the problem easier) introducing 100 copies of all records and randomizing their order. Inflating the datasets only makes the problem harder in terms of running time, does not change the problem in terms of accuracy if we restrict ourselves to use algorithms that just calculate statistics about the training sample. For instance, in our case, if one particular decision stump has an accuracy of 80% on the original training set, then this stump has the same accuracy in the inflated dataset. Thus, if that is the stump of choice in the small dataset it will also be in the inflated version. Moreover, it also does not affect the results concerning sampling, neither makes the problem easier nor more difficult. The reason is that all the statistical bounds used here to calculate necessary sample sizes provide results that are independent of the size of the probability space from where we are sampling that, in this case, is the training set. The necessary sample sizes depend on the probabilities on the training set and these are unchanged when we inflate the dataset. In other words, if in one particular situation the necessary sample size is 10000, this sample size will be the same independently of whether we are sampling from the original dataset or from the inflated version. We are aware that this is perhaps not the best method to test our algorithms and real large datasets would have been better; but we still believe that the results are informative enough to show the goodness of our method. Finally, we have chosen only datasets with 2 classes since, according to previous experiments on boosting stumps, we are not expecting our base learner to be able to find weak hypothesis with accuracy better than 50% for most problems with a large number of classes. Apart from this, the choice of datasets has been done so it reflects a variety of datasets sizes and combination of discrete and continuous attributes.

For every dataset, we had performed a 10-fold cross validation and for the algorithms using sampling (and thus, randomized) the results are averaged over 10 runs. All the experiments have been done in a computer with a CPU alpha 600Mhz using 256Mb of memory and a hard disk of 4.3Gb running under Linux. Since enough memory was available, all

Table 8.1 Accuracies of boosted decision stumps with and withouth sampling and that of Naive Bayes.

Name	Size	$||H_{DS}||$	DS	Exh.	AdaSel.	NB
agaricus	731100	296	88.68	97.74	97.84	98.82
kr-vs-kp	287600	222	68.24	93.19	92.89	88.05
hypothy.	284600	192	95.70	95.86	95.84	95.43
sick-euthy.	284600	192	90.74	91.02	90.93	90.26
german	90000	222	69.90	74.10	74.28	76.00
ionos	31590	408	76.18	90.26	89.53	89.14

the data has been loaded in a table in main memory and from there the algorithms have been run. Doing experiments with the data in external memory is part of our future plans. Loading the data took few seconds and since this time is the same for all the algorithms it has been omitted from the results. The time taken to construct a set with all possible decision stumps is included in the total time. As a test bed for comparing our boosting decision stumps algorithms we have chosen two well known learning algorithms, the decision tree inducer C4.5 of Quinlan (Quinlan, 1993) and the Naive Bayes classifier (both just used in a single run, not boosted). According to the discussion above about experiments with inflated datasets, for Naive Bayes classifier we had provided both, accuracy and running time results since this learning algorithm satisfies the requirements described before and therefore these results are meaningful. However, in the case of C4.5 we had only provided running time results, not accuracy results since this algorithm is not strictly based on probabilities over the sample. It makes, for instance, decisions about whether to split or not based on the actual number of instances following in one particular node and these numbers are obviously changed when we inflate the dataset. Thus, accuracy results of C4.5 in these sets cannot be used for comparison.

For comparison, we executed MadaBoost for boosting by re-weighting with the whole dataset; that is, MadaBoost with the base learner that selects the best decision stump by searching trough the whole dataset exhaustively. Below we use AdaSel., and Exh. respectively to denote the Decision Stump Selector using adaptive sampling and the exhaustive search selector. We also carried out the experiments with Exh. using the original AdaBoost and found the difference with MadaBoost with Exh. almost negligible. We have set the number of boosting rounds to be 10 which usually is enough to converge to a fixed training error (that is, although we keep obtaining hypothesis with accuracy slightly better than 50%, the training error does not get reduced anymore).

Table 8.2   Running times (in seconds) of MadaBoost with and without sampling, and that of Naive Bayes and C4.5.

Name	Exh.	AdaSel.	NB	C4.5
agaricus	892.31	2.07	16.34	21.65
kr-vs-kp	265.63	3.68	10.07	31.13
hypothyroid	233.24	5.82	7.14	67.40
sick-euthy.	232.05	6.84	7.08	162.76
german	80.75	16.96	1.08	20.34
ionos	56.95	6.29	0.85	29.47

Table 8.1 shows the accuracy obtained on these datasets by two combinations of selectors with MadaBoost. The columns "Size" and "$\|H_{DS}\|$" show, for each dataset, its size and the number of all possible decision stumps respectively. As we can see easily, there is no significant difference between the accuracies obtained by these three methods. These results indicate that our sampling method is accurate enough. Moreover, even though the hypotheses produced are very simple (a weighted majority of ten depth-1 decision trees), the accuracies are comparable to that obtained using Naive Bayes, a learning algorithm that has been reported to be competitive with more sophisticated methods like decision trees. Once we have established that there is no loss in accuracy due to the use of sampling, we should check whether there is any gain in efficiency. Table 8.2 shows the running times of MadaBoost combined with two selection algorithms, exhaustive one (Exh.), and the sampling (AdaSel.) version. As we mentioned, we have also provided the running time of Naive Bayes and C4.5 for those datasets. The reason for doing that is that boosting is usually a slow process and thus, even though we are reducing its running time by using sampling, we still want to know how is this running time compared to the running time of an state of the art learning method like C4.5 and to a very fast algorithm like Naive Bayes whose running time scales exactly linearly in the training set.

Let us discuss about the results. First, one can easily see that using all the dataset is a very slow process, particularly for large and complex datasets that need a large number of decision stumps. The running time of Exh. is a function of the dataset size, the number of decision stumps considered (which depends on how many attributes the dataset has and their range of values), and the number of boosting rounds.

For the algorithms using sampling, we can see that the running time has been all greatly reduced. For instance, for MadaBoost with sampling, the running time is, on average, reduced about by 1/42 of the Exh. case when including Agaricus, which is a particularly good case where the running time is reduced by 1/431; even if we do not count Agaricus, the average reduction of running time is about 1/22.

Surprisingly enough, for the sampling versions the fastest dataset becomes the largest one, Agaricus. It is due to the particular structure of this dataset. First, the decision stump size $\|H\|$ is of reasonable size. Second, during all the 10 boosting iterations one can find hypothesis with accuracy larger than 70% and thus, the sample sizes needed are very small. This contrasts with datasets like German where a similar number of decision stumps is considered and, even though the dataset is less than 1/8 of Agaricus, the running time on German is 8 times larger. This is because for this dataset, after the third boosting iteration, even the best stump has accuracy smaller than 60% and this affects the efficiency of the sampling method.

With respect to C4.5, we can see that our algorithm is faster in all datasets. More specifically, MadaBoost using sampling is around 8 times faster than C4.5 in average. Concerning Naive Bayes, the sampling version is faster in the three largest datasets, around 3 times in average and slower in the three smallest, again around 3 times in average although averaging over all datasets both algorithms take more or less the same time.

8.5. CONCLUDING REMARKS

We have presented a new methodology for sampling that, while keeping all the theoretical guarantees of previous ones, it is applicable in a wider setting and moreover, it is very likely that it is useful in practice. The key point is to perform sampling sequentially and determine the time to stop sampling by a carefully designed stopping condition. We theoretically give both justification and efficiency analysis of our algorithms, and then provide experimental evidence of the advantage of our method. In order to give some concrete example, we fix one specific problem, i.e., the design of a base learner for a boosting algorithm, and discuss how our method can be used. There are many other applications of our method where it seems very plausible that it will succeed and this are further discussed in (Domingo et al., 2000) where also improved versions of this method are presented.

Notes

1. The only assumption is that we can obtain examples independently under the same distribution, a natural assumption that holds for all problems considered here.

References

Domingo, C., Gavaldà, R., and Watanabe, R. (2000). Adaptive Sampling Methods for Scaling Up Knowledge Discovery Algorithms. *Journal of Knowledge Discovery and Data Mining*, to appear.

Domingo, C. and Watanabe, O. (2000). Madaboost: A modification of Adaboost. In *Thirteenth Annual Conference on Computational Learning Theory, COLT'2000*.

Dougherty, J., Kohavi, R., and Sahami, M. (1995). Supervised and unsupervised discretization of continuous features. In *Twelfth International Conference on Machine Learning*.

Fayad, U. and Irani, K. (1993). Multi-interval discretization of continuous-valued attributes for classification learning. In *13th International Joint Conference on Artificial Intelligence*, pages 1022–1027.

Feller, W. (1950). *An introduction to probability theory and its applications*. John Willey and Sons.

Freund, Y. and Schapire, R. (1997). A decision-theoretic generalization of on-line learning and an application to boosting. *Journal of Computer and System Sciences*, 55(1):119–139.

John, G. H. and Langley, P. (1996). Static versus dynamic sampling for data mining. In *Second International Conference on Knowledge Discovery and Data Mining*.

Kivinen, J. and Mannila, H. (1994). The power of sampling in knowledge discovery. In *ACM SIGACT-SIGMOD-SIGACT Symposium on Principles of Database Theory*, pages 77–85.

Lipton, R. and Naughton, J. (1995). Query size estimation by adaptive sampling. *Journal of Computer and System Science*, 51:18–25.

Lipton, R. J., Naughton, J. F., Schneider, D. A., and Seshadri, S. (1995). Efficient sampling strategies for relational database operations. *Theoretical Computer Science*, 116:195–226.

Provost, F., Jensen, D., and Oates, T. (1999). Efficient progressive sampling. In *5th Int. Conf. on Knowledge Discovery and Data Mining*.

Quinlan, J. R. (1993). *C4.5: Programs for machine learning*. Morgan Kaufmann, San Mateo, California.

Toivonen, H. (1996). Sampling large databases for association rules. In *22nd Int. Conference on Very Large Databases*, pages 134–145.

Wald, A. (1947). *Sequential Analysis*. Wiley Mathematical, Statistics Series.

Chapter 9

PROGRESSIVE SAMPLING

Foster Provost

New York University, New York, NY 10012

fprovost@stern.nyu.edu

David Jensen

University of Massachusetts, Amherst, MA 01003

jensen@cs.umass.edu

Tim Oates

Massachusetts Institute of Technology, Cambridge, MA 02139

oates@cs.umass.edu

Abstract   Training with too much data can lead to substantial computational cost. Furthermore, the creation, collection, or procurement of data may be expensive. Unfortunately, the minimum sufficient training-set size seldom can be known a priori. We describe and analyze several methods for *progressive sampling*—using progressively larger samples as long as model accuracy improves. We explore several notions of efficient progressive sampling, including both methods that are asymptotically optimal and those that take into account prior expectations of appropriate data size. We then show empirically that progressive sampling indeed can be remarkably efficient.

Keywords: Sampling, progressive sampling, iterative sampling, learning curves, machine learning, data mining, instance selection.

9.1.   INTRODUCTION

Algorithms that construct general knowledge from data—induction algorithms—have the opportunity to choose training instances more and more intelligently as they "learn." This notion has intrigued researchers for decades (Simon and Lea, 1973; Winston, 1975; Quinlan, 1983; Provost and Buchanan, 1995). But why is it an important notion *practically*? Why not simply give our induction algorithms all the data? One reason is that increasing the amount of data leads to greater computational cost. Depending on the situation, running an induction algorithm on all the data may be considerably less efficient than running on a small subset. Furthermore, the creation, collection, or procurement of data may be costly. Seldom are the best data freely and easily accessible in abundance.

Unfortunately, induction algorithms face competing requirements for accuracy and efficiency. The requirement for accurate models often demands the use of large data sets that allow algorithms to discover complex structure and make accurate parameter estimates. The requirement for efficient induction demands the use of small data sets, because the computational complexity of even the most efficient induction algorithms is linear in the number of instances, and most algorithms are considerably less efficient.

In this chapter[1] we consider automated instance selection that is limited to selecting training sets of the minimum sufficient size; the selection of individual instances is random. This problem seems simple, when compared to picking and choosing just the right instances. However, it is worthwhile to understand the simple before moving on to the complex.

Specifically, we study *progressive sampling* methods. Progressive sampling attempts to maximize accuracy as efficiently as possible, starting with a small sample and using progressively larger ones until model accuracy no longer improves. A central component of progressive sampling is a *sampling schedule* $S = \{n_0, n_1, n_2, \ldots, n_k\}$ where each n_i is an integer that specifies the size of a sample to be provided to an induction algorithm. For $i < j$, $n_i < n_j$. If the data set contains N instances in total, $n_i \leq N$ for all i. There are three fundamental questions regarding progressive sampling.

1. What is an efficient sampling schedule?

2. How can convergence (i.e., that model quality no longer increases) be detected effectively and efficiently?

3. As sampling progresses, can the schedule be adapted to be more efficient?

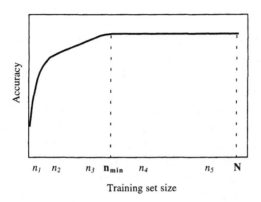

Figure 9.1   Learning curves and progressive samples

We discuss several ways to assess efficiency, and how various progressive sampling procedures fare with respect to each. Notably, we show that schedules in which the n_i increase geometrically are optimal in an asymptotic sense. We explore the question of optimal efficiency in an absolute sense: what is the most efficient schedule given one's prior expectations of convergence? Next, we address the crucial practical issue of convergence detection. We describe an interaction between the sampling schedule and the method of convergence detection, and we describe a practical alternative that avoids the worst aspects of the tradeoffs this interaction requires. We also discuss algorithms that schedule adaptively, based on knowledge of convergence and actual run-time complexity, obtained on the fly.

We then investigate empirically how a variety of schedules perform on large benchmark data sets. Finally, we discuss why progressive sampling is especially beneficial in cases where sampling from a large database is inefficient. We conclude that, in a wide variety of realistic circumstances, progressive sampling is preferable to analyzing all instances from a database. Surprisingly, it can be competitive even when the optimal sample size is known in advance.

9.2.   PROGRESSIVE SAMPLING

A *learning curve* (Figure 9.1) depicts the relationship between sample size and model accuracy. The horizontal axis represents n, the number of instances in a given training set, which can vary between zero and N, the total number of available instances. The vertical axis represents the accuracy of the model produced by an induction algorithm when given a training set of size n.

Learning curves typically have a steeply sloping portion early in the curve, a more gently sloping middle portion, and a plateau late in the curve. The middle portion can be extremely large in some curves (e.g., (Catlett, 1991a; Catlett, 1991b; Harris-Jones and Haines, 1997)) and almost entirely missing in others. The plateau occurs when adding additional data instances does not improve accuracy. The plateau, and even the entire middle portion, can be missing from curves when N is not sufficiently large. Conversely, the plateau region can constitute the majority of curves when N is very large. For example, in a recent study of two large business data sets, Harris-Jones and Haines (Harris-Jones and Haines, 1997) found that learning curves reach a plateau quickly for some algorithms, but small accuracy improvements continue up to N for other algorithms.

We assume that learning curves are well behaved. Specifically, we assume that the slope of a learning curve is monotonically non-increasing with n except for local variance. Locality is defined within a particular progressive sampling procedure.

Not all learning curves are well behaved. For example, theoretical analyses of learning curves based on statistical mechanics (Haussler et al., 1996; Watkin et al., 1993) have shown that sudden increases in accuracy are possible, particularly on small samples. However, empirical studies of the application of standard induction algorithms to large data sets—those of relevance to this chapter—have shown learning curves to be well behaved (Catlett, 1991b; Frey and Fisher, 1999; Harris-Jones and Haines, 1997; Oates and Jensen, 1997; Oates and Jensen, 1998). In addition, practical progressive sampling demands only that learning curves are well behaved at the level of granularity of the sampling schedule. Given the relatively course granularity of many schedules, sudden increases in accuracy can occur without impairing progressive sampling. Our empirical results in section 9.6 bear out these assumptions.

When a learning curve reaches its final plateau, we say it has *converged*. We denote the training set size at which convergence occurs as n_{min}.

Definition 1. Given a data set, a sampling procedure, and an induction algorithm, n_{min} is the size of the smallest sufficient training set. Models built with smaller training sets have lower accuracy than models built with from training sets of size n_{min}, and models built with larger training sets have no higher accuracy.

Figure 9.1 shows an example sampling schedule and its relation to a learning curve. Empirical estimates are necessary to determine n_{min} because the precise shape of a learning curve represents a complex in-

Compute schedule $S = \{n_0, n_1, n_2, \ldots, n_k\}$ of sample sizes
$n \leftarrow n_0$
$M \leftarrow$ model induced from n instances
while not *converged*
     recompute S if necessary
     $n \leftarrow$ next element of S larger than n
     $M \leftarrow$ model induced from n instances
end while
return M

Figure 9.2 Progressive sampling

teraction between the statistical regularities present in a given data set and the abilities of an induction algorithm to identify and represent those regularities. In general, these characteristics are not known in advance, nor is their interaction well understood. Thus, in many cases, n_{min} is nearly impossible to determine from theory. However, n_{min} can be approximated empirically by a progressive sampling procedure. We denote by $\hat{n}_{min}$ a procedure's approximation to n_{min}.

Figure 9.2 is a generic algorithm that defines the family of progressive sampling methods. An instance of this family has particular methods for selecting a schedule, for determining convergence, and for altering the schedule adaptively. The next three sections consider each of these methods in turn.

9.3. DETERMINING AN EFFICIENT SCHEDULE

We now discuss several alternative methods for selecting an efficient schedule. For now, we assume that progressive sampling is able to detect convergence and we assume that this detection can be performed efficiently (its worst-case run-time complexity is not worse than that of the underlying induction algorithm).

Several simple schedules are of interest. For example, later we will use for comparison the schedule composed of a single data set with all instances, $S_N = \{N\}$. We also will consider the simple schedule generated by an omniscient oracle, $S_O = \{n_{min}\}$.

John and Langley (John and Langley, 1996) define a progressive sampling approach we call *arithmetic sampling* using the schedule $S_a = n_0 + (i \cdot n_\delta) = \{n_0, n_0 + n_\delta, n_0 + 2n_\delta, \ldots, n_0 + k \cdot n_\delta\}$.[2] An example arithmetic schedule is $\{100, 200, 300, \ldots, n_k\}$. John and Langley compare arithmetic sampling with static sampling. *Static sampling* computes $\hat{n}_{min}$ without progressive sampling, based on a subsample's statistical similarity to the entire sample. For consistency, we consider static sampling as a degenerate progressive sampling procedure with schedule $S_{static} = \{\hat{n}_{min}\}$. John and Langley show that arithmetic sampling produces more accurate models than does static sampling. This is not surprising, given that n_{min} depends on the relationship between the data and the specific learning algorithm. In some cases, the difference between n_{min}'s for different learning algorithms is quite large (Harris-Jones and Haines, 1997).

Arithmetic sampling has an obvious drawback. If n_{min} is a large multiple of n_δ, then the approach will require many runs of the underlying induction algorithm. For example, if $n_{min} = 200,000$ and $n_0 = n_\delta = 100$, then 2000 runs will be necessary—more than half with $n > 100,000$ instances. John and Langley partially escape this difficulty by specifying the use of an incremental induction algorithm (e.g., a simple Bayesian classifier), whose run time depends only on the additional instances, rather than having to reprocess all prior instances. Unfortunately, S_a can be extremely inefficient for the vast majority of induction algorithms, which are not incremental. We compare arithmetic schedules to other approaches in section 9.6.

An alternative schedule escapes the limitations of arithmetic sampling. *Geometric sampling* uses the schedule S_g:

$$S_g = a^i \cdot n_0 = \{n_0, a \cdot n_0, a^2 \cdot n_0, a^3 \cdot n_0, \dots, a^k \cdot n_0\},$$

for some constants n_0 and a. An example geometric schedule is $\{100, 200, 400, 800, \dots\}$. The next section explains why geometric schedules are robust.

9.3.1.   Asymptotic Optimality

How do simple schedules compare to the schedule of an omniscient oracle, $S_O = \{n_{min}\}$? We now show that geometric sampling is an *asymptotically optimal* schedule. That is, in terms of worst-case time complexity, geometric sampling is equivalent to S_O.

For a given data set, let $f(n)$ be the (expected) run time of the underlying induction algorithm with n sampled instances. We assume that the asymptotic run-time complexity of the algorithm, $\Theta(f(n))$, is polynomial (no better than $O(n)$). Since most induction algorithms have $\Omega(n)$ run-time complexity, and many are strictly worse than $O(n)$, this assumption does not seem problematic. Under these assumptions, geometric sampling is asymptotically optimal.[3]

Theorem 1. For induction algorithms with polynomial time complexity $\Theta(f(n))$, no better than $O(n)$, if convergence also can be detected in $O(f(n))$, then geometric progressive sampling is asymptotically optimal among progressive sampling methods in terms of run time.

Proof: The run-time complexity of induction (with the learning algorithm in question) is $\Theta(f(n))$, where n is the number of instances in the data set. As specified in Definition 1, let n_{min} be the size of the smallest sufficient training set. The ideal progressive sampler has access to an oracle that reveals n_{min}, and therefore will use the simple, optimal schedule $S_O = \{n_{min}\}$. The run-time complexity of S_O is $\Theta(f(n_{min}))$. Notably, the run time of the optimal progressive sampling procedure grows with n_{min} rather than with the total number of instances N.

By the definition of S_g, geometric progressive sampling runs the induction algorithm on subsets of size $a^i \cdot n_0$ for $i = 0, 1, \ldots, b$, where $b+1$ is the number of samples processed before detecting convergence.

Now, we assume convergence is well detected, so

$$a^{b-1} \cdot n_0 < n_{min} \leq a^b \cdot n_0,$$

which means that

$$a^i \cdot n_0 < \frac{a^i}{a^{b-1}} \cdot n_{min},$$

for $i = 0, 1, \ldots, b$. Since $O(f(\cdot))$ is at best linear, the run time of S_g (on all subsets, including running the convergence-detection procedure) is

$$O(f(\sum_{i=0}^{b} \frac{a^i}{a^{b-1}} \cdot n_{min})).$$

This is

$$O(f(a \cdot n_{min} \cdot (1 + \frac{1}{a} + \frac{1}{a^2} + \cdots + \frac{1}{a^b}))).$$

The final, finite sum is less than the corresponding infinite sequence. Because $a > 1$ for S_g, this converges to a constant that is independent of n_{min}. Since $O(f(\cdot))$ is polynomial, the overall run time of S_g is $O(f(n_{min}))$. Therefore, progressive sampling asymptotically is no worse than the optimal progressive sampling procedure, S_O, which runs the induction algorithm only on the smallest sufficient training set. $\square$

Because it is simple and asymptotically optimal, we propose that geometric sampling, for example with $a = 2$, is a good default schedule for mining large data sets. We provide empirical results testing this proposition and discuss further reasons in section 9.6. First we explore further the topic of optimal schedules.

9.3.2.   Optimality with Respect to Expectations of Convergence

Comparisons with S_N and S_O represent two ends of a spectrum of expectations about n_{min}. Can optimal schedules be constructed given expectations between these two extremes?

Prior expectations of convergence can be represented as a probability distribution. For example, let $\Phi(n)$ be the (prior) probability that convergence requires more than n instances. By using $\Phi(n)$, and the run-time complexity of the learning algorithm, $f(n)$, we can compare the expected cost of different schedules. Thus we can make statements about *expectation-based optimality*. The schedule with minimum expected cost is optimal in a practical sense.

In many cases there may be no prior information about the likelihood of convergence occurring for any given n. Assuming a uniform prior over all n yields $\Phi(n) = (N - n)/N$.[4] At the other end of the spectrum, oracle optimality can be cast as a special case of expectation-based optimality, wherein $\Phi(n) = 1$ for $n < n_{min}$, and $\Phi(n) = 0$ for $n \geq n_{min}$ (i.e., n_{min} is known).

The better the information on the likely location of n_{min}, the lower the expected costs of the schedules produced. For example, we know that learning curves almost always have the characteristic shape shown in Figure 9.1, and that in many cases $n_{min} \ll N$. Rather than assume a uniform prior, it would be more reasonable to assume a more concentrated distribution, with low values for very small n, and low values for very large n. For example, data from Oates and Jensen (Oates and Jensen, 1997; Oates and Jensen, 1998) show that the distribution of the number of instances needed for convergence over a large set of the UCI databases (Blake et al., 1999) is roughly log-normal.

Given such expectations about n_{min}, is it possible to construct the schedule with the minimum expected cost of convergence? This seems a daunting task. For each value of n, from 1 to N, a model can either be built or not, leading to 2^N possible schedules. However, identification of the optimal schedule can be cast in terms of dynamic programming (Bellman, 1957), yielding an algorithm that requires $O(N^2)$ space and $O(N^3)$ time.

Let $f(n)$ be the cost of building a model with n instances and determining whether accuracy has converged. As described above, let $\Phi(n)$ be the probability that convergence requires more than n instances. Clearly, $\Phi(0) = 1$. If we let $n_0 = 0$, then the expected cost of convergence by following schedule S is given by the following equation:

Table 9.1 Expected costs of various schedules given $N = 10$, $f(n) = n^2$ and a uniform prior.

Schedule	Cost
$S_1 = \{1, 2, 3, 4, 5, 6, 7, 8, 9, 10\}$	121.0
$S_2 = \{10\}$	100.0
$S_3 = \{2, 6, 10\}$	72.8

$$C = \sum_{i=1}^{k} \Phi(n_{i-1}) f(n_i)$$

To better understand what this equation captures, consider a simple example in which $N = 10$ and $S = \{3, 7, 10\}$. The value of C is as follows:

$$C = \sum_{i=1}^{k} \Phi(n_{i-1}) f(n_i) = \Phi(0) f(3) + \Phi(3) f(7) + \Phi(7) f(10)$$

With probability 1 ($\Phi(0) = 1$), an initial model will be built with 3 instances at a cost of $f(3)$. If more than 3 instances are required for convergence, an event that occurs with probability $\Phi(3)$, a second model will be built with 7 instances at a cost of $f(7)$. Finally, with probability $\Phi(7)$ more than 7 instances are required for convergence and a third model will be built with all 10 instances at a cost of $f(10)$. The expected cost of a schedule is the sum of the cost of building each model times the probability that the model will actually need to be constructed.

Consider another example in which $N = 10$; the uniform prior is used ($\Phi(n) = (N - n)/N$), and $f(n) = n^2$. The costs for three different schedules are shown in Table 9.1. The first schedule, in which a model is constructed for each possible data set size, is the most expensive of the three shown. The second schedule, in which a single model is built with all of the instances, also is not optimal in the sense of expected cost given a uniform prior. The third schedule shown has the lowest cost of all $2^{10} = 1024$ possible schedules for this problem.

Given N, f and Φ, we want to determine the schedule S that minimizes C. That is, we want to find the optimal schedule. As noted previously, a brute force approach to this problem would explore exponentially many schedules. We take advantage of the fact that optimal schedules are composed of optimal sub-schedules to apply dynamic pro-

Table 9.2 Optimal schedules for $N = 500$ and various $f(n)$ given a uniform prior.

N = 500		
$f(n)$	Schedule	Cost
n	{500}	500
$n^{1.5}$	{27, 248, 500}	9,470
n^2	{10, 99, 312, 500}	181,775
n^3	{2, 17, 77, 204, 375, 500}	70,096,920

gramming to this problem, resulting in a polynomial time algorithm. Let $m[i, j]$ be the cost of the minimum expected-cost schedule of all samples in the size range $[i, j]$, with the requirement that samples of i and of j instances be included in $m[i, j]$. The cost of the optimal schedule given a data set containing N instances is then $m[0, N]$. The following recurrence can be used to compute $m[0, N]$:

$$m[i, j] = \min \begin{cases} \Phi(i)f(j) \\ \min_{i<k<j} m[i, k] + m[k, j] \end{cases}$$

Both the bottom-up table-based and top-down memoized implementations of this equation require $O(N^2)$ space to store the values of m and $O(N^3)$ time.

The results of applying dynamic programming to determine the optimal schedules for $N = 500$ and various $f(n)$ are shown in Table 9.2, which shows the optimal schedules for each $f(n)$ along with the associated costs. Note that the optimal schedule depends on $f(n)$. In general, the larger the complexity of $f(n)$, the more frequently the schedules indicate that models should be constructed. Second, although the n_i in any given schedule are quasi-geometrically increasing, the multiplicative factor is by no means a constant. In fact, the factor seems to decrease dramatically near the ends of the schedules.

As stated earlier, the running time of the algorithm that determines the optimal schedule is $O(N^3)$. This clearly is impractical for data sets with millions of instances. Fortunately, the running time actually is cubic in the number of data set sizes at which a model can be built, which can be a small fraction of N if one is willing to sacrifice precision in the placement of model construction points (for example, looking only at multiples of 100 or 1000 instances).

Non-uniform priors can strongly affect the schedules produced by dynamic programming. Table 9.3 shows optimal schedules based on a

Table 9.3 Optimal schedules for $N = 500$ and various $f(n)$ given a log-normal prior.

	$N = 500$	
$f(n)$	Schedule	Cost
n	$\{57, 143, 285, 500\}$	183
$n^{1.5}$	$\{36, 93, 180, 318, 500\}$	2,355
n^2	$\{16, 50, 108, 191, 318, 500\}$	33,473
n^3	$\{4, 23, 50, 93, 149, 231, 348, 500\}$	9,026,006

log-normal prior such that $\Phi(\log(x)) = 1 - N(x)$ where $N(x)$ is the cumulative density of a normal distribution with mean $\log(50)$ and standard deviation 1. Despite the fact that the schedules in this table call for more models to be constructed than schedules based on the uniform prior (see Table 9.2), the expected costs are lower because more precise information about the location of n_{min} is available.

We call progressive sampling based on the optimal schedule determined by this dynamic programming procedure, DP sampling.

9.4. DETECTING CONVERGENCE

A key assumption behind all the progressive sampling procedures discussed above is that convergence can be detected accurately and efficiently. We present some preliminary results below, and we believe that convergence detection remains an open problem on which significant research effort should be focused.

Convergence detection is fundamentally a statistical judgment, irrespective of the specific convergence criterion or the method to estimate whether that criterion has been met. In their paper on arithmetic sampling, John and Langley (John and Langley, 1996) model the learning curve as sampling progresses. They determine convergence using a stopping criterion modeled after the work of Valiant (Valiant, 1984). Specifically, convergence is reached when $Pr((acc(N) - acc(n_i)) > \epsilon) \leq \delta$, where $acc(x)$ is the accuracy of the model that an algorithm produces after seeing x instances, ϵ refers to the maximum acceptable decrease in accuracy, and δ is a probability that the maximum accuracy difference will be exceeded on any individual run. A model of the learning curve is used to estimate $acc(N)$.

Statistical estimates of convergence face several challenges. First, statistical estimation of complete learning curves is fraught with difficulties. Actual learning curves often require a complex functional form to estimate accurately. The curve shown in Figure 9.1 has three regions of behavior—a primary rise, a secondary rise, and a plateau. Most simple functional forms (e.g., the power laws used by Frey and Fisher (Frey and Fisher, 1999) and by John and Langley (John and Langley, 1996)) generally cannot capture all three regions of behavior, often causing the estimated curves to converge too quickly or never to converge. Estimating convergence is generally more challenging than fitting earlier parts of the curve, and even fairly small errors can mislead progressive sampling methods. For example, a power law may fit the early part of the curve well, but will represent a long, final plateau as a long (perhaps slight) incline.

More important, the need for accurate statistical estimates must be balanced against the goal of computational efficiency. Statistical estimates of convergence are aided by increasing the number of points in a schedule, but this directly impairs efficiency. Even worse, determining convergence is aided most by samples for which $n_i > n_{min}$, because these points will most assist the statistical determination that a plateau has been reached. Of course, these are the very sample sizes that, for efficiency reasons, progressive sampling schedules should avoid.

The most promising approach we have yet identified for convergence detection—*linear regression with local sampling* (LRLS) —begins at the latest scheduled sample size n_i and samples l additional points in the local neighborhood of n_i. These points are then used to estimate a linear regression line, whose slope is compared to zero. If the slope is sufficiently close to zero, convergence is detected. LRLS takes advantage of a common property of learning curves: the slope of the line tangent to the curve constantly decreases. If LRLS ever estimates that the slope is zero, it is unlikely to ever become non-zero for larger n.

LRLS increases progressive sampling's complexity by a constant factor. In section 9.6 we show that it approximates n_{min} with consistently high accuracy, though at the cost of significant increases in absolute run time.

9.5.   ADAPTIVE SCHEDULING

Determining optimal schedules for DP sampling requires a model of the probability of convergence and a model of the run-time complexity of the underlying induction algorithm. In the previous section we assumed that these were known in advance, but our prior knowledge may be less than perfect. We now show that a progressive sampling procedure can build both of these models adaptively. The key insight is that a progressive sampling algorithm can obtain substantial amounts of information cheaply by including small samples in its schedule.

As we argued above, the assumption of uniform probabilities of convergence for all n is probably incorrect. However, progressive sampling algorithms can model the convergence probability dynamically. For example, a progressive sampling algorithm might assume that the accuracy of a particular algorithm on a particular data set can be modeled by a power law. A simple power law is shown by Frey and Fisher (Frey and Fisher, 1999) to model learning curves better than a variety of alternatives, and a similar approach is used by John and Langley (John and Langley, 1996) to determine convergence (see section 9.4). Such a modeling approach could allow a progressive sampling procedure to improve the efficiency of its schedule adaptively during execution.

This enhances the benefits of quasi-geometric schedules that take many small samples early, when the learning curve has the most variation, but while running the induction algorithm is inexpensive and the impact of a suboptimal schedule is low. When the cost of running the induction algorithm increases, the model of the convergence probabilities will be much better, and thus the actual performance will be closer to the true optimal.

The second assumption of DP sampling is that we have an accurate model of the run-time complexity (in n) of the underlying induction algorithm. Run-time complexity models are not always easy to obtain. For example, our empirical results below use the decision-tree algorithm C4.5 (Quinlan, 1993), for which reported time complexity varies widely.

Moreover, DP sampling requires the *actual* run-time complexity for the problem in question, rather than a *worst-case* complexity. We obtained empirical estimates of the complexity of C4.5 on the data sets used below, and found $O(n^{1.22})$ for LED, $O(n^{1.37})$ for WAVEFORM, and $O(n^{1.38})$ for CENSUS.[5]

As with learning curve estimation, progressive sampling can determine the actual run-time complexity dynamically as the sampling progresses. As before, early in the schedule, with small samples, suboptimal scheduling due to an incorrect time-complexity model will have little overall effect. As the samples grow and bad estimates would be costly, the time-complexity model becomes more accurate.

9.6. EMPIRICAL COMPARISON OF SAMPLING SCHEDULES

We have shown that, in principle, progressive sampling can be efficient. We now evaluate whether progressive sampling can be used for practical scaling. We hypothesize that geometric sampling and DP sampling are considerably less expensive than using all the data when convergence is early, and not too much more expensive when convergence is late. We further hypothesize that, for large data sets and non-incremental algorithms, these versions of progressive sampling are significantly better than arithmetic progressive sampling. We also investigate how well simple geometric sampling compares with DP sampling.

We compare progressive sampling with several different schedules: $S_N = \{N\}$, a single sample with all the instances; $S_O = \{n_{min}\}$, the optimal schedule determined by an omniscient oracle; $S_a = 100 + (100 \cdot i)$, arithmetic sampling with $n_0 = n_\delta = 100$; $S_g = 100 \cdot 2^i$, geometric sampling with $n_0 = 100$ and $a = 2$; and S_{dp}, DP sampling with schedule recomputation after dynamic estimation of priors and run-time complexity.

Of the three progressive sampling methods, only DP sampling revises its schedules. In order to determine the probability of convergence, with S_{dp} progressive sampling estimates the learning curve dynamically by applying linear regression to estimate the parameters of a power law. We also dynamically estimate $f(n)$, the actual time complexity of the underlying induction algorithm, under the assumption that $f(n) = c_0 \cdot n^c$, based on the observed run time on the samples taken so far. Progressive sampling with S_{dp} first builds models on 100, 200, 300, 400 and 500 instances to get an initial estimate of the learning curve and actual runtime complexity. From these, the method estimates the convergence probability distribution and the complexity of the algorithm. Then, as described in Figure 9.2, it iteratively checks for convergence, rebuilds the schedule by recomputing the distribution and the complexity with the latest information, and then produces a new classifier.

For our first set of experiments, we assume that convergence can be detected accurately and without cost. The progressive sampling algorithms each had access to a function that returns FALSE if $n < n_{min}$ and TRUE if $n \geq n_{min}$. The function could only be accessed as a way of testing convergence, not as a method of schedule construction.

To instantiate the oracle and the costless convergence detection procedure, we determined n_{min} empirically by analyzing the full learning curve and applying a technique developed by Oates and Jensen (Oates and Jensen, 1997). The technique takes sets of three adjacent points on a learning curve (we used points from an arithmetic schedule with $n_\delta = 1000$), averages their accuracies, and then compares that accuracy with the accuracy on all N instances. The oracle chooses the first set (the set furthest to the left of the learning curve) for which average ac-

Table 9.4 Computation time for several progressive sampling methods

Data set	Full	Arith	Geo	DP	Oracle
LED	16.40	1.77	0.55	0.56	0.18
CENSUS	16.33	59.68	5.57	5.41	2.08
WAVEFORM	425.31	1230.00	41.84	50.57	22.12

curacy is within 1% of the accuracy on all instances. The middle point of that set is n_{min}.

For our experiments we used 3 large data sets from the UCI repository (Blake et al., 1999): LED, WAVEFORM (with 10% noise), and CENSUS (adult). For LED and WAVEFORM we used 100,000 instances, and for CENSUS 32,000. Based on the technique above, we found that for LED $n_{min} = 2,000$, for WAVEFORM $n_{min} = 12,000$, and for CENSUS $n_{min} = 8,000$.

Table 9.4 shows running times in seconds averaged over 10 runs on each of the three data sets. Before each run the order of the instances was randomized. All runs were performed on a 400MHz DEC alpha.

The results confirm our hypotheses regarding the relative efficiency of progressive sampling. Geometric progressive sampling is between three and thirty times faster than learning with all the data. It also is between three and thirty times faster than arithmetic progressive sampling. On the other hand, geometric progressive sampling is only two to three times slower than the oracle-optimal procedure, S_O.

Perhaps surprisingly, DP progressive sampling is not faster than geometric sampling. Constructing the DP schedule takes less than $\frac{1}{10}$th of a second, so the DP overhead is not responsible. DP would be faster with more precise knowledge of n_{min}. Indeed, non-adaptive DP with uniform priors consistently produces poor schedules on these data sets. Adaptive modeling of the probability of convergence is critical, but apparently we must devise a more accurate technique if we want to beat geometric sampling.[6]

Now we move on to the detection of convergence. We used LRLS to estimate convergence for adaptive DP sampling. For these experiments, at each schedule point, ten samples were taken for convergence estimation, $l = 10$, and convergence was indicated the first time that the 95% confidence interval on the slope of the regression line included zero. Table 9.5 compares the accuracy of the models built using LRLS convergence detection, with the accuracy using optimal convergence detection from above (i.e., it can query the oracle); each value represents

Table 9.5 Mean accuracy for DP with LRLS and with optimal (free) convergence detection

Data set	DP-LRLS	DP-free
LED	73.02	72.91
CENSUS	84.83	85.38
WAVEFORM	76.88	76.36

Table 9.6 Computation times with the full data set, with LRLS, with free convergence detection, and with the oracle-determined optimal schedule

Data set	Full	DP-LRLS	DP-Free	Oracle
LED	16.40	11.27	0.56	0.18
CENSUS	16.33	9.22	5.41	2.08
WAVEFORM	425.31	142.92	50.57	22.12

the average of 10 runs where instances were randomized prior to each run.

LRLS identifies convergence accurately. In most cases, LRLS correctly identifies n_b as the first schedule point after n_{min}, and in nearly all other cases convergence is identified in a sample for which $n_b > n_{min}$. In no data set was the mean accuracy at estimated convergence statistically distinguishable from the accuracy on n_{min} instances, the point of true convergence.

However, LRLS has a large effect on absolute computation time. The additional sampling and executions of the induction algorithm create a large (constant) factor increase in the total computational cost. Table 9.6 compares the run time with LRLS to running with "free" convergence detection, to running on the full data set, and to knowing n_{min} in advance.

These costs could be reduced by more efficient estimation of convergence. In addition, the costs of running on the full data set would increase dramatically if N increased, while this would not affect the computational cost of any form of progressive sampling. Thus, as the total size of data sets increases, progressive sampling becomes more attractive.

9.7. DISCUSSION

Our presentation does not highlight all the benefits of progressive sampling. For example, our analysis assumes that sampling from a large database is instantaneous. As this assumption is relaxed, the relative efficiency of progressive sampling becomes better. Progressive sampling can take advantage of data as they arrive, effectively creating a pipelined induction process. For standard induction based on slow data access, the CPU sits idle and waits for the sampling to complete, and then runs the induction algorithm on the resultant data set. Progressive sampling can immediately get started on the first sample points, computing its first estimates of the learning curve and $f(n)$. Thereafter, sampling first fills up a test-set buffer, so that when induction each subset is finished, the test-set buffer (containing data for the next subset) is used first to estimate the accuracy. Then the test set buffer can be shifted into the training buffer, and so on. Moreover, the slower the sampling, the more work can be done on convergence detection. With very slow sampling, the efficiency of progressive sampling will be the same as if n_{min} were known a priori.

The method of Musick et al. (Musick et al., 1993) for determining the best attribute at each decision-tree node can be seen as an instance of the generic progressive sampling algorithm shown in Figure 9.2, if we regard each node of the decision tree as an individual induced model. Specifically, based on an analysis of information gain and its statistical properties, they compute an estimate of the sample size needed to have less than a specified loss in information. However, because this estimate can overshoot n_{min} greatly, they then calculate n_δ for an efficient arithmetic schedule, and revise the estimate after executing each schedule point. Other sequential multi-sample learning methods (Provost and Kolluri, 1999) are degenerate instances of progressive sampling, typically using fixed arithmetic schedules and treating convergence detection simplistically, if at all.

For this chapter, we have considered only drawing random samples from the larger data set. We believe that the ideas of progressive sampling apply to other methods of selecting instances, but have not yet studied the general case. Methods for active sampling , choosing subsequent samples based upon the models learned previously, are of particular interest. A classic example of active sampling is windowing (Quinlan, 1983), wherein subsequent sampling chooses instances for which the current model makes errors. Active sampling changes the learning curve. For example, on noisy data, windowing learning curves are notoriously ill behaved: subsequent samples contain increasing amounts of noise, and performance often decreases as sampling progresses (Fürnkranz, 1998). It would be interesting to examine more closely the use of the techniques outlined above in the context of active sampling, and the potential synergies.

9.8.   CONCLUSION

With this work we have made substantial progress toward efficient progressive sampling. We have shown that if convergence detection can be done very efficiently, then progressive sampling can be far better than learning from all the data, and almost as efficient as being given the minimum sufficient training set by an oracle. We have shown that convergence detection can be done effectively and moderately efficiently. We also have shown that geometric sampling is remarkably robust: its efficiency is insensitive to the number of points in the schedule (unlike arithmetic sampling); it is asymptotically no worse than knowing the point of convergence in advance, and in practice it performs as well as much more complicated adaptive scheduling.

What is left are two well-defined challenges for future research on instance selection: increase the efficiency of convergence detection, and devise an accurate method for estimating points of convergence from partial learning curves. Note that although the methods may differ, these two tasks apply generally to any instance selection procedure that has as its goal minimizing the number of instances needed.

Acknowledgments

This research is supported in part by DARPA and AFOSRF under contract numbers DARPA/AFOSRF 49620-97-1-0485 and DARPAN66001-96-C-8504. The U.S. Government is authorized to reproduce and distribute reprints for governmental purposes notwithstanding any copyright notation hereon. The views and conclusions contained herein are those of the authors and should not be interpreted as necessarily representing the official policies or endorsements either expressed or implied, of DARPA, AFOSRF or the U.S. Government.

Notes

1. This chapter, except for cosmetic changes and the omission of one section, is identical to a prior paper (Provost et al., 1999).

2. The first and last points in the schedule, n_0 and n_k, depend on external factors such as N and the induction program's (fixed) computational overhead.

3. We follow the reasoning of Korf, who shows that progressive deepening is an optimal schedule for conducting depth-first search when the smallest sufficient depth is unknown (Korf, 1985) (Provost, 1993).

4. This implies that $\Phi(N) = 0$ and thus that $n_{min} \leq N$.

5. We followed a similar procedure to Frey and Fisher (Frey and Fisher, 1999) and assumed that the running time could be modeled by: $y = c_0 \cdot n^c$, gathered samples of CPU time required to build trees on 1,000 to 100,000 instances in increments of 1,000, then took the log of both the CPU time and the number of instances, ran linear regression, and used the resulting slope as an estimate of c.

6. Experiments with increasingly precise knowledge of n_{min} show DP sampling to be less efficient than geometric sampling even for prior distributions centered on n_{min}, if the distributions are wide. Of course, if the distributions are very narrow, DP sampling comes very close to S_O, the optimal schedule.

References

Bellman, R. E. (1957). *Dynamic programming.* Princeton University Press.

Blake, C., Keogh, E., and Merz, C. (1999). UCI repository of machine learning databases. http://www.ics.uci.edu/~mlearn/MLRepository.html.

Catlett, J. (1991a). Megainduction: A test flight. In *Proceedings of the Eighth International Workshop on Machine Learning*, pages 596–599. Morgan Kaufmann.

Catlett, J. (1991b). *Megainduction: Machine learning on very large databases.* PhD thesis, School of Computer Science, University of Technology, Sydney, Australia.

Frey, L. J. and Fisher, D. H. (1999). Modeling decision tree performance with the power law. In Heckerman, D. and Whittaker, J., editors, *Proceedings of the Seventh International Workshop on Artificial Intelligence and Statistics.* San Francisco, CA: Morgan Kaufmann.

Fürnkranz, J. (1998). Integrative windowing. *Journal of Artificial Intelligence Research*, 8:129–164.

Harris-Jones, C. and Haines, T. L. (1997). Sample size and misclassification: Is more always better? Working Paper AMSCAT-WP-97-118, AMS Center for Advanced Technologies.

Haussler, D., Kearns, M., Seung, H. S., and Tishby, N. (1996). Rigorous learning curve bounds from statistical mechanics. *Machine Learning*, 25:195–236.

John, G. and Langley, P. (1996). Static versus dynamic sampling for data mining. In *Proceedings of the Second International Conference on Knowledge Discovery and Data Mining*, pages 367–370. AAAI Press.

Korf, R. (1985). Depth-first iterative deepening: An optimal admissible tree search. *Artificial Intelligence*, 27:97–109.

Musick, R., Catlett, J., and Russell, S. (1993). Decision theoretic subsampling for induction on large databases. In *Proceedings of the Tenth International Conference on Machine Learning*, pages 212–219, San Mateo, CA. Morgan Kaufmann.

Oates, T. and Jensen, D. (1997). The effects of training set size on decision tree complexity. In Fisher, D., editor, *Machine Learning: Proceedings of the Fourteenth International Conference*, pages 254–262. Morgan Kaufmann.

Oates, T. and Jensen, D. (1998). Large data sets lead to overly complex models: an explanation and a solution. In Agrawal, R. and Stolorz, P., editors, *Proceedings of the Fourth International Conference on Knowledge Discovery and Data Mining (KDD-99)*, pages 294–298. Menlo Park, CA: AAAI Press.

Provost, F. and Buchanan, B. (1995). Inductive policy: The pragmatics of bias selection. *Machine Learning*, 20:35–61.

Provost, F., Jensen, D., and Oates, T. (1999). Efficient progressive sampling. In *Proceedings of the SIGKDD Fifth International Conference on Knowledge Discovery and Data Mining*.

Provost, F. and Kolluri, V. (1999). A survey of methods for scaling up inductive algorithms. *Data Mining and Knowledge Discovery*, 3(2):131–169.

Provost, F. J. (1993). Iterative weakening: Optimal and near-optimal policies for the selection of search bias. In *Proceedings of the Eleventh National Conference on Artificial Intelligence*, pages Menlo Park, CA, 749-755. AAAI Press.

Quinlan, J. (1983). Learning efficient classification procedures and their application to chess endgames. In Michalski, R., J., C., and Mitchell, T., editors, *Machine Learning: An AI approach*, pages 463–482. Morgan Kaufmann., Los Altos, CA.

Quinlan, J. R. (1993). *C4.5: Programs for Machine Learning*. Morgan Kaufmann, San Mateo, California.

Simon, H. and Lea, G. (1973). Problem solving and rule induction: A unified view. In Gregg, editor, *Knowledge and Cognition*, pages 105–127. Lawrence Erlbaum Associates, New Jersey.

Valiant, L. G. (1984). A theory of the learnable. *Communications of the ACM*, 27(11):1134–1142.

Watkin, T., Rau, A., and Biehl, M. (1993). The statistical mechanics of learning a rule. *Reviews of Modern Physics*, 65:499–556.

Winston, P. H. (1975). Learning structural descriptions from examples. In Winston, P. H., editor, *The Psychology of Computer Vision*, pages 157–209. New York: McGraw-Hill.

Chapter 10

SAMPLING STRATEGY FOR BUILDING DECISION TREES FROM VERY LARGE DATABASES COMPRISING MANY CONTINUOUS ATTRIBUTES

Jean-Hugues Chauchat and Ricco Rakotomalala

Laboratory ERIC - University Lumiére Lyon

5, av. Pierre Mendes-France

F-69676 Bron, France

{chauchat,rakotoma}@univ-lyon2.fr

Abstract   We propose a fast and efficient sampling strategy to build decision trees from a very large database, even when there are many continuous attributes which must be discretized at each step. Successive samples are used, one on each tree node. After a brief description of two fast sequential simple random sampling methods, we apply elements of statistical theory in order to determine the sample size that is sufficient at each step to obtain a decision tree as efficient as one built on the whole database. Applying the method to a simulated database (virtually infinite size), and to five usual benchmarks, confirms that when the database is large and contains many numerical attributes, our strategy of fast sampling on each node (with sample size about $n = 300$ or 500) speed up the mining process while maintaining the accuracy of the classifier.

Keywords: Decision tree, data mining, sampling, discretization.

10.1.   INTRODUCTION

In this chapter we propose a fast and efficient sampling strategy to build decision trees from a very large database, even when there are many continuous attributes which must be discretized at each step.

Decision trees, and more generally speaking decision graphs, are efficient and simple methods for supervised learning. Their "step by step" characteristic allows us to propose a strategy using successive samples, one on each tree node. In that way, one of the decision tree method most limiting aspects is overcome.

Working on samples is especially useful in order to analyze very large databases, in particular when these include a number of continuous attributes which must be discretized at each step. Since each discretization requires to sort the data set, this is very time consuming. Section 10.2 outlines the general decision tree method and the numerical attributes discretization problem.

In Section 10.3, we describe our new sampling strategy at each step, we give a brief description of two fast sequential random sampling methods and then we apply elements of statistical theory in order to determine the sample size that is sufficient at each step to obtain a decision tree as efficient as one built on the whole database.

In Section 10.4, we apply the whole method to a simulated database (virtually infinite size), and then to five classical benchmarks in machine learning. The results confirm that when the database is large and contains many numerical attributes, our strategy of fast sampling on each node (with sample size about $n = 300$ or 500) reduces drastically learning time while maintaining the generalization error rate. Conclusion and future work are in the fifth section.

10.2.   INDUCTION OF DECISIONR TREES

10.2.1.   Framework

Decision trees (Breiman et al., 1984), and more generally speaking decision graphs (Zighed and Rakotomalala, 2000), are efficient, step by step, and simple methods for supervised classification. Supervised classification means that a classification pre-exists and is known for each record in the (training) database we are working on: the patient has been cured, or not; the client has accepted a certain offer, or not; the machine breaks down, or not. Those situations have two values; sometimes there are three or more. The final objective is to learn how to assign a new record to its true class, knowing the available attributes (age, sex, examinations results, etc.) (Mitchell, 1997).

The wide utilization of decision tree method is based on its simplicity and ease of use. One is looking for a dataset partition represented by a lattice graph (Figure 10.1) which is easily interpretable. This partition must minimize a certain criterion. Generic algorithms (CART, ChAID, etc.) make local optimization. To start with, we are looking for one

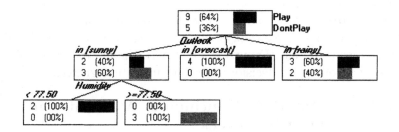

Figure 10.1 Decision tree on weather dataset

Table 10.1 Weather dataset

Outlook	Temp (^{o}F)	Humidity	Windy	Class
sunny	75	70	yes	play
sunny	80	90	yes	don't play
sunny	85	85	no	don't play
sunny	72	95	no	don't play
sunny	69	70	no	play
overcast	72	90	yes	play
overcast	83	78	no	play
overcast	64	65	yes	play
overcast	81	75	no	play
rain	71	80	yes	don't play
rain	65	70	yes	don't play
rain	75	80	no	play
rain	68	80	no	play
rain	70	96	no	play

attribute which allows us to divide the whole data set into subsets so as to link the qualitative variable (as much as possible) with the given classes. Thus the first "node" of the tree is constructed, followed by the same procedure for each subset, and so on. For example, let us consider a tiny dataset (Table 10.1) that we use again and again to illustrate decision tree method (Quinlan, 1993). Entirely fictitious, it supposedly concerns the conditions that are suitable for playing an unspecified game. A decision tree to predict the class attribute (play or not) may be built by recursive splitting (Figure 10.1).

Decision tree algorithms are very easy to implement, using a simple recursive procedure. In spite of their simplicity, they have a very good predictive power compared with more complex methods such as neural network (Michie et al., 1994). One can assert that, as with the linear model in regression, the decision tree is a reference method that enables one to get to know the complex links between attributes and classification we are learning. Nowadays, as Knowledge Discovery in Databases (KDD) is growing fast (Fayyad et al., 1996), the number of studies and software is growing on decision trees and induction graphs.

10.2.2.   Handling Continuous Attributes in Decision Trees

Most training-by-examples symbolic induction methods have been designed for categorical attributes, with finite value sets (Mitchell, 1997). For instance "sex" has two values: male or female. However, when we want to use continuous attributes (income, age, blood pressure, etc.), we must divide the value set in intervals so as to convert the continuous variable into a discrete one. This process is named "discretization" .

The first methods for discretization were relatively simple, and few papers have been published to evaluate their effects on machine learning results. Recently, much theoretical research has been done on this issue. The general problem has been clearly formulated and several discretization methods are now in use (Dougherty et al., 1995), (Zighed et al., 1999).

Initial algorithms processed "global discretization", during the pre-processing stage: each continuous attribute was converted to a discrete one; after which, a regular symbolic learning method was used (Frank and Witten, 1999). The process is as follows: each continuous variable is first of all sorted, then several cutting points are tested so as to find the subdivision which is the best according the class attribute, that is a subdivision which optimizes a splitting measure (entropy gain, chi-square, purity measure, etc.). During this pre-process, one is looking, at the same time, for the number of intervals and their boundaries. Working on large databases, these methods produce sometimes too many intervals. Using MDLPC algorithm (Fayyad and Irani, 1993) on "covtyp" database (Bay, 1999), for instance, we obtained as many as 96 intervals (!) for some attributes.

In the particular case of decision trees, a "local discretization" strategy is possible, whereby the optimal binary split on each node is sought. This strategy has two benefits: first, it is not necessary to determine how many intervals should be created as each split creates two intervals; the same variable can be later reintroduced in the tree with other discretization threshold; second, interaction among attributes is accounted for. Consider the case of the attribute "humidity" (Table 10.1): the

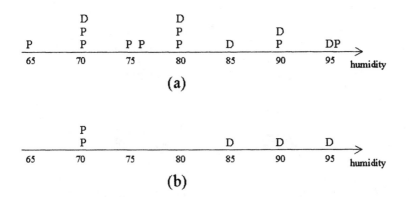

Figure 10.2 Discretization of "humidity" during pre-processing stage according to the class attribute (**Play**, **Don't play**) (a); discretization of the same attribute on a node, after filtering examples with the proposition "Outlook=Sunny" (b)

first splitting attempt proves unsatisfactory (Figure 10.2-a). Yet, after splitting the root node with "outlook", on the subset described by "outlook=sunny", a split on "humidity" produces an optimal solution: the groups are "pure" on either side of "humidity=77.5" (Figure 10.2-b).

Local discretization strategy thus makes it possible to compare the predictive capacity of all the attributes, whether continuous or not. In spite of their simplicity, we are facing here of one of the principal bottlenecks in the development of graphs. The discretization of the continuous variables requires initially a sorting of the values in ascending order which, in the best case, is of $O(n \log n)$. This is why we propose to use sampling by reducing the number n of records to which these calculations apply.

10.3. LOCAL SAMPLING STRATEGIES FOR DECISION TREES

The learning algorithm is based on a local sampling process. On each leaf, during construction, a sample is drawn from the part of the database that corresponds to the path associated to the leaf (Figure 10.3). Every time, the sample has full size, as long as the size of the database for that leaf is large enough; otherwise, all available examples are retained. This approach optimizes information use: at first it is overabundant, and a sample is enough to determine appropriate rules in a reasonable amount of time. By the end of the process, when information is rare, a larger fraction of the individuals is used, eventually all of them.

In practice, the process is as follows:

1. First, a complete list of individuals on the base is drawn;

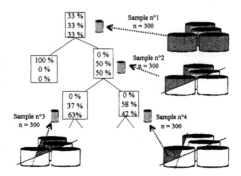

Figure 10.3 Steps of building decision tree using local sampling

2. The first sample is selected while the base is being read; an array of records associated with each attribute is kept in memory;

3. This sample is used to identify the best segmentation attribute, if it exists; otherwise, the stopping rule has played its role and the node becomes a terminal leaf;

4. If a segmentation is possible, then the list in Step 1. is broken up into sub-lists corresponding to the various leaves just obtained;

5. Step 4. requires passing through the database to update each example's leaf; this pass is an opportunity to select the samples that will be used in later computations.

Steps 3. to 5. are iterated until all nodes are converted to terminal leaves.

In order to save memory space, lists and samples associated to a given node are erased once it has been either segmented or determined to be a terminal leaf.

10.3.1.   Determining The Sample Size

If a useful split may be found on a given node *using the complete database*, then the size of the sample must be such that:

1) This split be recognized as such, that is the power of the test must be sufficient;

2) The discretization point be estimated as precisely as possible;

3) If, on the given node on the base, many splitting attributes are possible, the criterion for the optimal attribute remains maximal in the sample.

The first objective will be examined now. The two other objectives are rather complex and will be briefly reviewed in Section 10.4.1.

10.3.1.1 Testing Statistical Signification for A Link. In order to determine the sample size we need for each node, we use statistical tests concepts: probability of type I and type II errors (usually named α and β).

On the current node of the tree, we are looking for the (dichotomized) attribute which provides *the best split* (according to a criterion T which often is either the observed-χ^2 or the information gain (see below)).

The split is done if two conditions are met:

1) if this split is the best (according to the criterion T) among all the possible splits for all attributes;

2) if this split is possible, that is to say if the criterion $T(Sample$ $data)$ is so unlikely to occur if H_0 were true. The null hypothesis, H_0, states there is no link, *in this node in the whole database*, between the class attribute and the predictive attribute we are testing. The idea is that the criterion, calculated from the sample, should measure the evidence against H_0. Specifically, one computes the $p - value$, that is the probability of T being greater than or equal to $T(Sample\ data)$, given that H_0 is true. H_0 is rejected, so the split is possible, if the $p - value$ is less than a predetermined significance level, α.

Statisticians (Salzberg, 1999) often point out to dataminers that the standard hypothesis testing framework rests on the basic assumption that only two hypotheses are ever entertained on a given sample. In fact, we consider a very large number of possible models on a given node (numerous attributes and many discretizations in two subset for each attribute). As a consequence, the true significance level α', that is the probability of selecting an attribute without any usefulness, is larger than α, the nominal level. Fortunately, there is an upper bound for α': if there are q possible attributes, and if one computes $(p - value)_j$ independently for each attribute j ($j = 1, q$), then, the probability α' of observing at least one of them smaller than α is:

$$\begin{aligned}
\alpha' &= P[Min_{j=1,q}(p - value)_j < \alpha] = P[\cup_{j=1,q}\{(p - value)_j < \alpha\}] \\
&\leq \sum_{j=1,q} P[(p - value)_j < \alpha] = q \times \alpha
\end{aligned}$$

So, in order to get results that are truly significant at a reasonable level α', one must use a very small value for α (10^{-5} or 10^{-7}); as a result, it is very unlikely to get, from a sample set, a split without any relevance to the whole database.

The significance level α limits the *type I error probability*. Now we need to take account of β (the *type II error probability*). β is the probability to not selecting an attribute, even if there is a link, *in this node in the whole database*, between the class attribute and the predictive attribute we are testing (i.e. if H_0 is false). It is equivalent to deal with $(1 - \beta)$, the "power of the test".

In order to calculate the sample size we need to limit β, we have to study the probability distribution of our criterion when H_0 is false, that is to say when there is a link in the database.

10.3.1.2   Notations.   Let Y denote the class attribute, and let $Y_i (i = 1...p)$ be the categories to which cases are to be assigned, X any predictor attribute, $X_j (j = 1...q)$ one of the possible discrete values of X.

π_{ij} is the proportion of $(Y = Y_i$ and $X = X_j)$-*records in the subpopulation corresponding to the node we are working on, in the whole database*; $\pi_{i+} = \sum_j \pi_{ij}$ and $\pi_{+j} = \sum_i \pi_{ij}$ are the marginal proportions, and $\pi_{ij}^0 = \pi_{i+} \times \pi_{+j}$ are the products of marginal proportions.

Let n_{ij} denote the number of (ij) cell *in the sample tabulation*. Using simple random sampling, we have $E(n_{ij}) = n \times \pi_{ij}$, where $E(n_{ij})$ is the expected value of n_{ij} and n the number of records *in the sample*.

10.3.1.3   Probability Distribution of The Criterion.   In this paper we measure the link between the class attribute and the predictive attribute, either by the Pearson χ^2 statistic, or by the information gain which is based on Shannon's entropy. We do this because these are the two most used criteria in the context of induction trees (Zighed and Rakotomalala, 2000), and probability distributions of these measures are perfectly known as soon as the following information is given: - proportions π_{ij} in the whole database, - the sampling procedure, - the sample size n.

Under the hypothesis of simple random sampling, the probability distributions of these two statistics are identical (Agresti, 1990):

- When the null hypothesis H_0 is true and the sample size is moderately large $(E(n_{ij}) > 5, \forall ij)$, both have approximate chi-square distribution with degrees of freedom $\nu = (p - 1)(q - 1)$; its mean is ν and its variance is 2ν.

- When the null hypothesis H_0 is false, the distribution is approximately non-central chi-square; this distribution has two parameters: $\nu = (p - 1)(q - 1)$, as below, and λ, a non-centrality parameter. The mean is $\nu + \lambda$ and the variance is $2(\nu + 2\lambda)$.

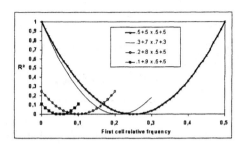

Figure 10.4 R^2 values for four kinds of cross tabulations (2x2), characterized by their margin distributions: $\pi_{1+}; \pi_{2+}$; $\pi_{+1}; \pi_{+2}$

The regular (central) chi-square distribution is the special case $\lambda = 0$. When H_0 is true, $\lambda = 0$; the further the truth is from H_0, the larger λ and the power of χ^2-test are.

The noncentral chi-square density function has no closed analytic formulation, but it is asymptotically normal for large values of λ (Johnson and Kotz, 1969). The parameter λ is a function of the sample size n and of the frequencies π_{ij} in the whole database (Agresti, 1990).

The value of λ for the information gain is:

$$\lambda_I = 2n \sum_{ij} \pi_{ij} \log \left(\frac{\pi_{ij}}{\pi_{ij}^0} \right) \simeq n \times \Phi^2$$

and for the Pearson χ^2 statistic, it is:

$$\lambda_K = n \sum_{ij} \frac{(\pi_{ij} - \pi_{ij}^0)^2}{\pi_{ij}^0} = n \times \Phi^2$$

For $p = 2$ classes and an attribute X with $q = 2$ values, it is well known that $\Phi^2 = \lambda_K/n$ becomes R^2, that is the squared Pearson coefficient of correlation for the two flag attributes.

Figure 10.4 presents R^2 values for four kinds of cross tabulations (2x2), characterized by their marginal distributions.

For example, for a table of type $\dfrac{\begin{array}{|c|c|} \hline 0.2 & 0.3 \\ \hline \end{array}}{\begin{array}{|c|c|} \hline 0.3 & 0.2 \\ \hline \end{array}}$ ($\pi_{1+} = \pi_{2+} = \pi_{+1} =$

$\pi_{+2} = 0.5$, and $\pi_{11} = 0.2$, and consequently $\pi_{12} = 0.3$; $\pi_{21} = 0.3$ and $\pi_{22} = 0.2$), we read $R^2 = 0.04$ (i.e. $R = 0.2$).

10.3.1.4   Equalization of Nominal Risk Probabilities α and β.   Now we are able to find the value of the minimum sample sizes

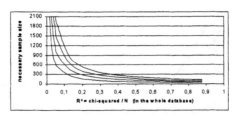

Figure 10.5 Sample Size that equalize α and β error probabilities: $\alpha = \beta = 10^{-5}$ (upper line) ; 10^{-4} ; 10^{-3} and 10^{-2} (lower line)

in order to get a given power $(1 - \beta)$, giving both the significance level α and the link (measured by R^2) on the whole database.

Let $T_{1-\alpha}$ denote the critical value, such as

$$P_{H_0}(T > T_{1-\alpha}) = P_{H_0}[p - value < \alpha] = \alpha$$

As the the criterion T (the observed-χ^2 or the information gain) has an asymptotic normal distribution when the null hypothesis H_0 is false (the mean is $\nu + \lambda$ and the variance is $2(\nu + 2\lambda)$),

$$\beta = P_{H_1}(T < T_{1-\alpha}) = P\left\{Z < \left[(T_{1-\alpha} - (\nu + \lambda))/\sqrt{2(\nu + 2\lambda)}\right]\right\}$$

H_1 denotes the alternative hypothesis and Z the standardized normal variable.

If $p = q = 2$, then $\nu = (p - 1)(q - 1) = 1$ and $\lambda = nR^2$ and:

$$\beta = P_{H_1}(T < T_{1-\alpha}) = P\left\{Z < \left[(T_{1-\alpha} - 1 - nR^2)/\sqrt{2 + 4nR^2}\right]\right\}$$

Figure 10.5 shows the necessary sample size n for $\alpha = \beta$, depending of the significance level α, and the link (measured by R^2) in the whole database.

Figure 10.5 deserves some comments:

- the necessary sample size n for α and β equalization decreases as the link R^2 increases on the whole database. When the link between X and Y is high, a few cases are sufficient to reveal it. Conversely, the weaker the link (R^2) is in the database, the larger the sample size must be to make evidence for it.

- the necessary sample size n for α and β equalization increases as the significance level α decreases: if one wants to reduce risk probabilities, a larger sample is needed.

- the gap between necessary sample sizes for risk probabilities varying from 10^{-2} to 10^{-5} varies inversely as R^2, and is not so large.

- generally speaking, Figure 10.5 shows that a sample size of about $n = 300$ or 500 cases is sufficient for R^2 of the order of 0.15 to

0.40, that is to say for proportion in the database as $\dfrac{\begin{array}{|c|c|} \hline 0.73 & 0.07 \\ \hline \end{array}}{\begin{array}{|c|c|} 0.07 & 0.13 \\ \hline \end{array}}$

or $\dfrac{\begin{array}{|c|c|} \hline 0.5 & 0.0 \\ \hline \end{array}}{\begin{array}{|c|c|} 0.3 & 0.2 \\ \hline \end{array}}$. Using the largest frequency class assignment rule,

the average misclassification rates in the database are 14% for the first example and 20% for the second. Those misclassification rates are like the rates we observe in real databases.

The calculations above are based on 2×2 cross-tables, which does not restrict the scope of the study. Learning $p > 2$ classes, the sample size should increase as p.

10.3.2.   Sampling Methods

We need an efficient toolkit to draw sample sets of size n from a database containing N records ($0 \leq n \leq N$), each of them having the same probability ($\frac{n}{N}$) to be chosen; in addition, each of the ($\frac{N!}{n!(N-n)!}$) possible sample sets must have the same probability.

Using simple random sampling without replacement, sampling variability is smaller than with replacement (Sollich and Krogh, 1996); it decreases to zero as, *in the node we are working on, in the whole database, the subpopulation size* tends to n the fixed sample size.

Vitter (Vitter, 1987) presents two sequential random sampling algorithms, which select the records in the same order as they appear in the database, so the disk access time is reduced. The sequential sampling methods iteratively select the next record for the sample in an efficient way. They respect the equiprobability conditions. They also have the advantage of being extremely fast and simple to implement.

The algorithm S sequentially processes the database records and determines whether each record should be included in the sample (Fan et al., 1962). The first record is selected with probability (n/N). If m records have already been selected from among the first t records in the base, the $(t+1)$st record is selected with probability ($n - m)/(N - t)$.When n records have been selected, the algorithm terminates. The average number of independent uniform variate generated by Algorithm S is approximately N, and the running time is of $O(N)$.

The algorithm D (Vitter, 1987) uses random jump between selected records of the database. Algorithm D does the sampling in $O(n)$ time, on the average; that is significantly faster than algorithm S, when $n \ll N$. In practice, one should use algorithm D when $n < N/7$ and algorithm S when $n \geq N/7$.

10.4.   EXPERIMENTS

Above, we studied some partial aspects of the sample size determination problem. As real data mining problems are very complex (many attributes, depending on each other, each one being discretized on each node), we have to check on real databases our assertion that a fix sample size of around 400 is sufficient.

10.4.1.   Objective and Context of The Experiments

Our first goal here is to show that a tree built with local sampling has a generalization error rate comparable to that of a tree built with the complete database. Performance stability does not convey the same meaning here as it does in other sample based learning strategies. With "association rules" (Kivinen and Manilla, 1994), the goal is to find the minimum sample size at which one finds the rules (with a given level of probability) that would have been established on the complete database. With a "decision tree" approach, the two trees need not be identical for two reasons.

First, induction using tree is by nature unstable; slight modifications to the learning set can cause radical changes in the shape of the tree because of alternative choices of segmentation attributes on the nodes (Dietterich and Kong, 1995). Of course, sampling on nodes introduces additional variability, but it is rather small compared to the natural variability of the method, and the sampling variability can be controlled by an adequate choice of the size of the sample.

Second, when the semantics of the results is important because they will be used to interpret causal phenomena (in health sciences, marketing, etc.) (Witten and Frank, 2000), the tree is often built by a subject matter expert hand: at each step, he chooses the splitting attributes among those selected as most relevant by the software. From this point of view, it does not matter that the variables be exactly classified in the same order using the complete base or a mere sample; yet it is important that the most relevant attributes be distinguished from the others. An appropriate choice of sample size should fulfil this requirement.

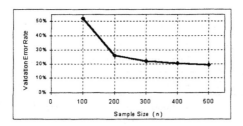

Figure 10.6 Average GENERALIZATION ERROR RATE according to the sample size drawn on each node (Breiman et al. Waves Dataset; Binary discretization of 21 continuous attributes at each node)

Our second goal is to show that sampling reduces computing time. The normal, "industrial" operative mode of our procedure is as follows: some DBMS manages the database and only the sample selected on a node is uploaded in memory. This is not how our experiments were set up. In effect, various factors such as access time to the hard drive, or server performances, would pollute the results. We chose to upload the complete base in memory to measure more precisely the time gains due to sampling. Thus, any change in computing time is solely due to sampling; to allow pure measurements, the automatic swap mechanism of the computer OS was disabled[1].

10.4.2.   Experiments on An Artificial Database

We will now apply the whole method (with sampling and binary discretization on each node of the tree) to a well known artificial problem: the "Breiman's et al. waves" (Breiman et al., 1984). In § 2.6 of the book, they pose this problem, now traditional: each of three classes is characterized by a specific weighting combination of 21 pseudo-random standardized normal variables.

We generated 100 times two files, one of 500, 000 records for the training, the other of 50, 000 records for the validation. Binary discretization was processed on each node for each attribute. Figure 10.6 show the error rate evolution of ChAID decision tree algorithm (Kass, 1980), according to the sample size drawn from the file on each node ($n = 100$, 200, 300, 400, 500).

The learning time is quasi null for $n = 100$ because the tree stops very quickly, even immediately: the χ^2-test has a low power (if n is too small, the *observed-*χ^2 is not significative, even if $R^2 > 0$). From $n = 200$ to 500, the run time increases from 43 s to 140 s. Using the whole 500, 000 records dataset, the run time is 1, 074 s.

Figure 10.6 shows how the error in generalization decreases as the sample size n increases. Even for this problem considered as a difficult

Table 10.2 Characteristics of the databases used for the comparisons

Name	Database Size	Learning Sample Size	Continuous / All attributes
adult	48,842	24,000	6/14
covtype	581,012	100,000	54/54
letter	20,000	16,000	16/16
nursery	12,960	10,000	0/7
shuttle	58,000	43,500	9/9

one, the marginal profit becomes weak starting from sample sizes of $n = 300$ records; one approaches then 18%, the minimum of generalization error rate obtained with trees using the entire database, with its $N = 500,000$ records.

10.4.3.   Results with Real Benchmark Databases

Now, we will check our preceding results by applying our strategy to real datasets. We chose large and various databases so as to carry out a convincing experimentation. The UCI Internet site (Bay, 1999) collects several databases which are classical benchmarks in machine learning. We chose five databases (Table 10.2) which contain more than $12,900$ individuals, and up to $580,000$ [2].

In order to evaluate our sampling strategy, we measure the generalization error rate and the running time. We repeated 10 times the following operations: we subdivide randomly the database in a training set and in a test set, as indicated in Table 10.2, then we build the trees on the training set and calculate the error rates on the test one. We choose various size of learning set according to the size of the whole database.

The tree built on the complete database is compared to the trees built with sampling on each node. Four sample sizes were tested ($n = 100$; 200 ; 300 ; 400 ; 500).

The average generalization error rate and the reduction in computing time due to sampling are presented in Table 10.3 and 10.4. ChAID algorithm, known for its speed, was used (Kass, 1980). ChAID uses pre-pruning that relies on a χ^2-test (we set $\alpha = 10^{-5}$).

The main conclusions are :

In Table 10.3, we see the influence of n (sample size on each node) on the generalization error rate. The size must not be too small. With $n = 100$, the machine stops very quickly without learning anything be-

Table 10.3 Error rate for various node sample sizes.

Base	E(ALL)	E(100)	E(200)	E(300)	E(400)	E(500)
adult	0.146	0.177	0.162	0.152	0.146	0.146
covtype	0.205	0.353	0.293	0.257	0.225	0.205
letter	0.197	0.212	0.205	0.204	0.201	0.197
nursery	0.023	0.124	0.039	0.026	0.024	0.024
shuttle	0.001	0.067	0.014	0.004	0.003	0.001

Table 10.4 Learning time on complete databases (milliseconds) and learning time reduction ratio with sampling on nodes.

Base	L(ALL)	R(100)	R(200)	R(300)	R(400)	R(500)
adult	11988	0.02	0.16	0.19	0.29	0.34
covtype	312354	0.02	0.04	0.09	0.18	0.27
letter	22827	0.52	0.64	0.67	0.80	0.82
nursery	704	0.29	0.75	0.77	0.80	0.98
shuttle	7185	0.08	0.08	0.08	0.09	0.12

cause the power of the χ^2-test is too low. In other words β (the probability to not selecting an attribute, even if there is a link in the whole database) is too large. On the other hand, from $n = 300$, one approaches the performances of the tree using the whole base. This confirms the theoretical and experimental results.

As expected, sampling drastically reduces computing time (see ratio $\frac{Time\_with\_sampling}{Time\_without\_sampling}$ in Table 10.4). As with the "Wave" example, learning time is strongly dependent on the size of the sample on each node. It is remarkable that there is little reduction in computing time when the file comprises only qualitative attributes, as with "Nursery". Conversely, "Shuttle" and "Covtype", combining quantitative attributes and large training file sizes (100,000 cases) allow substantial reductions. The "Adult" database, with its mix of continuous and discrete attributes, shows intermediate time reduction.

The case of "Letter" differs from the others, time reduction is not strong in spite of the presence of continuous attributes. This is explained by data fragmentation. It leads quickly to small subset on the nodes; the whole database contains less than n records complying with the rules defining this node; actually, in those cases, sampling is without effect, all individuals in those classes are used, our algorithm thus is working like the classical algorithm, hence the similarity in the results. This interpretation is confirmed by the evolution of generalization error rate according to the sample size. A small sample is enough to obtain a good error rate (e.g. $n = 100$).

All these evidences confirm the theoretical and experimental results: when the database is large and contains many numerical attributes, then our strategy of fast sampling on each node (with samples size about $n = 300$ or 500) reduced considerably running times while preserving the generalization error rate.

10.5.   CONCLUSION AND FUTURE WORK

Decision trees, and more generally speaking decision graphs, are efficient, step by step, and simple methods for supervised classification. However, mining on very large databases, in particular when these include a number of continuous attributes which must be discretized at each step, is very time consuming. In these cases, working on samples is especially useful. The decision tree "step by step" characteristic allows us to propose a strategy using successive samples, one on each tree node. Theoretical and empirical evidences show that our strategy of fast sampling on each node (with samples size about $n = 300$ or 500) reduces considerably learning time while preserving the accuracy.

This work opens on some new research areas. Should optimal sampling methods (stratified random sampling, selection with unequal probabilities, etc.) be used ? Those methods were developed for surveys on economic or sociological issues, when the cost of data collection is high compared to calculation time. They must be adapted for data mining: in our situation the information is already known, it is in the database. We have to check whether the gain in accuracy that would obtained by these methods, at fixed size, might also be obtained by sample size increase, at fixed learning time.

An interesting way, especially learning imbalanced classes, may be the equal-sized sampling strategy on each node (Chauchat et al., 1998): a random sample of size n_k is to be drawn from the original database for each class y_k. The size of the sample file is n ($n = \sum_{k=1}^{p} n_k$.). If the n_k are equal, the sample is said to be balanced, or equal-sized. We are currently working on a real life marketing application of targeting a rare specific customer group from a several hundred of thousands customers database for whom some 200 attributes are known, 95% of them continuous. The class attribute is whether a client connected to some remote service. Results show, once again, that local equal-sized sampling on the nodes while constructing the tree requires small samples to achieve the performance of processing the complete base, with dramatically reduced computing times (Chauchat et al., 2000).

An other question is the implementation of sampling in the core of the queries in databases.

Acknowledgements

We are very grateful to the two anonymous referees who provided helpful comments, and to M. Jean Dumais, Section Chief, Methodology Branch, Statistics Canada and guest lecturer at Université Lumière for his careful proof reading of the manuscript and his suggestions.

Notes

1. the computer is a Pentium II, 450MHz, 256Mb Ram, operating under Windows 98

2. for the requirements of the experimentation, binary attributes were treated like continuous attributes in covtype database

References

Agresti, A. (1990). *Categorical data analysis.* John Wiley, New York.

Bay, S. (1999). The uci kdd archive [http://kdd.ics.uci.edu]. Irvine, CA: University of California, Department of Computer Science.

Breiman, L., Friedman, J., Olshen, R., and Stone, C. (1984). *Classification and Regression Trees.* California : Wadsworth International.

Chauchat, J., Boussaid, O., and Amoura, L. (1998). Optimization sampling in a large database for induction trees. In *Proceedings of the JCIS'98-Association for Intelligent Machinery,* pages 28–31.

Chauchat, J., Rakotomalala, R., and Robert, D. (2000). Sampling strategies for targeting rare groups from a bank customer database. In *(To appear) Proceedings of the Fourth European Conference PKDD'2000,* Lyon, France. Springer Verlag.

Dietterich, T. and Kong, E. (1995). Machine learning bias, statistical bias and statistical variance of decision trees algorithms. In *Proceedings of the 12th International Conference on Machine Learning.*

Dougherty, J., Kohavi, R., and Sahami, M. (1995). Supervised and unsupervised discretization of continuous attributes. In Kaufmann, M., editor, *Machine Learning : Proceedings of the 12th International Conference (ICML-95),* pages 194–202.

Fan, C., Muller, M., and Rezucha, I. (1962). Development of sampling plans using sequential (item by item) selection techniques and digital computers. *Journal of American Statistics Association*, 57:387–402.

Fayyad, U. and Irani, K. (1993). Multi-interval discretization of continuous-valued attributes for classification learning. In *Proceedings of The 13th Int. Joint Conf. on Artificial Intelligence*, pages 1022–1027. Morgan Kaufmann.

Fayyad, U., Piatetsky-Shapiro, G., and Smyth, P. (1996). Knowledge discovey and data mining : Towards an unifying framework. In *Proceedings of the 2nd International Conference on Knowledge Discovery and Data Mining*.

Frank, E. and Witten, I. H. (1999). Making better use of global discretization. In *Proc. 16th International Conf. on Machine Learning*, pages 115–123. Morgan Kaufmann, San Francisco, CA.

Johnson, N. and Kotz, S. (1969). *Continuous Univariate Distributions*, volume 2. Wiley.

Kass, G. (1980). An exploratory technique for investigating large quantities of categorical data. *Applied Statistics*, 29(2):119–127.

Kivinen, J. and Manilla, H. (1994). The power of sampling in knowledge discovery. In *Proceedings of 13th SIGACT-SIGMOD-SIGART Symposium on Principles of Database Systems*, volume 13, pages 77–85.

Michie, D., Spiegelhalter, D., and Taylor, C. (1994). *Machine learning, neural and statistical classification*. Ellis Horwood, London.

Mitchell, T. (1997). *Machine learning*. McGraw Hill.

Quinlan, J. (1993). *C4.5: Programs for Machine Learning*. Morgan Kaufmann, San Mateo, CA.

Salzberg, S. (1999). On comparing classifiers : A critique of current research and methods. *Data mining and Knowledge discovery*, 1:1–12.

Sollich, P. and Krogh, A. (1996). Learning with ensembles : How overfitting can be usefull. In Touretzky, D., Mozer, M., and Hasselmo, M., editors, *Advances in Neural Information Processing Systems*, pages 190–196. MIT press.

Vitter, J. (1987). An efficient algorithm for sequential random sampling. *ACM Transactions on Mathematical Software*, 13(1):58–67.

Witten, I. and Frank, E. (2000). *Data Mining: practical machine learning tools and techniques with JAVA implementations*. Morgan Kaufmann.

Zighed, D., Rabaseda, S., Rakotomalala, R., and Feschet, F. (1999). Discretization methods in supervised learning. In Kent, A. and Williams, J., editors, *Encyclopedia of Computer Science and Technology*, volume 40, pages 35–50. Marcel Dekker, Inc.

Zighed, D. and Rakotomalala, R. (2000). *Graphes d'Induction - Apprentissage et Data Mining*. Hermes.

Chapter 11

INCREMENTAL CLASSIFICATION USING TREE-BASED SAMPLING FOR LARGE DATA

Hankil Yoon
CISE Department, University of Florida
Gainesville, FL 32611 U.S.A.
hyoon@cise.ufl.edu

Khaled AlSabti
Computer Science Department, King Saud University
P.O.Box 51178 Riyadh 11543, Kingdom of Saudi Arabia
alsabti@ccis.ksu.edu.sa

Sanjay Ranka
CISE Department, University of Florida
Gainesville, FL 32611 U.S.A.
ranka@cise.ufl.edu

Abstract   We present an efficient method called ICE for incremental classification that employs tree-based sampling techniques and is independent of data distribution. The basic idea is to represent the class distribution in the dataset by using the weighted samples. The weighted samples are extracted from the nodes of intermediate decision trees using a clustering technique. As the data grows, an intermediate classifier is built only on the incremental portion of the data. The weighted samples from the intermediate classifier are combined with the previously generated samples to obtain an up-to-date classifier for the current data in an efficient, incremental fashion.

Keywords: Incremental classification, decision tree, tree-based sampling, local clustering, weighted gini index.

11.1.   INTRODUCTION

Classification is an important, well-known problem in the field of data mining, and has remained an extensive research topic within several research communities. Over the years, classification has been successfully applied to diverse areas such as retail target marketing, medical diagnosis, weather prediction, credit approval, customer segmentation, and fraud detection (Mitchie et al., 1994).

Thanks to the advances in data collection/storage technologies, and large scale applications, the datasets for data mining applications have become large and may involve several millions of records. Each record typically consists of ten to hundreds attributes, some of them with large number of distinct values. In general, using large datasets results in the improvement in the accuracy of the classifier (Chan and Stolfo, 1997; Catlett, 1984), but the enormity and complexity of the data involved in these applications make the task of classification computationally very expensive. Building classifiers on such large datasets requires development of new techniques that minimize the overall execution time with comparable quality.

Another problem that makes classification more difficult is that most datasets are growing over time, continuously making the previously constructed classifier obsolete. Addition of new records (or deletion of old records) may potentially invalidate the existing classifier that has been built on the old dataset (which is still valid). However, constructing a classifier for large growing dataset from scratch would be extremely wasteful and computationally prohibitive.

In this chapter, we present a method that incrementally classifies a large database of records which grows in size over time, and that still results in a decision tree with comparable quality. The method, named ICE (*I*ncremental *C*lassification for *E*ver-growing large datasets), incrementally builds decision tree classifiers for large growing datasets using tree-based sampling techniques. The basic idea is to represent the class distribution in the dataset by weighted samples. The weighted samples are extracted from the nodes of intermediate decision trees using a clustering technique. As the data grows, an intermediate classifier is built only on the incremental portion of the data. The weighted samples from the intermediate classifier are combined with the previously generated samples to obtain an up-to-date classifier for the current data in an efficient manner.

Decision tree classification. Classification is a process of generating a concise description or model for each class of the *training dataset* in terms of the predictor attributes defined in the dataset. Each record of the training dataset consists of a set of predictor attributes and is tagged with a class label. Attributes with discrete domains are referred to as *categorical*, whereas those with ordered domains are referred to as *numerical*. The model may be subsequently tested with a *test dataset* for verification purpose. The model is then used to classify future records whose classes are unknown.

Among the classification models in the literature (Lim et al., 1997; Cheeseman et al., 1988; Ripley, 1996; James, 1985; Goldberg, 1989; Breiman et al., 1984; Quinlan, 1993; Shafer et al., 1996), decision trees are particularly suited for data mining for the following reasons. The decision tree is easily interpreted and comprehended by human beings, and can be constructed relatively fast (Breiman et al., 1984). While training neural networks is very expensive, inducing decision tree is efficient and thus suitable for large datasets. Also, decision tree generation algorithms do not require prior knowledge of domains or distributions on the data. Finally, decision trees result in good classification accuracy compared to the other models (Mitchie et al., 1994). For these reasons, we will focus on the decision tree model in this chapter.

Derivation of decision tree classifier typically consists of a *construction* phase and a *pruning* phase. In construction phase, a decision tree algorithm recursively partitions the training dataset until the *halting criterion* – each partition consists entirely or dominantly of records from one class – satisfies (Breiman et al., 1984; AlSabti et al., 1998; Quinlan, 1993; Shafer et al., 1996). Each non-leaf node of the tree contains a *splitting condition* which is a test on one or more attributes and determines how the data is partitioned. A split selection method such as **gini** index (Breiman et al., 1984) is computed in order to discover the splitting condition at each node. A partition which meets the halting criterion is not divided further, and the node that represents the partition is labeled with the dominant class.

In general, the decision tree generated in the construction phase is perfect for the known records and may be overly sensitive to statistical irregularities and idiosyncrasies of the training dataset. Thus, most algorithms perform a pruning phase, in which nodes are iteratively pruned to prevent *overfitting* and to obtain a tree with higher accuracy for unlabeled records. An algorithm based on the *minimum description length* (MDL) principle (Quinlan and Rivest, 1989) is used in this research to prune the decision tree. In addition, since the pruning phase can be generally executed in-memory and its cost is very small compared to that of the construction phase, we restrict our attention only to the construction phase in this chapter.

No	Age	Profession	Class
1	24	Clerk	C1
2	17	Business	C1
3	22	Teacher	C1
4	35	Professor	C1
5	40	Professor	C2
6	48	Clerk	C1
7	52	Teacher	C2
8	46	Clerk	C1

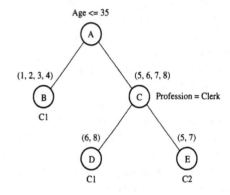

(a) Training data set   (b) Decision tree classifier induced

Figure 11.1   Example of a decision tree classifier

Figure 11.1 shows a sample decision tree for the given training dataset. Each record describes of a person, which is tagged with one of the two class labels – C1 and C2. Each record is characterized by two attributes, *age* and *profession*, the former being numerical and the latter being categorical. The splitting conditions (*age* $\leq$ 35) and (*profession* = clerk) partition the records into the corresponding classes. The classifiers shows a concise and meaningful description of the data involving *age* and *profession* by which a person is classified into one of the two classes.

Decision tree classifiers have different properties, which are used to compare classifiers. The *accuracy* of a classifier is the primary metric. It measures the predictive performance of the classifier and is determined by the percentage of the test dataset examples that are correctly classified (Weiss and Kulikowski, 1991). The *misclassification* metric has been defined to deal with different misclassification costs. The *risk* metric extends the misclassification metric to include the gain from correctly classified cases. The *comprehensibility* metric prefers classifiers that are easy to understand.

11.2.   RELATED WORK

Most algorithms in the machine learning and statistics are main memory algorithms whereas today's databases are typically much larger than main memory (Agrawal et al., 1993). The construction phase of the various decision tree classifiers differs in selecting the test criterion for partitioning a set of records. ID3 (Quinlan, 1986) and C4.5 (Quinlan, 1993) examine the solution space of all possible decision trees to a fixed depth, and employ a simple scheme that selects a test which minimizes the impurity of the partition. On the other hand, CART (Breiman et al., 1984), and SPRINT (Shafer et al., 1996) choose the test with the lowest **gini** index. A recently proposed algorithm, CLOUDS (AlSabti et al., 1998), samples the splitting points for numerical attributes followed by an estimation step to narrow the search space of the best split, reducing cost substantially while maintaining the quality of the generated trees.

Discretization and sampling are approximate techniques that can be used to scale up the classifier for large datasets. C4.5 uses *windowing* technique which repeatedly augments the samples by adding misclassified records (Quinlan, 1993). *Stratification* is a selective sampling method in which samples are drawn within the same classes (Catlett, 1984). The sampling techniques proposed by AlSabti (AlSabti, 1998) evaluate the **gini** index at only a subset of points along each numerical attributes.

A framework called Rainforest is recently proposed for developing fast and scalable algorithms for constructing decision trees that gracefully adapt to the amount of main memory available (Gehrke et al., 1998). The BOAT algorithm optimistically constructs the exactly same decision tree from samples extracted by using bootstrapping technique (Gehrke et al., 1999). With changes in data distribution, however, BOAT requires another pass over dataset and needs to keep temporary information to incrementally construct the tree.

11.3. INCREMENTAL CLASSIFICATION

Problem definition. A decision tree T has been built for a dataset D of size N by a decision tree classification algorithm. After a certain period of time, a new incremental dataset d of size n is collected, and a new decision tree T' needs to be built for the combined dataset $D + d$. This problem is nontrivial because addition of new records may change some splitting points, and consequently T' may be quite different from T. The goal is to build T' with minimal cost such that T' is comparably as accurate as the decision tree built for $D + d$ from scratch.

Impurity-based split selection methods. We use the **gini** index (Breiman et al., 1984) to illustrate impurity-based split selection methods for binary split. This class of methods is the most popular (Breiman et al., 1984; Quinlan, 1986), and it has been shown that they produce decision trees with high predictive accuracy (Lim et al., 1997). Most previous work in the database area uses this class (Shafer et al., 1996; Morimoto et al., 1998). Therefore, using this class of method will make comparisons easy and fair.

Let n and C_i be the total number of records in a set S and the number of records that belong to class i, respectively. Let m be the number of classes. Then, the **gini** index is defined as follows:

$$gini(S) = 1 - \sum_{i=1}^{m} (\frac{C_i}{n})^2 \tag{11.1}$$

When S is split into two subsets, S_1 and S_2, with n_1 and n_2 records, respectively, the corresponding **gini** index of the set S is calculated in terms of the **gini** index of the subsets as follows:

$$gini_{split}(S) = \frac{n_1}{n} gini(S_1) + \frac{n_2}{n} gini(S_2) \tag{11.2}$$

The **gini** index defined above does not lend itself for reuse in accordance with the changes in the number of records (and hence the class distribution). Impurity-based split methods such as **gini** index cause such instability (Gehrke et al., 1999), which makes the induction of T' from T difficult.

11.3.1.   Weighted Gini Index

A weight can be assigned to a record when the record represents a number of other records. Since the **gini** index works only on the records with equal weight, however, it is inappropriate to handle such weighted records. Therefore, the **gini** index must change accordingly as each record carries different weight. In this section, we introduce the weighted **gini** index as reasoning for incremental classification. The motivation for the weighted **gini** index is to replace similar records by a weighted representative and use them in an incremental batch learning.

Suppose that there exists a set S' of n sorted numerical points in nondecreasing order, with each point (call it w-point) associated with a weight k for any positive integer $k \geq 1$. Let w_j be the weight of j-th w-point in S'. Then, the total weight sum W of S' is:

$$W = \sum_{j=1}^{n} w_j \geq n \tag{11.3}$$

The set S' is now divided into two subsets, S'_1 and S'_2, with n_1 and n_2 w-points, respectively, such that $n = n_1 + n_2$ and all the w-points in S'_1 are less than any one in S'_2. Now the question is how we take the weights into account in calculating the impurity of a set. Recall that the class distribution C_i in Equation 11.1 is the probability for a point in S to be classified into class i. Let Q_i be the weight sum of the w-points that belong to class i in S'. Then, Equation 11.1 is to be re-written as following:

$$gini^w(S') = 1 - \sum_{i=1}^{m} (\frac{Q_i}{W})^2 \qquad (11.4)$$

We now consider the distribution of w-points between S'_1 and S'_2 in computing the **gini** index in Equation 11.2. In Equation 11.2, the terms $\frac{n_1}{n}$ and $\frac{n_2}{n}$ represent the *portions* of the subsets S_1 and S_2, respectively, in the set S. In other words, these simple fractions mean that the impurity of a set tends to increase as the *number* of records contained in the set increases. When the points are associated with weights, this statement illustrates that the impurity of a set tends to increase as the *weight sum* of points contained in the set increases.

Let W_1 and W_2 be the weight sums of subsets S'_1 and S'_2, respectively. Then, we can modify the Equation 11.2 based on Equations 11.3 and 11.4 as follows:

$$gini^w_{split}(S') = \frac{W_1}{W} gini^w(S'_1) + \frac{W_2}{W} gini^w(S'_2) \qquad (11.5)$$

Now let us check if $gini^w_{split}$ is *consistent* with the original $gini_{split}$ index, i.e., $gini^w_{split} = gini_{split}$ at a common splitting point x. Suppose that w-points are chosen from S such that for any p represented by s_{i-1} and q represented by s_i for two adjacent w-points s_{i-1} and s_i, $p \leq q$ holds as depicted in Figure 11.2. Then there always exists a point x such that $p \leq x \leq q$ and x does not cut into any w-point. In other words, x is a clear-cut point on the dimension Z.

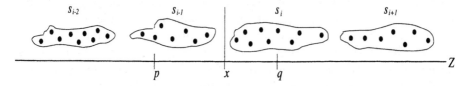

Figure 11.2 The set S' of w-points along dimension Z

Lemma 1. The $gini^w_{split}$ index value at a clear-cut splitting point x on a dimension is consistent with the original **gini** index value $gini_{split}$ at the same splitting point.

Proof Suppose a point x that clear-cuts the set S' of size n into two subsets S'_1 and S'_2, with n_1 and n_2 w-points, respectively, where $n = n_1 + n_2$. Since there is no w-point in which contains x, the number of points before (after) x remains to be n_1 (n_2). This yields that $W_1 = n_1$ and $W_2 = n_2$, thus making $W = W_1 + W_2 = n_1 + n_2 = n$. Therefore, with respect to x, $gini^w(S') = gini(S)$ implying that the class distribution Q_i remains unchanged for subsets S'_1 and S'_2, and thus $gini^w(S')_{split} = gini_{split}(S)$ at any such point x.   □

Now, we extend this property of w-point to a multi-dimensional space. Note that the decision tree is constructed such that the splitting condition at each node *clearly* divides the data partition of the node by the splitting value along the dimension used in the condition. Each divided subpartition is recursively used at the next level. The entire data can be projected on to a multi-dimensional hyperspace, and it can be said that each node represents a hyperbox in a multi-dimensional space composed of the attributes that participate in the splitting conditions along the root to the node.

Therefore, assuming that a single w-point represents all the records of a node, we come to the following lemma.

Lemma 2. The $gini^w_{split}$ index value is consistent with the original **gini** index value $gini_{split}$ with respect to the splitting conditions.

Proof It directly follows from Lemma 1 and from the fact that the splitting conditions along the path from the root to any child node clearly divide each dimension of the hyperspace.   □

When more than one w-point are chosen to represent a node, Lemma 2 may not hold. Note, however, that the weighted **gini** index will always be computed between two adjacent w-points when constructing a decision tree on a set of w-points. If each w-point is chosen from around the center of a set of original (unweighted) records in each node, $gini^w_{split}$ would not be very different from $gini_{split}$. Therefore, weighted samples can be used to incrementally construct a decision tree for large growing data such that weighted samples are extracted from the decision trees built on the incremental partitions. Each w-point is a *clustered* representation of a number of records, which implies that clustering techniques can be used to select w-points from the original dataset. We describe how samples can be extracted in Section 11.4

11.3.2.   ICE: A Scalable Method for Incremental Classification

An efficient method called ICE for incremental classification on large growing datasets (Yoon, 2000) is described in Figure 11.3. D_i denotes a partition of the entire dataset that is collected during i-th time period (*epoch i*). The partition may have been collected since the last partition D_{i-1} was classified, or that it is partitioned simply because the dataset is too large to be processed at once. A decision tree T_i is built for D_i, and a set S_i of samples representing D_i is extracted from T_i. The set S_i is combined together with the previous sets of samples (denoted $U_i = S_1 \bigcup \cdots \bigcup S_i$). U_i is then used as the training set for the eventual decision tree classifier C_i at epoch i, after which U_i is preserved for the next epoch.

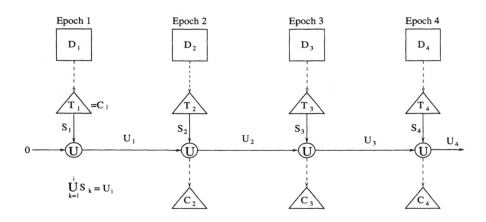

Figure 11.3 The description of ICE over four epochs

The method requires building two separate decision trees, T_i and then C_i, at epoch i. However, the cost for running a decision tree algorithm on two comparably small datasets (D_i and U_i) separately is relatively minor. The *current* entire dataset ($\bigcup D_i$) is presumably very large compared to D_i and U_i, making construction of the decision tree from scratch inefficient and impractical. The dataset U_i is the space overhead of ICE. This overhead saves *at least* one pass over the entire dataset, and makes the overall process very efficient and practical.

ICE not only is suitable for incremental classification when the dataset is ever-growing, but can also be easily extended to diverse applications (Yoon, 2000). The characteristics of ICE include size scalability, easy parallelization, inherent migration into distributed environment, temporal scalability, algorithm independence, minimal data access, and flexibility in dealing with addition and deletion of records as partition.

11.4. SAMPLING FOR INCREMENTAL CLASSIFICATION

There is supposedly no superior sampling method that consistently outperforms the others, nor a classification method that works well for all applications. The quality of the resulting classifier depends primarily upon the characteristics of the dataset. The goal pursued in this chapter is to identify sampling techniques that are better suited for the incremental classification in terms of accuracy and performance.

11.4.1. Sampling from Decision Tree

One of the problems with random sampling is that it is *sensitive* to the distribution of data: when the dataset is skewed, the randomly chosen samples are not likely to be a good representation of the dataset. Due to this, the resulting decision tree is generated *nondeterministically*, depending on what samples are chosen. We observe that the decision tree built on a dataset can be used to generate a good sample of the dataset. Since the skewedness of distribution is already reflected to the tree itself, we can reliably generate good samples in a more consistent manner from the tree. Such a sampling method is expected to result in a new decision tree with better or comparable quality than random sampling does, let alone efficiency.

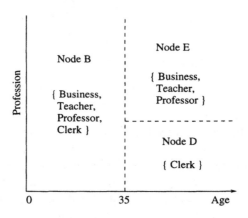

Figure 11.4 The hyperbox representation of a decision tree classifier

Each node in a decision tree includes a set of records. One can view each node of a decision tree as a hyperbox in a multi-dimensional hyperspace composed of the attributes of the dataset. Figure 11.4 shows such hyperboxes of the decision tree derived from the dataset in Figure 11.1(a). Each segment of region corresponds to a leaf node of the decision tree. During construction phase, weighted samples from the nodes of the tree can be extracted easily. If the samples are to be from the leaf nodes, for example, we extract samples from nodes B, D, and E.

11.4.2.   Sampling Techniques for Incremental Classification

The weighted samples extracted from the decision tree are expected to show better characteristics than the ones drawn directly from the dataset. These samples will enable a better decision tree classifier to be constructed in an incremental fashion. We discuss a couple of sampling techniques that can be used within ICE for incremental classification in this section.

Random sampling.   Weighted samples are chosen randomly from each node of the decision tree built for the current incremental partition. We denote this method as ICE-random. While this method is simple and easy, the probability of right samples to be selected is still low. Therefore, the classifier built on random weighted samples may be still inconsistent in terms of quality. The accuracy of the resulting classifier, for instance, may fluctuate depending on the samples chosen.

An alternative to random sampling is *stratified sampling* by which samples are randomly chosen based on the class labels in proportion to the size of the class in each node (Catlett, 1984). Choosing samples in a class is random. This method may somewhat alleviate the drawback of random sampling. However, the samples chosen randomly within the same class may still be insufficient to represent the records at the node from which the samples are chosen.

Local clustering.   Clustering operation discovers collections of points in the data space so that the points in a cluster are *closer* to each other than to the points in other clusters (Kaufman and Rousseeuw, 1990). Classification is somewhat similar to clustering except that it deals with class labels to find clusters of records with the same class label. If clustering is applied such that it identifies clusters of the points with the *same* class label, the cluster representatives may be a good representation of the dataset that can be used for incremental classification.

Therefore, a clustering algorithm can be applied to the records in each node of an intermediate decision tree to obtain good weighted samples. Samples are the cluster centers found by the clustering algorithm with the weight being the number of records in the cluster. Since clustering is executed against the local partition in each node, this method is called local clustering, and denoted ICE-local when used within ICE. The cost of clustering decreases dramatically as the number of records becomes smaller. Since it is applied to a supposedly small set of records in each node, its cost can be deemed minor compared to the cost on the entire dataset. We can also adopt one of the well-known efficient clustering algorithms such as BIRCH (Zhang et al., 1996) and DBSCAN (Ester et al., 1996).

Conversely, a clustering technique can be applied to each incremental portion to extract samples for incremental classification (**global clustering**, denoted ICE-global). This sampling method can become an option if incremental portions are small enough to run this method. However, the cost of this method could be very expensive because of the assumed large size of the dataset. For this reason, it will be used only for comparison.

11.4.3.   Cost of Incremental Classification Using ICE

It is obvious that sampling adds an extra cost to the task of classification. Rather than discussing the cost of sampling itself, we put more emphasis on the cost for obtaining the final decision tree classifier using the sampling techniques presented in the previous section.

Let $T(A, D)$ be the cost function for building a decision tree classifier for a dataset D using an algorithm A. In general, $T(A, D)$ takes the form of $|D| \cdot f(A, D)$, where $f(A, D)$ depends primarily on the size of D. When $|D|$ exceeds some threshold so that A must run out-of-core, or A involves expensive operations such as sorting, $f(A, D)$ will become asymptotically large, elevating the total cost higher.

For a constant r (sampling rate) such that $0 < r \le 1$, the cost of ICE at epoch i using an algorithm A is composed of three components: building a tree t_i for D_i, extracting samples from t_i, and constructing the final classifier C_i for U_i. Therefore, the cost for ICE is defined:

$$T_{ICE}(A, D) = T(A, D_i) + r \cdot |D_i| + T(A, U_i)$$

The ratio R of the cost of ICE to the cost of building a decision tree from scratch is:

$$R = \frac{T(A, D_i) + r \cdot |D_i| + T(A, U_i)}{T(A, D)}$$

Assume that each partition D_i is of the same size, namely n. Then, at any epoch i, $|D| = i \cdot n$ and $|U_i| = r \cdot i \cdot n$. Assuming $f(A, D)$ is a constant, R becomes approximately $\frac{1 + i \cdot r}{i}$, which converges to r for a large i. Note that the cost of extracting samples from t_i $(r \cdot |D_i|)$ can be ignored compared to $T(A, D)$. In practice, however, since $|D_i|, |U_i| \ll |D|$, the functions $f(A, D_i)$ and $f(A, U_i)$ are likely much smaller than $f(A, D)$. This implies that $\frac{1+i \cdot r}{i}$ is the upper bound of R. Therefore, it is clear that as the dataset grows over time, building a decision tree for the entire dataset becomes more expensive, whereas the cost of running ICE remains very low, regardless of the size of the entire dataset.

11.5. EMPIRICAL RESULTS

In the experiments, we used a simple greedy search as the base decision tree classifier algorithm used in ICE. At each node, the algorithm sorts *all* the numerical attributes of the records to calculate the **gini** index. The histograms of categorical attributes are discovered at the root node and passed down the tree iteratively. The attribute with the lowest **gini** index is chosen to split the records at each node. Note, however, that any decision tree algorithm can be used in ICE.

The experimental results of ICE are compared against that of random sampling (from the entire dataset). The sampling techniques used are **ICE-local** and **ICE-global**. **ICE-random** is omitted on purpose since its result is not comparably better than random sampling from the entire dataset. In the experiments, a simple k-means clustering algorithm (Kaufman and Rousseeuw, 1990) was used with the cluster seeds chosen randomly. The value k varies depending on the size of the node and the sampling rate. The experiments were performed on a DEC Alpha workstation with Digital Unix OSF 1 V4.0B and 128 MB main memory. More complete experimental results can be found in (Yoon, 2000).

11.5.1. Experimental Results with Real Datasets

The **letter** dataset (Blake et al., 1998) contains records that represent characteristics of hand-written alphabet letters The dataset contains 15,000 and 5,000 records of 16 attributes for training and test datasets, respectively, with 3 classes and 5 epochs.

Figure 11.5 shows the predictive accuracy of each C_i after the first epoch (where $T_1 = C_1$) for this dataset. The straight line (denoted **Cumulative**) denotes the accuracy of the decision tree for the cumulative dataset $\bigcup D_i$ at epoch i, using the simple greedy search algorithm. The dotted straight line (denoted **Partition** i) represents the accuracy of the decision tree T_i built on the current partition D_i. $x\%$ records of D_i are extracted from T_i as samples, where x varies within 1, 2, 5, 10, 20, 30, and 50. The random samples are chosen from the *cumulative* dataset

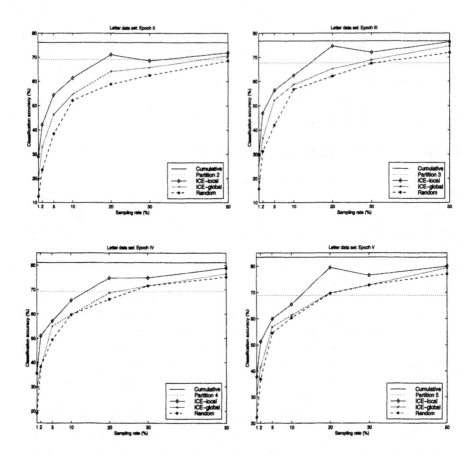

Figure 11.5 Experimental results for letter dataset with five epochs

$\bigcup D_i$ at epoch i, to which the simple greedy search method is applied (denoted **Random**).

The results of this experiment indicate that **ICE-local** consistently results in a better predictive accuracy at any sampling rate at any epoch. **ICE-global** is slightly better than random sampling. One could speculate that the impurity of the samples increases as epochs evolve and samples are chosen repeatedly. However, such effect of cumulating impurity seems either non-existent or very minor for this dataset. At about 15% sampling rate or higher, ICE becomes better than T_i and gets close to the accuracy of the decision tree built on the cumulative dataset.

Table 11.1 collectively shows the measures of decision tree quality for 15% sampling rate. Although 15% of the dataset cannot be regarded small enough, the purpose of this experiment is to prove the weighted sampling of ICE outperforms random sampling. Each number in the

Table 11.1 The quality measures for letter dataset at 15% sampling rate

Measure	Method	Epoch 3	Epoch 4	Epoch 5
Accuracy (%)	Cumulative	78.8	81.1	83.4
	Random	60.3 (1.12)	63.4 (0.90)	66.2 (0.99)
	ICE-local	65.1 (0.95)	66.8 (0.76)	69.1 (0.50)
	ICE-global	63.2	63.7	66.6
Number of leaves	Cumulative	1358	1621	1849
	Random	370 (5.59)	453 (12.15)	527 (17.92)
	ICE-local	224 (4.34)	274 (5.86)	390 (6.24)
	ICE-global	615	783	927
Execution time (secs)	Cumulative	6.02	8.08	10.37
	Random	0.78	1.06	1.29
	ICE-local	2.46	2.77	2.95
	ICE-global	0.83	1.12	1.42

table is an average over 10 executions and the numbers in parentheses are standard deviation that shows how the results vary on different runs for the same dataset. Other parameters remain unchanged.

In terms of accuracy, it is obvious that ICE-local *consistently* results in better decision trees than random sampling does. Note that the standard deviation in the table reveals how the results vary in different runs. In terms of tree size, it also indicates that the tree generated by ICE-local is more compact than that by random sampling, hence making it more comprehensive. The results also conform to the cost model given in the previous section. With more realistically large datasets, the speedup will be more evident.

11.5.2.   Experimental Results with Large Synthetic Datasets

In this section, ICE is tested for larger datasets with more complicated rulesets. The datasets used for experiments are synthetically created using the data generator described in (Yoon, 2000), where the detailed configurations of the datasets used in the experiments can be found.

The experiments were conducted on a Sun Ultra Enterprise 10000 with Solaris OS 5.6, 8 processors, and 2 GB main memory. The size of the dataset used is approximately 72 MB with 1 million records. When the benchmarking algorithm (a simple greedy search) was applied to such dataset, we had to give up collecting the results because the algo-

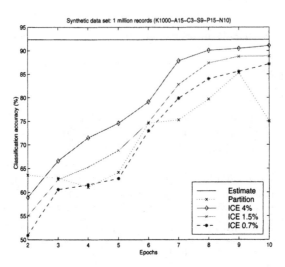

Figure 11.6　Experimental results for synthetic dataset (1 million records, 3 classes)

rithm was still executing after 36 hours even in such a powerful machine. Instead, a smaller dataset of 100,000 records with the same configuration was created and used to approximate the accuracy of the dataset for comparison purpose.

The experimental results are plotted in Figure 11.6. The dataset is divided into 10 equal-sized partitions (i.e., epochs), each with 100,000 records. The straight line on the top (denoted **Estimate**) is the estimated accuracy using the smaller dataset with the same configuration. The dotted line (denoted **Partition**) indicates the accuracy of each partition by the benchmarking algorithm, and each of the lines denoted **ICE** x% shows the accuracy of ICE at x% sampling rate.

This experimental result shows some important implications of ICE. First of all, ICE can be applied to large datasets that cannot otherwise be directly used to construct classifiers from. The total execution time of ICE (up to the 10th epoch) is approximately 90 minutes with a slight storage overhead for samples and additional time to build a decision tree for each partition. The maximum memory required is about 25 MB. Such overhead can be well justified by the achieved accuracy and the overall execution time because the cost of building a classifier from the entire dataset cannot even be measured. Secondly, the way the partitions are handled by ICE also hides the details of computing. It means that ICE can run on any of various computing configurations: a single processor (serial computing), a processor cluster (parallel computing), a network cluster (distributed computing), or any combination of them. Thirdly, it suggests how ICE be utilized for dynamically growing large datasets. The figure shows that the accuracy results of ICE up to the 6th epoch are comparably lower than the overall estimate. Since the entire dataset at an early epoch is presumably small, it would make sense to build a classifier using the entire dataset at early epochs while still generating samples for later epochs. This hybrid approach would result in more accurate classifiers at early epochs. When the accuracy converges to a satisfactory level after a few early epochs, ICE is applied alone to quickly construct a prototype classifier (which will be used until the next epoch). Finally, the results imply that larger sample size may improve the overall accuracy results of ICE. However, sampling rate need not be significantly high. As is shown by other experiments presented in (Yoon, 2000), only a marginal portion of the entire data is required to build a classifier of comparable accuracy with much less cost.

References

Agrawal, R., Imielinski, T., and Swami, A. (1993). Database Mining: A Performance Perspective. *IEEE Transactions on Knowledge and Data Engineering*, 5(6):914–925.

AlSabti, K. (1998). *Efficient Algorithms for Data Mining*. PhD thesis, Syracuse University.

AlSabti, K., Ranka, S., and Singh, V. (1998). Coulds: A decision tree classifier for large datasets. In *International Conference on Knowledge Discovery and Data Mining*, pages 2–8, New York, NY.

Blake, C., Keogh, E., and Merz, C. J. (1998). Uci repository of machine learning databases. The URL is http://www.ics.uci.edu/~mlearn/ML-Repository.html.

Breiman, L., Friedman, J. H., Olshen, R. A., and Stone, C. J. (1984). *Classification and Regression Trees*. Wadsworth, Belmont.

Catlett, J. (1984). *Megainduction: Machine Learning on Very Large Databases*. PhD thesis, University of Sydney.

Chan, P. K. and Stolfo, S. J. (1997). On the accuracy of meta-learning for scalable data mining. *Intelligent Information Systems*, 8:5–28.

Cheeseman, P., Kelly, J., Self, M., Stutz, J., and Taylor, W. (1988). Autoclass: A bayesian classification system. In *The 5th Internaltion Conference on Machine Learning*, pages 54–64, San Francisco, CA.

Ester, M., Kriegel, H., Sander, J., and Xu, X. (1996). A density-based algorithm for discovering clusters in large spatial databases with noise. In *International Conference on Knowledge Discovery and Data Mining*, pages 226–231, Portland, OR.

Gehrke, J., Ganti, V., Ramakrishnan, R., and Loh, W.-Y. (1999). Boat: Optimistic decision tree construction. In *ACM SIGMOD Conference*, pages 169–180, Philadelphia, PA.

Gehrke, J., Ramakrishinan, R., and Ganti, V. (1998). Rainforest: A framework for fast decision tree classification of large datasets. In *Internation Conference on Very Large Databases*, pages 416–427, New York, NY.

Goldberg, D. E. (1989). *Genetic Algorithms in Search, Optimization and Machine Learning*. Morgan Kaufman, San Francisco, CA.

James, M. (1985). *Classification Algorithms*. Wiley and Sons, New York, NY.

Kaufman, L. and Rousseeuw, P. (1990). *Finding Groups in Data: An Introduction to Cluster Analysis*. Wiley and Sons, New York, NY.

Lim, T.-S., Loh, W.-Y., and Shih, Y.-S. (1997). An Emperical Comparison of Decision Trees and Other Classification Methods. Technical Report TR 979, Department of Statistics, University of Wisconsin, Madison.

Mitchie, D., Spiegelhalter, D. J., and Taylor, C. C. (1994). *Machine Learning, Neural and Statistical Classification*. Ellis Horwood, New York, NY.

Morimoto, Y., Fukuta, T., Matsuzawa, H., Tokuyama, T., and Yoda, K. (1998). Algorithms for mining association rules for binary segmentations of huge categorical databases. In *International Conference on Very Large Databases*, pages 380–391, New York, NY.

Quinlan, J. R. (1986). Induction of decision trees. *Machine Learning*, 1:81–106.

Quinlan, J. R. (1993). *C4.5: Programs for Machine Learning*. Morgan Kaufman, San Francisco, CA.

Quinlan, J. R. and Rivest, R. L. (1989). Inferring decision trees using minimum description length principle. *Information and Computation*, 80:227–248.

Ripley, B. D. (1996). *Pattern Recognition and Neural Networks*. Cambridge University Press, Cambridge, UK.

Shafer, J., Agrawal, R., and Mehta, M. (1996). Sprint: A scalable parallel classifier for data mining. In *International Conference on Very Large Databases*, pages 544–555, Bombay, India.

Weiss, S. M. and Kulikowski, C. A. (1991). *Computer Systems that Learn: Classification and Prediction Methods from Statistics, Neural Nets, Machine Learning, and Expert Systems*. Morgan Kaufman, San Mateo, CA.

Yoon, H. (2000). *Efficient Algorithms and Software for Mining Sparse, High-dimensional Data*. PhD thesis, University of Florida.

Zhang, T., Ramakrishinan, R., and Livny, M. (1996). Birch: An efficient data clustering method for very large databases. In *ACM SIGMOD Conference*, pages 103–114, Montreal, Canada.

IV

UNCONVENTIONAL METHODS

Chapter 12

INSTANCE CONSTRUCTION VIA LIKELIHOOD-BASED DATA SQUASHING

David Madigan
Soliloquy, Inc., New York City, USA.
dmadigan@soliloquy.com

Nandini Raghavan
Technology Research & Development, DoubleClick, Inc., New York City, USA.
nraghavan@doubleclick.net

William DuMouchel
AT&T Shannon Labs, Florham Park, New Jersey, USA.
dumouchel@research.att.com

Martha Nason
Department of Biostatistics, University of Washington, Seattle, USA.
mnason@biostat.washington.edu

Christian Posse
Talaria, Inc., Seattle, USA.
posse@talarianic.com

Greg Ridgeway
RAND, Santa Monica, California, USA.
gregr@rand.org

Abstract   Squashing is a lossy data compression technique that preserves statistical information. Specifically, squashing compresses a massive dataset to a much smaller one so that outputs from statistical analyses carried out on the smaller (squashed) dataset reproduce outputs from the same statistical analyses carried out on the original dataset. Likelihood-based data squashing (LDS) differs from a previously published squashing algorithm insofar as it uses a statistical model to squash the data. The results show that LDS provides excellent squashing performance even when the target statistical analysis departs from the model used to squash the data.

Keywords: Instance construction, instance construction, data squashing.

12.1.   INTRODUCTION

Massive datasets containing millions or even billions of observations are increasingly common. Such data arise, for instance, in large-scale retailing, telecommunications, astronomy, computational biology, and internet logging. Statistical analyses of data on this scale present new computational and statistical challenges. The computational challenges derive in large part from the multiple passes through the data required by many statistical algorithms. When data are too large to fit in memory, this becomes especially pressing. A typical disk drive is a factor of $10^5 - 10^6$ times slower in performing a random access than is the main memory of a computer system (Gibson et al., 1996). Furthermore, the costs associated with transmitting the data may be prohibitive. The statistical challenges are many: what constitutes "statistical significance" when there are 100 million observations? how do we deal with the dynamic nature of most massive datasets? how can we best visualize data on this scale?

Much of the current research on massive datasets concerns itself with *scaling up* existing algorithms - see, for example, (Bradley et al., 1998) or (Provost and Kolluri, 1999). In this paper we focus on the alternative approach of *scaling down* the data. Most of the previous work in this direction has focused on sampling methods such as random sampling, stratified sampling, duplicate compaction (Catlett, 1991), and boundary sampling (Aha et al., 1991; Syed et al., 1999). Recently DuMouchel et al. (DuMouchel et al., 1999) [DVJCP] proposed an approach that instead *constructs* a reduced dataset. Specifically their data squashing algorithm seeks to compress (or "squash") the data in such a way that a statistical analysis carried out on the squashed data provides the same outputs that would have resulted from analyzing the entire dataset. Success with respect to this goal would deal very effectively with the computational challenges mentioned above - the entire armory of statistical tools could then work with massive datasets in a routine fashion and using commonplace hardware.

DVJCP's approach to squashing is model-free and relies on moment-matching. The squashed dataset consists of a set of pseudo data points chosen to replicate the moments of the "mother-data" within subsets of a partition of the mother-data. DVJCP explore various approaches to partitioning and also experiment with the order of the moments. On a logistic regression example where the mother-data contains 750,000 observations, a squashed dataset of 8,443 points outperformed a simple random sample of 7,543 points by a factor of almost 500 in terms of mean square error with respect to the regression coefficients from the mother-data. DVJCP provide a theoretical justification of their method by considering a Taylor series expansion of an arbitrary likelihood function. Since this depends on the moments of the data, their method should work well for any application in which the likelihood is well-approximated by the first few terms of a Taylor series, at least within subsets of the partitioned data. The empirical evidence provided to date is limited to logistic regression.

In this paper we consider the following variant of the squashing idea: suppose we declare a statistical model in advance. That is, suppose we use a particular statistical model to squash the data. Can we thus improve squashing performance? Will this improvement extend to models other than that used for the squashing? We refer to this approach as "likelihood-based data squashing" or LDS.

LDSis similar to DVJCP's original algorithm (or DS) insofar as it first partitions the dataset and then chooses pseudo data points corresponding to each subset of the partition. However the two algorithms differ in how they create the partition and how they create the pseudo data points. For instance, in the context of logistic regression with two continuous predictors, Figure 12.1 shows the partitions of the two-dimensional predictor space generated by the two algorithms for a single value of the dichotomous response variable. The DS algorithm partitions the data along certain marginal quantiles, and then matches moments. The LDS algorithm partitions the data using a likelihood-based clustering and then selects pseudo data points so as to mimic the target sampling or posterior distribution. Section 12.2 describes the algorithm in detail.

In what follows, we explore the application of LDS to logistic regression and neural networks.

Note that both the DS and LDS algorithms produce pseudo data points with associated weights. Use of the squashed data requires software that can use these weights appropriately.

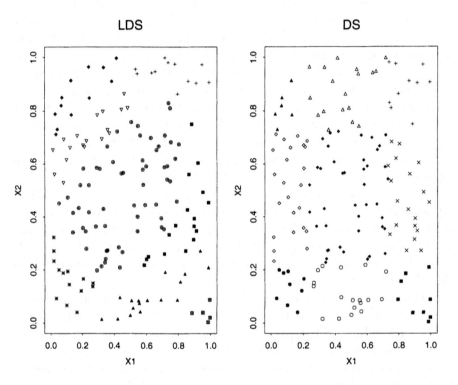

Figure 12.1 Data partitions created by LDS and DS. Note that the partition created by DS is a rectangular grid whereas the partition created by LDS depends on the likelihood function and can be irregular.

12.2.   THE LDS ALGORITHM

We motivate the LDS algorithm from a Bayesian perspective. Suppose we are computing the distribution of some parameter θ posterior to three data points d_1, d_2, and d_3 (the mother-data). We have:

$$Pr(\theta \mid d_1, d_2, d_3) \propto Pr(d_1 \mid \theta) Pr(d_2 \mid \theta) Pr(d_3 \mid \theta) Pr(\theta).$$

Now suppose $Pr(d_1 \mid \theta) \approx Pr(d_2 \mid \theta)$, at least for the values of θ with non-trivial posterior mass. Then one can *construct* a pseudo data point d^* such that

$$(Pr(d^* \mid \theta))^2 \approx Pr(d_1 \mid \theta) Pr(d_2 \mid \theta).$$

A squashed dataset comprising d^* with a weight of 2 and d_3 with a weight of 1 (see Table 12.1) will approximate the analysis posterior to the entire mother-data.

Table 12.1 Simple example of squashing when $Pr(d_1 \mid \theta) \approx Pr(d_2 \mid \theta)$. LDS constructs the pseudo data point d^* so that $Pr(d_1 \mid \theta) Pr(d_2 \mid \theta) Pr(d_3 \mid \theta) \approx (Pr(d^* \mid \theta))^2 Pr(d_3 \mid \theta)$.

Mother-data		Squashed-data	
Instance	*Weight*	*Instance*	*Weight*
d_1	1	d^*	2
d_2	1		
d_3	1	d_3	1

In practice, for every mother-data point d_i, LDS first evaluates $Pr(d_i \mid \theta)$ at a set of k values of θ, $\{\theta_1, \ldots, \theta_k\}$ to generate a *likelihood profile* $(Pr(d_i \mid \theta_1), \ldots, Pr(d_i \mid \theta_k))$ for each d_i. Then LDS clusters the mother-data points according to these likelihood profiles. Finally LDS constructs one or more pseudo data points from each cluster and assigns weights to the pseudo data points that are functions of the cluster sizes.

Note that since LDS clusters the mother data points according to their likelihood profiles, the resultant clusters typically bear no relationship to the kinds of clusters that would result from a traditional clustering of the data points. Figure 12.1, for example, shows LDS constructing several clusters containing data points with disparate (x_1, x_2) coordinates.

12.2.1.   Detailed Description

Let observations $y = (y_1, \ldots, y_n)$ be realized values of random variables $Y = (Y_1, \ldots, Y_n)$. Suppose that the functional form of the probability density function $f(y; \theta)$ of Y is specified up to a finite vector of unknown parameters $\theta \in \Re^p$. Denote by $l(\theta; y)$ the log likelihood of θ, that is, $l(\theta; y) = \log f(y; \theta)$ and denote by $\hat{\theta}$ the value of θ that maximizes $l(\theta; y)$.

The base version of LDS (base-LDS) proceeds as follows:

[SELECT] *Select Values of θ.* Select a set of k values of θ according to a central composite design centered on $\breve{\theta}$. $\breve{\theta}$ is an estimate of $\hat{\theta}$ generally based on at most one pass through the mother-data. A central composite design (Box et al., 1978) chooses $k = 1 + 2p + 2^p$ values of θ: one central point ($\breve{\theta}$), $2p$ "star" points along the axes of θ, and 2^p "factorial" points at the corners of a cube centered on $\breve{\theta}$. Figure 12.2 illustrates the design for $p = 3$. This design is a basic standard in response surface mapping (Box and Draper, 1987). Section 12.3 below addresses the exact locations of the star and factorial points.

[PROFILE] *Evaluate the Likelihood Profiles.* Evaluate $l(\theta_j; y_i)$ for $i = 1, \ldots, n$ and $j = 1, \ldots, k$. In a single pass through the mother-data, this creates a likelihood profile for each observation.

[CLUSTER] *Cluster the Mother-Data in a Single Pass.* Select a random sample of $n' < n$ datapoints from the mother-data to form the initial cluster centers. For the remaining $n - n'$ datapoints, assign each datapoint y_i to the cluster c that minimizes:

$$\sum_{j=1}^{k} \left(l(\theta_j; y_i) - \bar{l}_c(\theta_j;) \right)^2$$

where $\bar{l}_c(\theta_j;)$ denotes the average of the log likelihoods at θ_j for those data points in cluster c.

[CONSTRUCT] *Construct the Pseudo Data.* For each of the n' clusters, construct a single pseudo datapoint. Consider a cluster containing m datapoints, $(y_{i_1}, \ldots, y_{i_m})$. Let y_i^* denote the corresponding pseudo datapoint. The algorithm initializes y_i^* to $\frac{1}{m} \sum_k y_{i_k}$, i.e., the cluster mean, and then optionally refines y_i^* by numerically minimizing:

$$\sum_{j=1}^{k} \left((m \times l(\theta_j; y_i^*)) - \sum_{k=1}^{m} l(\theta_j; y_{i_k}) \right)^2 .$$

The idea here is to chose the pseudo datapoint so that its (weighted) contribution to the likelihood matches the total contribution of the datapoints in the cluster as closely as possible. In practice refining the cluster mean seems to provide little or no improvement in squashing performance so the results reported in this paper do not include the optional step.

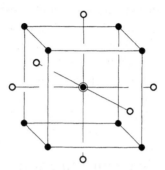

Figure 12.2 Central composite design for three variables. The open circles represent "star" points. The filled circles represent the "factorial" points.

As described, the algorithm requires two passes over the mother-data: one to estimate $\check{\theta}$, and one to evaluate the likelihood profiles and perform the clustering. The first pass can be omitted in favor of an estimate of $\check{\theta}$ based on a random sample, although this can adversely affect squashing performance - see Section 12.5 below.

There exist a variety of elaborations of the base algorithm, some of which we discuss in what follows. For large p, the central composite design will choose an unnecessarily large set of values of θ at the SELECT phase. The literature on experimental design (see, for example, (Box et al., 1978)) provide a rich array of fractional factorial designs that efficiently scale with p. The clustering algorithm in base-LDS can also be improved; Zhang et al. (Zhang et al., 1996) describe an alternative that could readily provide a replacement for the CLUSTER phase. Other elaborations include using alternative clustering metrics at the CLUSTER phase, varying both the number of pseudo points and the construction algorithm at the CONSTRUCT phase, and iterating the entire LDS algorithm. Some but not all of these elaborations require extra passes over the mother-data.

12.3.   EVALUATION: LOGISTIC REGRESSION

To evaluate the performance of LDS we conducted a variety of experiments with datasets of various sizes. In each case our primary goal was to compare the parameter estimates based on the mother-data with the corresponding estimates based on the squashed data. To provide a baseline we also computed estimates based on a simple random sample. We provide results both for simulated data and for the AT&T data from DVJCP. Following DVJCP we report results in the form of residuals from the mother-data parameter estimates, that is, (reduced-data parameter estimate - mother-data parameter estimate). The residuals are standardized by the standard errors estimated from the mother-data and are averaged over all the parameters in the pertinent model.

Note that reproducing parameter estimates represents a more challenging target than reproducing predictions since the former requires that we obtain high quality estimates for *all* the parameters. Section 12.3.4 below shows that accurate parameter estimate replication does result in high quality prediction replication.

12.3.1. Small-Scale Simulations

Implementation of base-LDS requires an initial estimate $\check{\theta}$ of $\hat{\theta}$ and a choice of locations for the k values of θ used in the central composite design. We carried out extensive experimentation with small-scale simulated mother-data in order to understand the effects of various possible choices on squashing performance.

For the initial estimate $\check{\theta}$ of $\hat{\theta}$ we considered three possibilities: $\hat{\theta}_{SRS}$, $\hat{\theta}_{ONE}$, and $\hat{\theta}$. $\hat{\theta}_{SRS}$ is a maximum likelihood estimator of θ based on a 10% random sample, $\hat{\theta}_{ONE}$ is an approximate maximum likelihood estimator of θ based on a single step of the standard logistic regression Newton-Raphson algorithm (this requires a single pass through the mother-data), and $\hat{\theta}$ is the maximum likelihood estimator of θ based on the mother-data.

In the central composite design, let d_F denote the distance of the 2^p "factorial points" from $\check{\theta}$ and let d_S denote the distance of the $2p$ "star" points from $\check{\theta}$, both distances in standard error units. Here we considered $d_F = \{0.1, 0.5, 1, 3\}$ and $d_S = \{0.1, 0.5, 1, 3\}$.

In each case, the mother-data consisted of 1000 observations generated from the following logistic regression model with $\theta = (\beta_1, \ldots, \beta_5)$:

$$\log \frac{Pr(Y = 1)}{1 - Pr(Y = 1)} = \beta_1 X_1 + \beta_2 X_2 + \beta_3 X_3 + \beta_4 X_4 + \beta_5 X_5 \quad (12.1)$$

with $X_1 \equiv 1$, $X_2, X_3, X_4, X_5 \sim U(0, 1)$ and $\beta_1, \ldots, \beta_5 \sim U(0, 0.5)$.

For each of 100 simulated mother-datasets from this model, LDS generated 48 squashed datasets corresponding to the 48 ($3 \times 4 \times 4$) design settings. Parameter estimates based on each of these, as well as on an SRS sample were computed. The LDS and SRS datasets were of size 100.

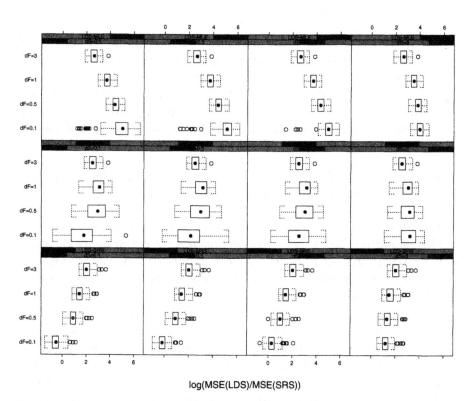

Figure 12.3 Small Scale Simulation Results. Each boxplot shows a particular setting of $\breve{\theta}$, d_F, and d_S. The horizontal axes show the log-ratio of the mean square error from random sampling to the mean square error from LDS.

Figure 12.3 shows boxplots of the standardized residuals of the parameter estimates. Several features are immediately apparent:

- With appropriate choices for d_F, LDS outperforms random sampling for all three settings of $\breve{\theta}$. Note that the results are shown on a $\log_{10}$ scale; for instance, for LDS-MLE with $d_S = 0.1$ and $d_F = 0.1$, LDS outperforms SRS by a factor of about 10^5.

- Squashing performance improves as the quality of $\breve{\theta}$ improves from $\hat{\theta}_{\mathrm{SRS}}$ to $\hat{\theta}_{\mathrm{ONE}}$ to $\hat{\theta}$.

- There is a dependence between the size of d_F and the quality of $\breve{\theta}$. For $\breve{\theta} = \hat{\theta}_{\mathrm{SRS}}$, $d_F = 3$ is the optimal setting amongst the four choices. For $\breve{\theta} = \hat{\theta}_{\mathrm{ONE}}$, several choices of d_F yield equivalent performance. For $\breve{\theta} = \hat{\theta}$, $d_F = 0.1$ is the optimal setting amongst the four choices. In practice, the following suggested values for d_F yield reasonable performance: for $\breve{\theta} = \hat{\theta}_{\mathrm{SRS}}, \hat{\theta}_{\mathrm{ONE}}$, and $\hat{\theta}$ respectively, set $d_F = 3, 1$, and 0.1.

- The choice of d_S has a relatively small effect on squashing performance. Setting $d_S \equiv d_F$ is a reasonable default.

Since $\breve{\theta}$ defines the center of the design matrix where LDS evaluates the likelihood profiles, it is hardly surprising that performance degrades as $\breve{\theta}$ departs from $\hat{\theta}$. It is evidently more important to cluster datapoints that have similar likelihoods in the region of the maximum likelihood estimator (which with large datasets will be close to the posterior mean) than to cluster datapoints that have similar likelihoods in regions of negligible posterior mass. What is perhaps somewhat surprising is the extent to which the design points need to depart from $\breve{\theta}$ when $\breve{\theta} \neq \hat{\theta}$. In that case it is best to evaluate the likelihood profiles at a diffuse set of values of θ most of which are far out in the tails of θ's posterior distribution. In fact, choosing d_S and d_F as large as 10 still gives acceptable performance when $\breve{\theta} \neq \hat{\theta}$. This implies that when LDS doesn't have a very good estimate of $\hat{\theta}$, it needs to ensure a very broad coverage of the likelihood surface.

12.3.2.   Medium-Scale Simulations

Here we consider the performance of LDS in a somewhat larger-scale setting. In particular, we simulated mother-datasets of size 100,000 from the logistic regression model specified by (12.1) again with $X_1 \equiv 1$, $X_2, X_3, X_4, X_5 \sim U(0,1)$ and $\beta_1, \ldots, \beta_5 \sim U(0, 0.5)$. Figure 12.4 shows the results for different choices of $\breve{\theta}$.

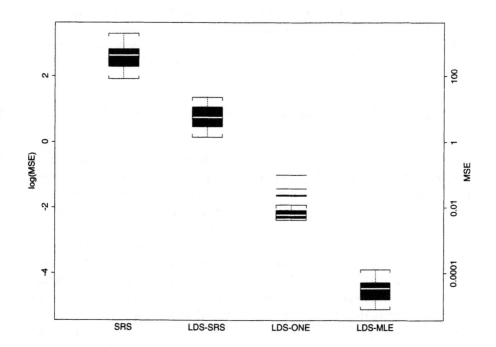

Figure 12.4 Performance of Base-LDS for 30 repetitions of the medium-scale simulated data. "SRS" refers to the performance of a 1% random sample. "LDS-SRS" refers to base-LDS with $\check{\theta} = \hat{\theta}_{\text{SRS}}$ (i.e., a maximum likelihood estimator of θ based on a 1% random sample), "LDS-ONE" refers to base-LDS with $\check{\theta} = \hat{\theta}_{\text{ONE}}$ (i.e., a maximum likelihood estimator of θ based on a single pass through the mother-data), and "LDS-MLE" refers to base-LDS with $\check{\theta} = \hat{\theta}$ (i.e., the maximum likelihood estimator of θ based on the mother-data). For LDS-SRS and LDS-ONE we set $d_F \equiv d_S \equiv 3$ whereas for LDS-MLE we set $d_F \equiv d_S \equiv 0.25$. Note that the vertical axis is on the log scale.

Clearly setting $\check{\theta} = \hat{\theta}_{\mathrm{SRS}}$ yields substantially poorer squashing performance than either $\check{\theta} = \hat{\theta}_{\mathrm{ONE}}$ or $\check{\theta} = \hat{\theta}$. However, Section 12.5 below describes how this can be alleviated with an iterative version of LDS that achieves squashing performance comparable to that for $\check{\theta} = \hat{\theta}$, but starting with $\check{\theta} = \hat{\theta}_{\mathrm{SRS}}$.

Note that even with 100,000 observations the five parameters in the model specified by (12.1) are often not all significantly different from zero. Experiments with models in which either all of the parameters are indistinguishable from zero or all of the parameters are significantly different from zero yielded LDS performance results that are similar to those reported here. For simplicity we only report the results from model (12.1).

12.3.3.   Larger-Scale Application: The AT&T Data

DVJCP describe a dataset of 744,963 customer records. The binary response variable identifies customers who have switched to another long-distance carrier. There are seven predictor variables. Five of these are continuous and two are 3-level categorical variables. Thus for logistic regression there are 10 parameters. As before we consider 1% random and squashed samples. With 10 parameters, the central composite design requires 1,024 factorial points, 20 star points, and 1 central point for a total of 1,045 points. This would incur a significant computational effort. In place of the fully factorial component of the central composite design, we evaluated two fractional factorial designs, a resolution V design requiring 128 factorial points and a resolution IV design requiring 64 points (Box et al., 1978, p.410). In brief, a Resolution V design does not confound main effects or two-factor interactions with each other, but does confound two-factor interactions with three-factor interaction, and so on. A Resolution IV design does not confound main effects and two-factor interactions but does confound two-factor interactions with other two-factor interactions. Table 12.2 describes the results.

LDS outperforms SRS by a wide margin and also provides better squashing performance than DS in this case.

If the actual parameter estimates from the mother-data are used for $\check{\theta}$ in the first step of the algorithm (i.e. setting $\check{\theta} = \hat{\theta}$), then it is possible to reduce the MSE to 0.01 (k=149). At the other extreme setting $\check{\theta} = \hat{\theta}_{\mathrm{SRS}}$ increases the MSE to 1.04 (k=149).

Table 12.2 Performance of Base-LDS for the AT&T data. k is the number of evaluations of the likelihood per data point. $\frac{SRS}{LDS}$ is the average MSE for simple random sampling (154.04 in this case) divided by the MSE for LDS (i.e., the improvement factor over simple random sampling). HypRect($\frac{1}{2}$) shows the most comparable results from DVJCP (Note that HypRect($\frac{1}{2}$) uses 8,373 observations as compared with 7,450 observations in the other rows).

k	$\check{\theta}$	d_F	d_S	MSE	$\frac{SRS}{LDS}$
85	$\hat{\theta}_{\text{ONE}}$	5	5	0.023	6697
149	$\hat{\theta}_{\text{ONE}}$	**5**	**5**	**0.019**	**8107**
DS HypRect($\frac{1}{2}$)				0.24	642
SRS (10 replications)				154.04	1

12.3.4. Prediction

Our primary goal so far has been to emulate the mother-data parameter estimates. A coarser goal is to see how well squashing emulates the mother-data predictions. Following DVJCP we consider the AT&T data where each observation in the dataset is assigned a probability of being a *Defector*. We used the parameter estimates from a 1% random sample and from a 1% squashed dataset to assign this probability and then compared these with the "true" probability of being a *Defector* from the mother-data model. For each observation in the mother-data, we compute (Probability based on reduced dataset) - (Probability based on the mother-data), multiplied by 10000 for descriptive purposes. Table 12.3 describes the results. LDS performs about two orders of magnitude better than simple random sampling and also outperforms the comparable model-free HypRect($\frac{1}{2}$) method from DVJCP.

12.4. EVALUATION: NEURAL NETWORKS

The evaluations thus far have focused on logistic regression. Here we consider the application of LDS (still using a logistic regression model to perform the squashing) to neural networks. We simulated data from a feed-forward neural network with two input units, one hidden layer with three units, and a single dichotomous output unit (Venables and Ripley, 1997). The left-hand panel of Figure 12.5 compares the test-data misclassification rate using a neural network model based on the mother-data (10,000 points) with the test-data misclassification rate based on either a simple random sample of size 1,000 (indicated by "x") or an

Table 12.3 Comparison of predictions for the AT&T data using logistic regression with all 10 main effects. For each reduced dataset the $N = 744,963$ predictive residuals are defined as (Probability based on reduced dataset) - (Probability based on the mother-data) $\times$ 10,000. Each row of the table describes the distribution of the corresponding residuals for a given reduction method.

Method	Mean	StDev	Min	Max
Random Sample	-41	193	-870	679
LDS	**0.4**	**2**	**-5**	**11**
HypRect($\frac{1}{2}$)	-2	9	-37	34

LDS squashed dataset of size 1,000 (indicated by "o"). In either case, predictions are based on a holdout sample of 1,000 generated from the same neural network model that generated the mother-data. The results are for 30 replications. It is apparent that LDS consistently reproduces the misclassification rate of the mother-data. The right-hand panel of Figure 12.5 compares the predictive residuals (i.e., (Probability based on reduced dataset) - (Probability based on the mother-data)) for the two methods. Table 12.4 shows the results in a format comparable with Table 12.3. These predictive results are not as good as those for the logistic regression analysis of the AT&T data (Table 12.3), but here the application is to a different model class to that used for the squashing and LDS substantially outperforms simple random sampling nonetheless.

Table 12.4 Comparison of neural network predictions for random sampling and LDS. For each reduced dataset the 1,000 residuals from the hold-out data are defined as (Probability based on reduced dataset) - (Probability based on the mother-data). Each row of the table describes the distribution of the corresponding residuals for a given reduction method. The results are averaged over 30 replications.

Method	Mean	StDev	Min	Max
Random Sample	-0.005	0.08	-0.29	0.25
LDS	0.0002	0.02	-0.06	0.07

12.5.   ITERATIVE LDS

Except where noted, the evaluations reported thus far utilize a single pass through the mother-data to compute $\breve{\theta}$. In the case of logistic regression, $\breve{\theta}$ is the output of the first step of the standard Newton-

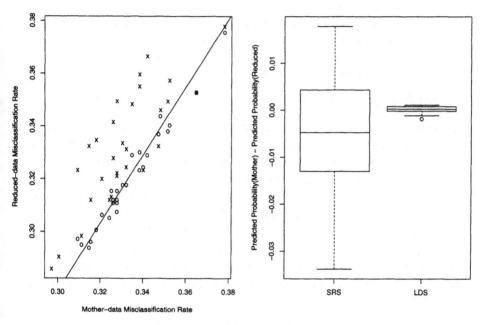

Figure 12.5 Comparison of neural network predictions for random sampling ("x") and LDS ("o"). The left-hand panel shows the misclassification rates for the mother-data predictions versus the reduced-data predictions. The right-hand panel shows the predictive residuals. Both panels reflect performance on 1,000 hold-out datapoints generated from from the same neural network model that generated the mother-data. The figure is based on 30 replications.

Raphson algorithm for estimating $\hat{\theta}$. In fact, this provides a remarkably accurate estimate of $\hat{\theta}$ and results in squashing performance close to that provided by setting $\check{\theta} = \hat{\theta}$.

For those cases where there does not exist a high-quality, one-pass estimate of $\hat{\theta}$, and furthermore many passes through the data are required for an exact estimate of $\hat{\theta}$, iterative LDS (ILDS) provides an alternative approach. ILDS works as follows:

1 Set $\check{\theta} = \hat{\theta}_{SRS}$, an estimate of $\hat{\theta}$ based on a simple random sample from the mother data.

2 Squash the mother-data using LDS (this requires one pass through the motherdata).

3 Use the squashed data to estimate $\hat{\theta}_{LDS}$.

4 Set $\check{\theta} = \hat{\theta}_{LDS}$ and go to (2).

In practice, this procedure requires three or four iterations to achieve squashing performance similar to the performance achievable when $\check{\theta} = \hat{\theta}$ with each iteration requiring a pass through the mother data.

Figure 12.6 shows the MSE reduction achievable with seven iterations. This is based on a 1% squashed sample from mother-data generated from model (12.1) with N=100,000 and 30 repetitions. Based on the experiments reported in Section 12.3.1, we reduced d_F and d_S as the iterations proceeded. Table 12.5 shows the schedule for results in Figure 12.6. Generally the performance is not sensitive to the particular schedule although it is important not to reduce d_F and d_S too quickly.

Table 12.5   "Cooling" schedule for ILDS.

Iteration	d_F	d_S
1	3	3
2	3	3
3	2	2
4	0.5	0.5
$>= 5$	0.25	0.25

12.6.   DISCUSSION

Madigan et al. (Madigan et al., 2000) discuss several possible refinements to the base-LDS algorithm as well as potential limitations. In

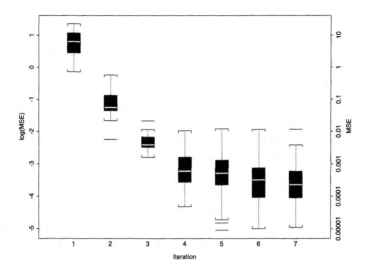

Figure 12.6 Squashing performance of ILDS. The first iteration sets $\check{\theta}$ equal to a maximum likelihood estimator of θ based on a 1% random sample. Subsequent iterations set $\check{\theta}$ to the maximum likelihood estimator based on the squashed 1% sample from the previous iteration.

particular, statistical methods that depend strongly on local data characteristics such as trees and non-parametric regression may be particularly challenging for squashing algorithms. A concern is that minor deviations in the location of the squashed data points may result in substantial changes to the fitted model. In this case, a constructive approach to squashing may be more promising than methods based on partitioning.

Madigan et al. (Madigan et al., 2000) present results that suggest that it is possible to use LDS to achieve a 100-fold reduction in computational effort for variable selection for certain model classes.

We have yet to evaluate LDS with a large number of input variables (i.e., large p). In the neural network context, preliminary experiments suggest that the squashing performance of base-LDS for neural networks does degrade as the number of units in the input layer increases. Including interaction terms in the logistic regression model used for the squashing alleviates the problem somewhat.

References

Aha, D., Kilber, W., and Albert, M. (1991). Instance-based learning algorithms. *Machine Learning*, 6:37–66.

Box, G. and Draper, N. (1987). *Empirical Model Building and Response Surfaces.* John Wiley & Sons.

Box, G., Hunter, W., and Hunter, J. (1978). *Statistics for Experimenters:An Introduction to Design, Data Analysis, and Model Building.* John Wiley & Sons.

Bradley, P., Fayyad, U., and Reina, C. (1998). Scaling clustering algorithms to large databases. In *Proceedings of the Fourth International Conference on Knowledge Discovery and Data Mining,* pages 9–15.

Catlett, J. (1991). Megainduction: A test flight. In *Proceedings of the Eighth International Workshop on Machine Learning,* pages 596–599.

DuMouchel, W., Volinsky, C., Johnson, T., Cortes, C., and Pregibon, D. (1999). Squashing flat files flatter. In *Proceedings of the Fifth ACM Conference on Knowledge Discovery and Data Mining,* pages 6–15.

Gibson, G., Vitter, J., and Wilkes, J. (1996). Report of the working group on storage I/O issues in large-scale computing. *ACM Computing Surveys,* 28.

Madigan, D., Raghavan, N., DuMouchel, W., Nason, M., Posse, C., and Ridgeway, G. (2000). Likelihood-based data squashing: A modeling approach to instance construction. *Journal of Data Mining and Knowledge Discovery.*

Provost, F. and Kolluri, V. (1999). A survey of methods for scaling up inductive algorithms. *Journal of Data Mining and Knowledge Discovery,* 3:131–169.

Syed, N., Liu, H., and Sung, K. (1999). A study of support vectors on model independent example selection. In *Proceedings of the Fifth ACM Conference on Knowledge Discovery and Data Mining,* pages 272–276.

Venables, W. and Ripley, B. (1997). *Modern Applied Statistics with S-PLUS.* Springer-Verlag.

Zhang, T., Ramakrishnan, R., and Livny, M. (1996). Birch: An efficient data clustering method for large databases. In *Proceedings of the 1996 ACM SIGMOD International Conference on Management of Data,* pages 103–114.

Chapter 13

LEARNING VIA PROTOTYPE GENERATION AND FILTERING

Wai Lam
Department of Systems Engineering and Engineering Management
The Chinese University of Hong Kong
Shatin, Hong Kong
wlam@se.cuhk.edu.hk

Chi-Kin Keung
Department of Systems Engineering and Engineering Management
The Chinese University of Hong Kong
Shatin, Hong Kong
ckkeung@se.cuhk.edu.hk

Charles X. Ling
Department of Computer Science
The University of Western Ontario
London, Ontario, Canada N6A 5B7
ling@csd.uwo.ca

Abstract   The family of instance-based learning algorithms have been shown to be effective for learning classification schemes in many domains. However, it demands high data retention rate and is sensitive to noise. We investigate an integration of instance-filtering and instance-averaging techniques to solve the problem. We compare different variants of integration as well as existing learning algorithms such as C4.5 and KNN. Our new framework shows good performance in data reduction while maintaining or even improving classification accuracy in 19 real data sets.

Keywords: Instance-based learning, prototype generation, instance-filtering, machine learning.

13.1.   INTRODUCTION

The family of instance-based learning algorithms are found to be effective in learning a classification scheme in different domains (Dasarathy, B. V., 1990). Unfortunately, they suffer from high data retention rate, computational cost and noise sensitivity. Many researchers try to solve this problem by either scaling up the search algorithm or reducing the data set size. However, scaling up the search algorithm does not reduce data retention rates and does not provide noise tolerance. On the other hand, data reduction can improve the algorithm if representative instances are stored and noisy instances are removed. Traditional data reduction methods adopt instance filtering or prototype averaging. Instance filtering focuses on selecting representative instances while prototype averaging generates artificial prototypes by generalizing or summarizing instances. In this paper, we investigate these two methods and propose an integration of them. Our algorithm, called Prototype Generation and Filtering (PGF), possesses several characteristics. It retains selected instances and learns generalized prototypes as the output. It is effective in data reduction while maintaining or even improving the classification accuracy. It is also insensitive to the order of presentation of instances and provides noise tolerance. Empirical results in 19 real data sets show that our new framework achieves good performance compared with existing algorithms such as KNN and C4.5.

This chapter is organized as follows. Section 13.2 presents related work on instance-filtering and instance-averaging. Section 13.3 describes our new proposed framework. Section 13.4 shows the experimental results and discussion on the results. Conclusions and future research directions are provided in Section 13.5.

13.2.   RELATED WORK

Different communities including machine learning and pattern recognition have proposed different methods for prototype learning in the past few decades. We review related research on instance selection approaches including instance-filtering and instance-averaging. We briefly describe major work in these approaches and point out the strengths and weaknesses of each of them.

13.2.1. Instance-Filtering Methods

In instance-filtering methods, editing rules are used to determine whether an instance should be retained as a prototype or not. Prototype descriptions are represented by those selected instances. Previous proposed methods differ from search direction and locations of instances retained. Some of the filtering techniques are described below.

The earliest editing rule called *Condensed Nearest Neighbor* (CNN) is introduced by Hart (Hart, P. E., 1968). This bottom-up algorithm randomly selects one instance for each class to form the initial prototype set. An instance is retained if it is misclassified by the current prototype set so that a *consistent subset* is obtained. Gates proposes an iterative, top-down variant of CNN called *Reduced Nearest Neighbor* (Gates, G. W., 1972). Starting with all instances as the prototype set, this algorithm deletes an instance if its deletion does not result in any misclassification of other instances. This algorithm also intends to find a consistent subset by retaining instance in the class boundaries. Swonger proposes another variant of CNN called *Iterative Condensation Algorithm* (Swonger, C. W., 1972). ICA allows both deletion and selection of instances within an iteration so that it is not sensitive to the initial state of the algorithm. Several other variants of CNN have been proposed including (Gowda, K. C. and Krisha, G., 1979) and (Ullmann, J. R., 1974).

The *Edited Nearest Neighbor* (ENN) algorithm proposed by Wilson is a top-down algorithm retaining central instances (Wilson, D. L., 1972). Opposed to CNN, ENN eliminates those instances misclassified by its K nearest neighbors. Tomek extends ENN to *unlimited editing* and *all K-NN* algorithms (Tomek, I., 1976). *Unlimited editing* continuously repeats ENN and stops after the subset of prototype becomes stable. In *all K-NN*, ENN is also repeated but different number of nearest neighbors are used in determining the acquisition of instances. The two variants of ENN give better performance in terms of classification accuracy and data reduction, but it requires a higher computational cost than pure ENN.

Ritter et al. introduce a top-down *Selective Nearest Neighbor* (SNN) algorithm to determine a possible minimal subset approximating the decision boundaries (Ritter et al, 1975). It ensures each instance in the original data set is closer to an instance in the selected set of the same class than any other instance of other classes.

Classification accuracy can be improved by removing noisy instances and irrelevant features in instance-based learning. Aha et al. have investigated these two factors in instance-filtering intensively. A noise-tolerant instance filtering called NTGrowth is proposed by Aha and Kibler (Aha, D. W. and Kibler, D., 1989). NTGrowth retains misclassified instances and keeps track of the classification performance of each of them. Noisy data, usually have low accuracy, will be discarded. Later, Aha et al. formalize NTGrowth to the well-known IB2 and IB3 algorithms retaining border instances incrementally (Aha et al., 1991). IB2 is similar to CNN saving misclassified instances except that instances are normalized by the range of attributes and missing values are tolerated while IB3 accepts instances with classification accuracy significantly greater than the frequency of the observed class. However, both methods do not retain correctly classified instances which may be useful in classification. Apart from noise tolerance, three incremental edited NN algorithms are proposed to determine the relative attribute relevance and handle novel attributes (Aha, D. W., 1992). Cost and Salzberg also propose a method called PEBLS using a weighted modified Value Difference Metric (MVDM) to weight symbolic features (Cost, S and Salzberg, S., 1993). A detailed review and evaluation of several major feature weighting methods in lazy learning has been done by Wettschereck et al. (Wettschereck et al., 1997).

Zhang proposes three more instance-filtering techniques, namely, Storage Reduction Instance Based Learning (SRIBL) storing misclassified instances, Typical Instance-Based Learning (TIBL) storing typical instances and Boundary Instance-Based Learning (BIBL) storing atypical instances (Zhang, J., 1992). Similar to CNN, SRIBL retains misclassified instances until all instances can be correctly classified. TIBL identifies new typical instances by considering their performance on previously misclassified instances and BIBL retains least typical instances including boundary and exceptional instances.

Minimum description length (MDL) principle is first applied in instance-filtering by Cameron-Jones (Cameron-Jones, R. M., 1992). Choice of instances constituting the prototype set is encoded in a bit string and a function is designed to calculate the coding cost of a bit string. MDL retains instances which can additionally decrease the coding cost of the current bit string. The classification accuracy of MDL is found to be slightly superior to IB3 with a significant improvement in data reduction rate.

Dasarathy presents a new editing rule to produce a Minimal Consistent Set (MCS) by considering the distance between nearest neighbors of the same class and that between *nearest unlike neighbor* (Dasarathy, B. V., 1994). This method achieves a minimal consistent subset and is insensitive to the initial order of presentation of instances. Instances are also selected in the order of their consistency property.

Recently, Wilson and Martinez propose three top-down data reduction techniques called RT1-RT3 (Wilson, D. R. and Martinez T. R., 1997). RT1 removes an instance if most of its *associate*, instances in the current subset having it as one of their *k*-nearest neighbors (KNN), are classified correctly without it. Noisy instances are usually removed as they can hardly classify their associates correctly while border instances will be retained as their associates tend to be classified correctly with their contribution in KNN classification. RT2 improves RT1 by considering associates of the original data set instead of the selected subset. It also changes the order of instance removal by sorting the instances in the subset by the distance to their nearest unlike neighbor first. This technique makes RT2 insensitive to the order of data presentation. In RT3, ENN is applied before sorting to filter out noise. It obtains a greater data reduction than RT2. RT3 has a high classification accuracy comparable to those of pure KNN with an average of around 14% data retention rate.

13.2.1.1 Advantages of Instance-Filtering Methods.

Firstly, filtering methods are usually simpler and faster. The behavior such as the convergence of simple instance-filtering algorithms can be investigated (Kibler, D. and Aha, D. W., 1988). Despite their simplicity, most filtering methods can gain accuracy comparable or even superior to pure NN classifier and also retain a consistent subset. As filtering techniques select representative instances from the original instance set, they can truly represent the original concept and the learned concept is easily comprehended. Lastly, since filtering rules can be designed to filter out or retain different instances such as border (Aha et al., 1991; Gates, G. W., 1972; Hart, P. E., 1968; Ritter et al, 1975; Wilson, D. R. and Martinez T. R., 1997), central points (Tomek, I., 1976; Wilson, D. L., 1972) and even noise and outliers (Aha et al., 1991; Hart, P. E., 1968; Wilson, D. R. and Martinez T. R., 1997). One can easily apply different rules simultaneously or separately to filter instances. This flexibility can also be extended to other instance selection methods such as instance-averaging.

13.2.1.2 Disadvantages of Instance-Filtering Methods.

However, instance-filtering techniques assume that ideal examples can be found in the original data set. This assumption limits the representation power of filtering methods. The generalization power is also limited by only selecting the original data. Some methods enforcing the edited subset to be consistent may result in overfitting (Kuncheva, L. I., Bezdek, J. C., 1998). Noisy instances at the border are usually selected by boundary instance retaining methods. For methods retaining central points, data reduction rate is usually low compared with instance-averaging (Tomek, I., 1976; Wilson, D. L., 1972). Some methods are sensitive to the order of presentation of instances (Aha et al., 1991; Gates, G. W., 1972; Hart, P. E., 1968).

13.2.2. Instance-Averaging Methods

Instance-averaging methods generate prototype description by averaging or summarizing some representative instances. Unlike instance-filtering, it involves generating artificial prototypes rather than just storing instances appearing in the original data set. We review some of the major previously proposed algorithms below.

An early instance-averaging algorithm is a top-down approach proposed by Chang (Chang, C. L., 1974). The algorithm merges two nearest instances by weighted averaging the two instances if the classification accuracy is not degraded after the merge. The merging process stops when the number of incorrect classifications starts to increase. Therefore the number of prototypes need not be specified before. It is found to be very effective to reduce the size of data set.

Bradshaw introduces the *Disjunctive Spanning* (DS) algorithm which uses weighted averaging on an selected prototype with instances correctly classified by it (Bradshaw, G., 1987). A prototype containing more instances has a higher weight so that its influence on the next merge will be larger. However, the author mentions that the learned prototypes may be non-prototypical.

Kibler and Aha improve DS by adding an *adaptive threshold* (AT) to limit the distance between averaged instances (Kibler, D. and Aha, D. W., 1988). Distance threshold in instance-averaging is first introduced by Sebestyen to prevent creation of non-prototypical instances (Sebestyen, G. S., 1962). Kibler and Aha report that AT improves both classification accuracy and data reduction rate of DS. DS is compared with an instance-filtering method called Growth and it is found that the two methods achieve similar results in both accuracy and reduction rate. However, the authors notice that instance-averaging techniques may result in misclassified instances in the prototype set and are unable to represent concave concepts.

An algorithm called Nested Generalized Exemplar (NGE) storing instances as hyperrectangles is proposed by Salzberg (Salzberg, S., 1991). Instances are generalized to form and resize hyperrectangles after correct predictions. The nested hyperrectangles allow the representation of nested concepts. A variant partitioning the feature space is also considered. However, Wettschereck and Dietterich show that NGE is significantly inferior to KNN on some real data sets (Wettschereck, D. and Dietterich, T. G., 1995). NGE is then improved by avoiding creating overlapping rectangles and using a feature weighted distance metric. The improved NGE is still inferior to KNN. As discussed by the authors, NGE is very sensitive to the shape of decision boundaries of the learned concept. Wettschereck also combines the nearest neighbor and hyperrectangle concepts to form a hybrid algorithm that classifies unseen cases by hyperrectangles if they are enclosed by any of them and by KNN otherwise (Wettschereck, D., 1994). In order to perform KNN, the hybrid method has to store all the instances. Despite its inability to reduce storage, it achieves a high accuracy comparable to KNN with a lower computational cost as hyperrectangles are used.

Datta and Kibler introduce a top-down splitting algorithm called *Prototype Learner* (PL) learning prototypes for each concept by generalization of high quality example partitions (Datta, P. and Kibler, D., 1995). This algorithm works on nominal attributes only and partitions instances of the same class by values of some features. Later they propose a *Symbolic Nearest Mean Classifiers* (SNMC) trying to learn a single symbolic prototype for each concept description (Datta, P. and Kibler, D., 1997a; Datta, P. and Kibler, D., 1997b). SNMC uses the MVDM metric to weigh and define distance of symbolic attributes. It begins by clustering homogeneous instances into different groups using k-means clustering. Cluster means will then become the learned prototypes. Different number of clusters (prototypes) are tried to obtain the best results.

Bezdek at el. propose a modified Chang's averaging method (MCA) learning multiple prototypes (Bezdek et al., 1998). MCA averages instances using simple means instead of weighted means and restricts the merging of instances with the same class label only. MCA is compared with three sequential competitive learning methods including *learning vector quantization* (LVQ), fuzzy LVQ (GLVQ-F) and *dog-rabbit* (DR) models as well as the original Chang's method and a fuzzy $\hat{c}$-means algorithm using resubstitution error rate. Experiments on the iris data show that MCA gains a zero resubstitution error and DR achieves a minimum number of prototypes. Kuncheva and Bezdek further compare the resubstitution error of above algorithms with that of genetic algorithm and a random search approach finding a subset of instance (Kuncheva, L. I., Bezdek, J. C., 1998). They find that generated prototypes usually gain higher resubstitution error than subset prototypes because the former may not match with real instances well.

13.2.2.1 Advantages of Instance-Averaging Methods.

In instance-filtering techniques, it is assumed that the most representative instance can be found in the training set. However, this may not be true and artificial prototypes generated by instance-averaging techniques can be more representative. Instance-averaging technique usually remove border points so that smoother decision boundaries can be obtained (Datta, P. and Kibler, D., 1997a; Datta, P. and Kibler, D., 1997b). They usually gain a larger data reduction rate than instance-filtering techniques. Instance-averaging techniques typically has a greater generalizing power. Through averaging instances, the most common characteristic of a concept can be learned and the learned concept has less risk to overfit the original data. As for noise tolerance, averaging methods can generalize away mislabeled instances in compact region by merging them with other instances rather than just selecting them as prototypes.

13.2.2.2 Disadvantages of Instance-Averaging Methods.

Instance-averaging methods are usually more complex than instance-filtering ones. This limits the study of the behaviors of instance-averaging techniques such as their generality and limitations in terms of the concept descriptions (Kibler, D. and Aha, D. W., 1988). Artificial instances by averaging and relabeling may be non-prototypical (Bradshaw, G., 1987). Therefore, learned prototypes are less reasonable and comprehensible than those selected by instance-filtering methods. A good resubstitution performance is also not guaranteed by relabeling merged instances (Kibler, D. and Aha, D. W., 1988; Kuncheva, L. I., Bezdek, J. C., 1998). The generated subset may even contain mislabeled instance so that unseen cases will be misclassified (Kibler, D. and Aha, D. W., 1988). Besides, simple averaging also fails to describe concave concepts (Kibler, D. and Aha, D. W., 1988). It leads to an inferior classification accuracy in some domains. Though averaging methods can generalize away mislabeled instances in compact regions, they cannot do so on outliers. Also, it is not easy to find an interpretation for the average value of discrete features computed.

13.2.3. Other Methods

Some researchers use stochastic approach to find a subset of real instances as prototypes. Instead of filtering instances one by one by editing rules, stochastic techniques search the space of sets of prototypes to determine the best prototype set. Skalak proposes two stochastic methods using random sampling (RS) and iterative random mutation hill climbing (RMHC) to select a set of prototypes (Skalak, D. B., 1994). In RS, random sampling is repeated and samples with maximum classification accuracy become the prototype set. In RMHC, a prototype set is represented by a binary string and a random bit is mutated in each iteration. The result prototype is the binary string with maximum fitness. RMHC can also be designed to select prototypes and features simultaneously by changing the bit coding. Cameron-Jones adds this search heuristic to the MDL approach to form the Explore algorithm (Cameron-Jones, R. M., 1995). Explore outperforms the original MDL algorithm in both classification accuracy and data reduction rate.

Tirri et al. introduce a probabilistic instanced-based learning algorithm learning predictive distributions of attributes by Bayesian inference to form prototypes instead of using feature values (Trri et al, 1996). Overfitting and sensitivity to the choice of distance metric in feature value prototypes can be avoided by this algorithm. A local metric for nearest neighbor is learned using a prototype subset is proposed by Ricci and Avesani (LASM) (Ricci, F. and Avesani, P., 1999). Given a prototype subset, LASM tries to find a system of asymmetric weights for each prototype. In the learning state, weights of prototypes will be updated by reinforcement step and punishment step for correct classification and misclassification of other instances respectively. The learned weights, as well as the prototypes itself, will be used to classify unseen cases.

13.3.   OUR PROPOSED ALGORITHM

13.3.1.   Motivation

From the perspective of filtering, averaging methods can contribute to the performance of the filtering process. As filtering methods do not conduct generalization on instances, they usually cannot gain a satisfactory level of data reduction. However, with the help of averaging methods, instances in compact regions can be generalized to a few or single prototypes so that the reduction rate can be significantly improved. On the other hand, filtering can assist averaging. For example, filtering can be done after or in the middle of the averaging process so that the filtered prototypes are no longer restricted to original instances which may be less representative than the generalized prototypes. Alternatively, we can design a filtering rule to filter out outliers before applying averaging techniques. In fact, most distant instances are exceptions or outliers. Artificial prototypes formed by merging and averaging distant instances are usually non-prototypical (Bradshaw, G., 1987). Furthermore, averaging methods suffer from their inability to represent concave concepts. Combining averaging with filtering techniques can provide a richer representation power for complex concepts. Kibler and Aha observe that averaging technique may retain misclassified prototypes (Kibler, D. and Aha, D. W., 1988). These prototypes can be removed by specifically designed filtering rules.

In view of the above motivation, we propose a framework, called *Prototype Generation and Filtering* (PGF), which combines the strength of instance averaging and instance filtering methods to generate high quality prototypes. Our objective is to significantly reduce the data set via prototypes while maintaining the same level of, or even better classification performance.

Only few previous works investigate the integration of these two methods in learning a classifier. A similar work is done by Wettschereck using a hybrid nearest-neighbor and nearest-hyperrectangle algorithm which attempts to combine the benefits of generalized and original instances (Wettschereck, D., 1994). However, no filtering technique is used to select instances in this algorithm and all instances are required to perform KNN when testing cases are not in any hyperrectangle. Thus no data reduction is obtained.

13.3.2.   The PGF Algorithm

As shown in Figure 13.1, PGF consists of two components, namely a basic instance generation method and a prototype filtering method. The prototype generation method is derived from a greedy instance-averaging technique. We first describe the basic prototype generation method.

13.3.2.1   The Basic Prototype Generation Method.   Our basic prototype generation method is based on an agglomerative clustering technique. A prototype is represented by a set of data instances together with the sufficient statistics, namely, the total number, mean and standard standard deviation of the instances. Statements 3–8 in Figure 13.1 comprise the basic prototype generation method. It learns prototypes in a top-down fashion. At the beginning, each instance is considered as a prototype. Let P be the current prototype set. At each iteration, all pairwise prototype distance are calculated. Two prototypes with the shortest distance are merged and a new prototype is generated replacing these two prototypes. The new prototype essentially contains all those instances in the original two prototypes. We calculate the mean and standard deviation of all the instances in the new prototype as the generalized representation of that prototype. After this merging process, a new prototype set P is formed. The prototype set P is evaluated by a prototype set score function (PROT_SET_SCORE) to predict the quality of the prototypes. If the prototype set is good, it is stored.

The merging process continues until the number of prototypes decreases to the number of classes in the data set. This method can return the prototype set which attains the highest value in the prototype set score function. The class of all the instances in a prototype is also stored. The majority class of all the instances in a prototype becomes the class of the prototype. After the algorithm terminates, the output prototype set will be used for classifying unseen cases. Suppose an unseen case needs to be classified, its distance between all the learned prototypes are calculated to find the nearest prototype. The class label of the nearest prototype is then assigned to the unseen case.

1	$P = $ *Training Set.*
2	FILTER(P).
3	*max_score* = PROT_SET_SCORE(P).
4	$P' = P$.
5	while (no. of prototypes in P > no. of class)
6	Find two prototypes, x and y with shortest distance in P.
7	MERGE(P, x, y).
8	If (PROT_SET_SCORE(P) >= *max_score*)
9	$P' = P$.
10	*max_score* = PROT_SET_SCORE(P).
11	Return P'.

Figure 13.1 The PGF algorithm.

The prototype set score function is based on classification accuracy. As our objective is to learn prototypes to classify unlabeled instances, classification accuracy on unseen cases is a reasonable indicator to predict the performance of prototypes. We divide the training set into a sub-training set and a tuning set. Prototypes are generated using the sub-training set and the tuning set is used for calculating the prototype set score using classification accuracy. The prototype set with the highest classification accuracy is the output.

The prototype generation method attempts to find common characteristics including the class label information by merging instances. It is reasonable to consider the information about the class labels during merging. In order to find homogeneous prototypes, some previous works just split the training set by each class and learn prototypes for each of them separately (Bezdek et al., 1998; Datta, P. and Kibler, D., 1995; Datta, P. and Kibler, D., 1997a; Datta, P. and Kibler, D., 1997b). This method guarantees fully homogeneous prototypes but the entire data distribution is distorted. Besides, the strength of averaging method to generalize away mislabeled instances of other labels in compact regions is disabled by the splitting the entire data set by classes. In view of this, we design a simple distance measure which considers both the Euclidean distance and the class labels. Basically we add a penalty value to penalize instances of different classes on top of the Euclidean distance. This distance measure encourages homogeneous instances to merge while preserving the original data distribution.

We integrate instance filtering methods into the basic prototype generation framework. It first applies prototype filtering method before prototype generation. Statement 2, namely, FILTER(P), is the instance filtering method. In prototype generation, grouping of outliers leads to the creation of poor prototypes. These poor prototypes will likely result in degradation in classification accuracy. If outliers or exceptions can be removed before the prototype generation is applied, the result prototypes will have a better quality. Moreover, the computational cost of prototype generation can be significantly reduced as the size of original data set becomes smaller after filtering.

13.3.2.2   Instance Filtering Methods Employed.   Different instance filtering techniques target at retaining instances in different locations such as at class boundaries or region centers. As described above, our PGF framework requires a instance filtering method which attempts to filter out less representative instances, outliers and hopefully noise. To this end, we have tried two different filtering methods which retain representative instances and are insensitive to the order of presentation of prototypes or instances. The first one is the ENN method proposed by Wilson which removes instances misclassified by its k nearest neighbors (Wilson, D. L., 1972). ENN can remove outliers and noise effectively since these instances can hardly be classified by its k nearest neighbors correctly.

Our second filtering method considers the classification accuracy of each prototype in the prototype set. The training set is splitted into a sub-training and a tuning set. In PGF, the accuracy performance of each prototype in the sub-training set is found by applying the prototype on the tuning set. Prototypes with accuracy higher than a certain threshold Q will be retained to perform prototype generation. This kind of accuracy measure has been used for prototype learning in some previous work. For example, IB3 uses historical performance on accuracy of each prototype to decide whether or not it is retained (Aha et al., 1991). NGE uses this information to learn weights for prototypes (Salzberg, S., 1991).

13.4. EMPIRICAL EVALUATION

13.4.1. Experimental Setup

We have conducted a series of experiments to investigate the performance of our PGF framework. Nineteen real-world data sets from the widely used UCI Repository (Blake, C. L. and Merz, C. J., 1998) were tested in the experiments. Table 13.1 shows the data sets we used.

Table 13.1 Data sets and their codes.

Data Set	Code	Data Set	Code	Data Set	Code
Balance-scale	Ba	New-thyroid	Ne	Vowel	Vw
Breast-cancer-w	Bc	Optdigits	Op	Wdbc	Wd
Glass1	Gl	Pendigits	Pe	Wine	Wi
Ionosphere	Io	Pima	Pi	Wpbc	Wp
Iris	Ir	Segmentation	Se	Yeast	Ye
Letter	Le	Shuttle	Sh		
Liver	Li	Sonar	Sn		

For each data set, we randomly partitioned the data into ten portions. Ten trials derived from 10-fold cross-validation were conducted for every set of experiments. The mean of the data retention rate (size) and classification accuracy of the ten trials were calculated to measure the performance for a particular data set. Note that the smaller the size and the higher the classification accuracy, the better is the performance. Euclidean distance metric with attribute values normalized by the range of features in the training set was used. Missing values were replaced by the mean value of the feature.

13.4.2. Experimental Results

Table 13.2 shows the average classification accuracy and data retention rate (size) of 10-fold cross-validation of each data set for different variants of the PGF algorithm. We also obtained the performance of pure filtering and pure abstraction methods so that comparative analysis can be conducted. The average classification accuracy of existing learning algorithms, namely, C4.5 and KNN, are also shown. A range of parameters for those algorithms are tested and the best performance of each algorithm is presented. The average performance across all data set is also given in each table. We first analyze the behavior of PGF by comparing its performance with pure filtering and abstraction methods. Then we compare the classification accuracy of PGF with that of C4.5 and KNN.

Table 13.2 The average generalization accuracy (acc) and data retention rate (size) of ten trials for PGF, pure instance filtering, pure prototype abstraction, C4.5 and KNN.

| | PGF | | | | Pure Filtering | | | | Pure Abstraction | | C4.5 | KNN |
| | PGF-ENN | | PGF-ACC | | ENN | | ACC | | | | | |
Data	acc	size	acc	size	acc	size	acc	size	acc	size	acc	acc
Ba	0.861	0.152	0.829	0.013	0.864	0.784	0.824	0.087	0.779	0.103	0.792	0.822
Bc	0.960	0.121	0.960	0.023	0.967	0.953	0.968	0.119	0.964	0.091	0.939	0.967
Gl	0.588	0.058	0.514	0.033	0.719	0.695	0.527	0.058	0.584	0.068	0.666	0.695
Io	0.880	0.109	0.855	0.022	0.846	0.868	0.866	0.099	0.892	0.196	0.900	0.861
Ir	0.913	0.038	0.933	0.068	0.953	0.954	0.933	0.111	0.927	0.097	0.953	0.947
Le	0.710	0.335	0.523	0.082	0.740	0.815	0.526	0.097	0.774	0.456	0.692	0.759
Li	0.559	0.152	0.557	0.063	0.597	0.623	0.568	0.077	0.571	0.153	0.642	0.641
Ne	0.926	0.060	0.852	0.028	0.963	0.967	0.833	0.115	0.944	0.153	0.921	0.948
Op	0.958	0.328	0.919	0.056	0.959	0.981	0.920	0.103	0.956	0.254	0.824	0.964
Pe	0.973	0.234	0.925	0.066	0.985	0.987	0.927	0.109	0.977	0.279	0.914	0.981
Pi	0.722	0.209	0.706	0.058	0.753	0.704	0.707	0.075	0.730	0.050	0.694	0.743
Se	0.948	0.236	0.911	0.076	0.956	0.967	0.919	0.146	0.965	0.325	0.951	0.956
Sh	0.984	0.209	0.981	0.061	0.985	0.996	0.982	0.115	0.987	0.214	0.989	0.987
Sn	0.833	0.472	0.716	0.051	0.833	0.860	0.716	0.095	0.866	0.552	0.706	0.837
Vw	0.959	0.252	0.632	0.096	0.987	0.989	0.636	0.110	0.971	0.252	0.779	0.970
Wd	0.954	0.195	0.935	0.038	0.958	0.954	0.950	0.106	0.938	0.191	0.944	0.970
Wi	0.938	0.112	0.921	0.023	0.954	0.951	0.882	0.095	0.938	0.112	0.888	0.960
Wp	0.747	0.033	0.733	0.018	0.733	0.712	0.738	0.063	0.747	0.018	0.676	0.732
Ye	0.560	0.067	0.512	0.064	0.562	0.528	0.518	0.071	0.505	0.404	0.545	0.545
Ave.	0.841	0.177	0.785	0.049	0.859	0.857	0.786	0.097	0.843	0.209	0.811	0.857

PGF-ENN. We investigate ENN and PGF-ENN to analyze how the abstraction method can help ENN in PGF. From Table 13.2, it is found that the data retention rate of ENN is dramatically improved from 85.7% to 17.7% with about 2.1% degradation in classification accuracy. ENN retains instances which can be correctly classified by their k nearest neighbors. We can imagine that if most of the instances are closely and homogeneously packed, a large portion of data will be retained as they are usually correctly classified. This accounts for the large data retention rate in ENN. On the contrary, our prototype abstraction method is strong in generalizing data sets with this kind of structure. Instances in closely packed regions will be generalized to few representative prototypes resulting in the significant reduction in data retention rate.

ENN can assist the abstraction method in PGF too. If ENN is performed before abstraction, noise, outliers and exceptions can be removed first. The removal of these instances can avoid the formation of non-representative prototypes in abstraction. Furthermore, a smoother decision boundary can also be obtained by the removal of border points. This may help the generalization of instances in abstraction. We can see from Table 13.2 that the data retention rate of pure prototype abstraction method is improved from 20.9% to 17.7% while keeping a similar classification accuracy.

PGF-ACC. ACC retains instances with classification accuracy higher than a certain threshold. As center instances usually gain high accuracy, they will be retained. When comparing pure abstraction and PGF-ACC, we find that data retention rate of abstraction is improved from 20.9% to 4.9%. Despite the significant improvement in data retention rate, the classification of abstraction is degraded from 84.3% to 78.5%. We know that pure abstraction discovers representative instances by generalizing the common characteristics of similar instances. However, in PGF-ACC, about 90% of instances are discarded by ACC before abstraction is applied. Therefore the prototypes generated in abstraction will be less representative leading to the degradation in classification accuracy. We suggest that filtering methods retaining center instances should not be used in PGF if classification accuracy is the main objective.

On the contrary, abstraction can help ACC in PGF. From Table 13.2, we can see that data retention rate of ACC is improved from 9.7% to 4.9% while maintaining similar classification accuracy. We know that ACC retains center instances. The above result shows those center instances selected by ACC can be further refined by abstraction to form representative prototypes without sacrificing classification accuracy.

In the second set of experiments, we compare PGF with C4.5 and KNN. In KNN, a range of k is tested and the best results are output. Table 13.2 shows the average classification accuracy and data retention rate of 10-fold cross-validation of these algorithms on all the data sets. PGF performs slightly better than C4.5 in the average classification accuracy across most of the data sets. When compared with KNN, PGF stores only 17.7% of the total data and gains a comparable accuracy. Hence, PGF achieves comparable classification performance with state-of-the-art learning algorithms such as C4.5 and KNN. More importantly PGF can drastically reduce the data size to less than 18% of the original size on average.

13.5. CONCLUSIONS AND FUTURE WORK

We have developed a new prototype learning framework, called Prototype Generation and Filtering (PGF), which integrates the strength of instance filtering and prototype averaging methods. Our PGF possesses several characteristics. It retains selected instances and learns generalized prototypes. It can effectively reduce the data set size while maintaining or even improving the classification accuracy. It also provides noise tolerance and is insensitive to the order of presentation of instances. We have conducted sets of experiments to demonstrate the effectiveness of our framework.

In the future, we intend to investigate the effect of the distance measure on our PGF framework. The current design makes use of a simple distance metric considering both the Euclidean distance and class label distribution. More sophisticated schemes can be employed. Another extension is to explore different prototype generation methods. We can adopt different clustering techniques in the generation process.

References

Aha, D. W. (1992). Tolerating Noisy, Irrelevant, and Novel Attributes in Instance-Based Learning Algorithms. *International Journal of Man-Machine Studies*, 36:267–287.

Aha, D. W. and Kibler, D. (1989). Noise-Tolerant Instance-Based Learning Algorithms. *Proceedings of the Eleventh International Joint Conference on Artificial Intelligence*, pages 794–799.

Aha, D. W., Kibler, D. and Albert, M. K. (1991). Instance-Based Learning Algorithms. *Machine Learning*, 6:37–66.

Bezdek, J. C., Reichherzer, T. R., Lim, G. S. and Attikiouzel, Y. (1998). Multiple-Prototype Classifier Design. *IEEE Transactions on Systems, Man, and Cyberneics*, 28(1):67–79.

Blake, C.L. and Merz, C.J. (1998). UCI Repository of Machine Learning Database. Irvine, CA: University of California Irvine, Department of Information and Computer Science.
http://www.ics.uci.edu/~mlearn/MLRepository.html.

Bradshaw, G. (1987). Learning about Speech Sounds: The NEXUS project. *Proceedings of the Fourth International Workshop on Machine Learning*, pages 1–11.

Cameron-Jones, R. M. (1992). Minimum Description Length Instance-Based Learning. *Proceedings of the Fifth Australian Joint Conference on Artificial Intelligence*, pages 368–373.

Cameron-Jones, R. M. (1995). Instance Selection by Encoding Length Heuristic with Random Mutation Hill Climbing. *Proceedings of the Eighth Australian Joint Conference on Artificial Intelligence*, pages 293–301.

Chang, C. L. (1974). Finding Prototypes for Nearest Neighbor Classifiers. *IEEE Transactions on Computers*, 23(11):1179–1184.

Cost, S and Salzberg, S. (1993). A Weighted Nearest Neighbor Algorithm for Learning with Symbolic Feature. *Machine Learning*, 10:57–78.

Dasarathy, B. V. (1990). *Nearest Neighbor (NN) Norms: NN Pattern Classification Technique*. IEEE Computer Society Press.

Dasarathy, B. V. (1994). Minimal Consistent Set (MCS) Identification for Optimal Nearest Neighbor Decision Systems Design. IEEE Transactions on Systems, Man, and Cyberneics, 24(3):511–517.

Datta, P. and Kibler, D. (1997). Learning Symbolic Prototypes. *Proceedings of the Fourteenth International Conference on Machine Learning*, pages 75–82.

Datta, P. and Kibler, D. (1997). Symbolic Nearest Mean Classifier. *Proceedings of the Fourteenth National Conference of Artificial Intelligence*, pages 82–87.

Datta, P. and Kibler, D. (1995). Learning Prototypical Concept Description. *Proceedings of the Twelfth International Conference on Machine Learning*, pages 158–166.

Gates, G. W. (1972). The Reduced Nearest Neighbor Rule. *IEEE Transactions on Information Theory*, 18(3):431–433.

Gowda, K. C. and Krisha, G. (1979). The Condensed Nearest Neighbor Rule Using the Concept of Mutual Nearest Neighborhood. *IEEE Transactions on Information Theory*, 25(4):488–490.

Hart, P. E. (1968). The Condensed Nearest Neighbor Rule. *IEEE Transactions on Information Theory*, 14(3):515–516.

Kibler, D. and Aha, D. W. (1988). Comparing Instance-Averaging with Instance-Filtering Learning Algorithms. *Proceedings of the Third European Working Session on Learning*, pages 63–80.

Kuncheva, L. I., Bezdek, J. C. (1998). Nearest Prototype Classification: Clustering, Genetic Algorithms, or Random Search? *IEEE Transactions on Systems, Man, and Cyberneics*, 28(1):160–164.

Ricci, F. and Avesani, P. (1999). Date Compression and Local Metrics for Nearest Neighbor Classification. *IEEE Transactions on Pattern Analysis and Machine Intelligence*, 21(4):380–384.

Ritter, G. L, Woodruff, H. B. and Lowry, S. R. (1975). An Algorithm for a Selective Nearest Neighbor Decision Rule. *IEEE Transactions on Information Theory*, 21(6):665–669.

Salzberg, S. (1991). A Nearest Hyperrectangle Learning Method. *Machine Learning*, 6:251–276.

Sebestyen, G. S. (1962). *Decision-Making Process in Pattern Recognition*. New York: The Macmillan Company.

Skalak, D. B. (1994). Prototype and Feature Selection by Sampling and Random Mutation Hill Climbing Algorithms. *Proceedings of the Eleventh International Conference on Machine Learning*, pages 293–301.

Swonger, C. W. (1972). Sample Set Condensation for a Condensed Nearest Neighbor Decision Rule for Pattern Recognition. In Watanabe, S., editor, *Frontiers of Pattern Recognition*. Academic Press, New York, NY, pages 511–519.

Tomek, I. (1976). An Experiment with the Edited Nearest–Neighbor Rule. *IEEE Transactions on Systems, Man, and Cyberneics*, 6(6):448–452.

Trri, H., Knotkanen, P. and Myllymäki, P. (1996). Probabilistic Instance-Based Learning. *Proceedings of the Thirteenth International Conference on Machine Learning*, pages 158–166.

Ullmann, J. R. (1974). Automatic Selection of Reference Data for Use in a Nearest Neighbor Method of Pattern Classification. *IEEE Transactions on Information Theory*, 20(4):431–433.

Wettschereck, D. (1994). A Hybrid Nearest–Neighbor and Nearest–Hyperrectangle Algorithm. *Proceedings of the Seventh European Conference on Machine Learning*, pages 323–335.

Wettschereck, D., Aha, D. W. and Mohri, T. (1997). A Review and Empirical Evaluation of Feature Weighting Methods for a Class of Lazy Learning Algorithms. *Artificial Intelligence Review*, 11:273–314.

Wettschereck, D. and Dietterich, T. G. (1995). An Experimental Comparison of the Nearest–Neighbor and Nearest–Hyperrectangle Algorithms. *Machine Learning*, 19:5–27.

Wilson, D. L. (1972). Asymptotic Properties of Nearest Neighbor Rules Using Edited Data. *IEEE Transactions on Systems, Man, and Cyberneics*, 2:431–433.

Wilson, D. R. and Martinez T. R. (1997). Instance Pruning Techniques. *Proceedings of the Fourteenth International Conference on Machine Learning*, pages 403–411.

Zhang, J. (1992). Selecting Typical Instances in Instance-Based Learning. *Proceedings of International Conference on Machine Learning*, pages 470–479.

Chapter 14

INSTANCE SELECTION BASED ON HYPERTUPLES

Hui Wang

School of Information and Software Engineering

University of Ulster

Newtownabbey, BT 37 0QB, N.Ireland

H.Wang@ulst.ac.uk

Abstract   Instance selection aims to search for a representative portion of data
that serves the same purpose as the whole data. In this chapter we
propose a novel procedure for instance selection based on *hypertuples*,
a generalization of traditional database tuples. This procedure has two
tasks: building a model and selecting instances based on the model.
For the first task, we propose to merge data tuples while ensuring some
criteria are satisfied. This merge operation results in a set of hypertuples which, under certain conditions, serves as a model of the original
data. We identify two types of criteria for the task of instance selection: preserving classification structures and maximizing density of
hypertuples. For the first criterion we propose a formalism that leads
to a unique solution – the *least E-set*. We then propose algorithms for
finding this unique solution and for finding a compromised solution efficiently. For the second criterion we propose a new measure of density,
which is normalized and quantized, and which applies to both numerical
and categorical data. Using this measure of density, we then propose
a hill-climbing algorithm that can efficiently find a quasi-optimal set of
hypertuples, which is "quasi-densest".

Having a model of data, we can generate a set of representative
instances – the second task of the procedure. We propose to calculate
the centers of the hypertuples in the model and take these centers as
the representative instances of the original data. To use the selected
instances for classification, we propose to use a nearest neighbor (NN)
approach.

Experiments using real world public data show that, when used with
the proposed NN classifier, the selected instances are not only representative but even outperform C5 in some cases.

Keywords: Representative instances, hypertuples, hypertuple hyperrelations, hyperrelation density.

14.1. INTRODUCTION

Instance selection aims to search for a representative portion of data that serves the same purpose as the whole data. Instance selection is becoming increasingly important in many KDD applications due to the need for efficiency in speed and storage.

Sampling is a major method for instance selection, where randomness is a key issue and significantly affects the quality of the selected instances. Other methods include density estimation – finding the representative instances for each cluster, and boundary hunting – finding the critical instances to form boundaries to differentiate data points of different classes. These methods are meant to be general and they take instance selection as a process separate from model building[1]. Therefore the justification of the selected instances has to be done separately.

In this chapter we view instance selection as an integral part of the model building process. With this view in mind, we propose a procedure for model building with instance selection as an integrated process. This procedure has two integral tasks: building an initial model, and selecting instances as a way to simplify the initial model. The advantage of this procedure is, the validity of the selected instances is dependent upon the initial model. If the initial model is justified and valid, so are the selected instances.

We propose to use data reduction for the task of building initial models. Data reduction is to reduce the size of data while preserving decision related information. It is expected that the reduced data should give the same decision as the original data. Clearly the way in which data reduction is carried out is closely tied to the type of decision. For classification type of decision data should be reduced in such a way that classification information is preserved. In other words, the reduced data and the original data should produce the same (or at least similar in performance) classifiers. For clustering type of decision, neighborhood information should be preserved.

In this chapter we present a novel approach to data reduction for classification. The basic idea of the approach is, we merge pairs of data tuples using columnwise set union operation while preserving the classification structure hidden in the data. The result of each merge operation is a *hypertuple*, a generalization of the traditional database tuple in the sense that each field is a set of values rather than a single value. Once the merge operation can not proceed any further, we get a model of the data, which is a set of hypertuples and is called a *hyperrelation*. The model has significantly less number of tuples than the original data, though the hypertuples are *coarser* than the tuples in the original data. This model can be used as a classifier (see, for example, Wang et al. (1998)). Alternatively we can go a step further to select a set of simpler tuples to represent this model. This has at least two potential benefits: (1) significantly less amount of space is used to store the model, and (2) existing classifiers (e.g., C5, NN) can be used. For this task of instance selection, we propose to calculate the centers of the hypertuples in the model and use them to represent the model. Since a model is a generalized representation of data, these centers hence become a representation of the data.

14.2. DEFINITIONS AND NOTATION

To present our findings concisely and within the given page limit, we briefly introduce some notational conventions and definitions that are used throughout the chapter.

14.2.1. Order and Lattices

A *partial order* on a set U is a binary relation $\leq$ which is reflexive, antisymmetric, and transitive. Suppose that $\langle U, \leq \rangle$ is a partially ordered set and $T \subseteq U$. We let $\downarrow T \stackrel{\text{def}}{=} \{y \in U : (\exists x \in T)\ y \leq x\}$. If $T = \{a\}$, we will write $\downarrow a$ instead of $\downarrow \{a\}$; more generally, if no confusion can arise, we shall identify singleton sets with the element they contain.

A *semilattice* $\mathcal{L}$ is a nonempty partially ordered set such that for each $x, y \in \mathcal{L}$ the least upper bound $x + y$ exists. For $A \subseteq \mathcal{L}$, we denote the least upper bound of X by $\text{lub}(X)$. The greatest element of $\mathcal{L}$, if it exists, is denoted by 1; if $\mathcal{L}$ is finite then 1 exists, and it is equal to $\text{lub}(\mathcal{L})$.

If $A \subseteq \mathcal{L}$, we denote by $[A]$ the *subsemilattice* of $\mathcal{L}$ generated by A [2], i.e. $[A] = \{t \in \mathcal{L} : \exists X \subseteq A \text{ such that } t = \text{lub}(X)\}$. The greatest element in $[A]$ is $\text{lub}(A)$. If A is finite, $[A]$ is also finite.

An element $a \in A$, $a \neq 1$, is called *maximal* in A, if for all $x \in A$, $a \leq x$ implies $x = a$. If $A = \mathcal{L}$, we just speak of maximal elements.

For $A, B \subseteq \mathcal{L}$, we say that B *covers* A, written as $A \preceq B$ if for each $s \in A$ there is some $t \in B$ such that $s \leq t$. We also set $A + B \stackrel{\text{def}}{=} \{a + b : a \in A, b \in B\}$.

A comprehensive discussion on lattice theory can be found in (Grätzer, 1978).

14.2.2.   Decision Systems

An *information system* is a tuple $\mathcal{I} = \langle U, \Omega, V_x \rangle_{x \in \Omega}$, where $U = \{a_0, \ldots, a_N\}$ is a nonempty finite set and $\Omega = \{x_0, \ldots, x_T\}$ is a nonempty finite set of mappings $x_i : U \to V_{x_i}$ [3].

We interpret U as a set of objects and Ω as a set of attributes or features, each of which assigns to an object a its value under the respective attribute. Let $V \stackrel{\text{def}}{=} \prod_{x \in \Omega} V_x$. For $a \in U$, we let $\Omega(a) \stackrel{\text{def}}{=} \langle x(a) \rangle_{x \in \Omega} \in V$. Each $\Omega(a)$ is called a *tuple*, and the collection of all tuples is denoted by D. Thus, for each $t \in D$, there is at least one $a \in U$ such that $\Omega(a) = t$.

A *decision system* $\mathcal{D}$ is a pair $\langle \mathcal{I}, d \rangle$, where $\mathcal{I}$ is an information system as above, and $d : D \twoheadrightarrow V_d = \{d_0, \ldots, d_K\}$ is an onto mapping, called a *labeling* of D; the value $d(t)$ is called the *label* of t.

We will also refer to d as the *decision attribute* or *function*, and interpret $d(t)$ as follows:

$$\text{If } a \in U \text{ and } \Omega(a) = t, \text{ then decide } d(t).$$

The mapping d induces a partition $\mathcal{P}_d$ of D with the classes $\{D_0, \ldots, D_K\}$, where $t \in D_i \iff d(t) = d_i$.

In this chapter we consider a dataset represented as a decision system $\mathcal{D}$. Then D is the set of data tuples, d is the decision (classification) function, and V is the set of all possible data tuples in a problem domain. Therefore each $t \in D$ is associated with a class label $d(t)$. In practice, however, we are only given the decision system surrounding a dataset and the information system is usually unknown. So in the sections below our discussion focuses mainly on a subset of the domain lattice (sublattice) [D] generated from the dataset D.

14.2.3.   Hypertuples and Domain Lattice

Given a dataset expressed as a decision system, an elegant mathematical structure (lattice) is implied. This structure makes it possible to investigate data reduction, machine learning, as well the relationship between the two from an algebraic perspective.

Let $\mathcal{L} \stackrel{\text{def}}{=} \prod_{x \in \Omega} 2^{V_x}$. Then $t \in \mathcal{L}$ is a vector $\langle t(x) \rangle_{x \in \Omega}$, where $t(x) \subseteq V_x$ are sets of values [4]. The elements of $\mathcal{L}$ are called *hypertuples*; the elements t of $\mathcal{L}$ with $|t(x)| = 1$ for all $x \in \Omega$ are called *simple tuples*. Any set of hypertuples is called a *hyperrelation*. Note that V is a set of *all* simple tuples for a given problem domain, and D is the set of simple tuples described in the decision system $\mathcal{D}$.

$\mathcal{L}$ is a lattice under the ordering

$$t \leq s \iff t(x) \subseteq s(x) \tag{14.1}$$

with the sum and product operations, and the maximal element (i.e., 1) given by

$$t + s = \langle t(x) \cup s(x) \rangle_{x \in \Omega}, \tag{14.2}$$

$$t \times s = \langle t(x) \cap s(x) \rangle_{x \in \Omega}, \tag{14.3}$$

$$1 = \langle V_x \rangle_{x \in \Omega}. \tag{14.4}$$

$\mathcal{L}$ is called *domain lattice* for $\mathcal{D}$.

Table 14.1(a) is a dataset consisting of three simple tuples, where $V_{X_1} = \{a, b\}$ and $V_{X_2} = \{0, 1\}$. Table 14.1(b) and (c) are sets of hypertuples, which are the least and greatest E-sets respectively for the dataset, to be defined later.

Table 14.1 (a) *A set of simple tuples in a decision system.* (b) *A set of hypertuples as the least E-set.* (c) *A set of hypertuples as the greatest E-set.*

U	X_1	X_2	d
u_0	a	0	α
u_1	a	1	α
u_2	b	0	β

(a)

U	2^{X_1}	2^{X_2}	d
u'_0	$\{a\}$	$\{0,1\}$	α
u'_1	$\{b\}$	$\{0\}$	β

(b)

U	2^{X_1}	2^{X_2}	d
u'_0	$\{a\}$	V_{X_2}	α
u'_1	$\{b\}$	V_{X_2}	β

(c)

14.3. MERGING HYPERTUPLES WHILE PRESERVING CLASSIFICATION STRUCTURE

In this section we consider how to merge hypertuples while preserving classification structure in the hope that, as such, the final set of hypertuples should be representative of the original dataset, and it does not lose information relevant to classification.

14.3.1. Equilabeledness and E-Set

Here we analyze the task of data reduction for classification in order to introduce some concepts and to identify our objectives.

First of all we take a look at the *decision relevant* information for classification. Let the dataset be D coupled with a decision function d. D generates a lattice [D], which is a sublattice of $\mathcal{L}$. Data reduction is to find $S \subseteq [D]$ with a new decision function d'. Note that the definition of d is on D, and the definition of d' is on $D' \overset{\text{def}}{=} \downarrow S \cap V$. It is necessary that $D \subseteq D'$ and that d' is consistent with d on D – that is, $d(x) = d'(x)$ for $x \in D$. Thus d' can be understood as a generalization of d; or, equivalently, S is a generalization of D. We also call S a model or hypothesis of D. For data reduction purpose it is also expected $|S| \ll |D|$. Therefore, loosely speaking, *data reduction is to find a concise generalization of the dataset.*

Clearly there are many different generalizations of a dataset. Then which one do we select? Note that all tuples in D are labeled by d while elements in $V \setminus D$ are not. Our solution is a set of equilabeled hypertuples. Intuitively, an equilabeled hypertuple is $t \in [D]$ which covers at least one labeled element and all labeled elements covered by t have the same label. The reason for selecting equilabeled hypertuples is twofold. First, an equilabeled hypertuple covers at least one labeled tuple, and this significantly reduces the search space. Most importantly, if the generalization S of D includes one hypertuple which doesn't cover any labeled tuple at all, this generalization is deemed to have generalized beyond what is given in the dataset. Second, an equilabeled hypertuple does not cover tuples with different labels, and this guarantees the consistency of the generalization with the original data. As a result, the label of the tuples covered can be extended to the equilabeled hypertuple itself. For an illustrative example, readers are invited to consult Wang et al. (1999).

Formally, we call an element $r \in [D]$ *equilabeled* with respect to D_q, if $\emptyset \neq \downarrow r \cap D \subseteq D_q$. In other words, r is equilabeled if $\downarrow r$ intersects D, and every element in this intersection is labeled d_q for some $q \leq K$. Recall that K is the number of classes. In this case, we say that r *G-belongs to* D_q. We denote the set of all equilabeled elements G-belonging to D_q by $\mathcal{E}_q$, and let $\mathcal{E}$ be the set of all equilabeled elements. Note that $D \subseteq \mathcal{E}$, and that $q, r \leq K, q \neq r$ implies $\mathcal{E}_q \cap \mathcal{E}_r = \emptyset$.

We will now extend d over all of $[D]$ by setting

$$d(r) = \begin{cases} d_q, & \text{if } r \in \mathcal{E}_q, \\ unknown, & \text{otherwise.} \end{cases} \tag{14.5}$$

Now $\mathcal{E}$, along with the extended labeling, is the space where our expected generalization will come from. This space, however, is still too large. Since the elements in $\mathcal{E}$ are partially ordered – some are covered by some others – we need only look at those which are not covered by any; they are *maximal*. Our wish to find maximal elements in some context leads to the following notions.

Definition 1. Let P be such that $D \subseteq P \subseteq V$. We let

$$\omega(P) \stackrel{\text{def}}{=} \{h \in \mathcal{E} : \exists X \subseteq P, h = \text{lub}(X)\}$$

$$\epsilon(P) \stackrel{\text{def}}{=} \{t : t \text{ is maximal in } \omega(P)\}.$$

$\epsilon(P)$ is called the *E–set for P*, and $t \in [D]$ is said to be *in context P* if $t \in \omega(P)$.

It can be easily shown that $\epsilon(D) \preccurlyeq \epsilon(P) \preccurlyeq \epsilon(V)$ for $D \subseteq P \subseteq V$. We then call $\epsilon(D)$ *least E-set* and $\epsilon(V)$ *greatest E-set*, which are denoted by E and $\mathbb{E}$ respectively. Clearly both E-sets are unique. It is not hard to see that $\mathbb{E}$ is the set of all maximal elements in $\mathcal{E}$.

Our approach to data reduction aims to find the least and greatest E-sets. The following section will explain why we are interested in these E-sets.

14.3.2. Data Reduction as A Search for Hypotheses

Here we set out to characterize the least and greatest E-sets in the context of concept learning, in order to justify our selection of the two E-sets.

Concept learning is to approximate a function (called target function) from examples (Mitchell, 1997) by a hypothesis. A hypothesis is a function, which can be Boolean-valued, discrete-valued, or continuous-valued. Since our discussion is limited to decision systems, our hypotheses are discrete-valued. As discussed earlier, a decision system comes with a decision function d which can be extended (generalized) to the set of equilabeled hypertuples $\mathcal{E}$ (See Eq.14.5). Then any subset H of $\mathcal{E}$ is a qualified hypothesis for the dataset D, provided that H is consistent with D. This can be guaranteed by the condition $D \preccurlyeq H$.

Definition 2. Let P be such that $D \subseteq P \subseteq V$. A *hypothesis for* D *in context P* is a $H \subseteq \epsilon(P)$ such that $D \preccurlyeq H$. We use $\text{GEN}(D|P)$ to denote the set of all hypotheses for D in P, and let $\text{GEN}(D) \overset{\text{def}}{=} \{\text{GEN}(D|P) : D \subseteq P \subseteq V\}$.

Similarly, we define a *hypothesis for* D_q. Note that for a hypothesis H for D, $H \cap \mathcal{E}_q$ is a hypothesis for D_q. Conversely, if H_q is a hypothesis for D_q for each $q \leq K$, then $H \overset{\text{def}}{=} \bigcup_{q \leq K} H_q$ is a hypothesis for D.

Definition 3. Let $H_j, H_k \in \text{GEN}(D)$ be two hypotheses for D. Then H_j is *more general than* H_k if and only if $H_k \preccurlyeq H_j$. H_j is *(strictly) more general than* H_k, written $(H_k \prec H_j)$, if and only if $(H_k \preccurlyeq H_j) \wedge (H_j \not\preccurlyeq H_k)$.

A hypothesis H for D is *maximally specific* if and only if $H \in \text{GEN}(D)$ and there is no $H' \in \text{GEN}(D)$ such that $H' \prec H$; similarly, H is *maximally general* if and only if $H \in \text{GEN}(D)$ and there is no $H' \in \text{GEN}(D)$ such that $H \prec H'$. We denote by SSET the set of all maximally specific hypotheses for D, and by GSET the set of all maximally general hypotheses for D.

Mitchell calls SSET the *specific boundary* and GSET the *general boundary* for D (Mitchell, 1997). It is shown (Wang et al., 1999) that SSET is the least E-set and GSET is the greatest E-set.

14.3.3.   Extraction of Hyperrelations

In this section we discuss how to find the least E-sets. Note that the least E-set is the set of all maximal elements in $\omega(D)$. Clearly if we have an algorithm which is able to find all maximal elements in a set, we can find the least and greatest E-sets in a same way by fixing different contexts. The SUMUP algorithm (Wang et al., 1998) is designed for this purpose. Given a dataset D, we consider a class D_q of D. Let P be such that $D \subseteq P \subseteq V$. The algorithm iteratively examines $H_i \subseteq \mathcal{E}_q$ which covers D_q. This iteration stops when H_i generalizes to a point where generalizing any further assumes extra information outside D_q.

Algorithm 14.3.1 (Sumup algorithm).   *1 $H_0 \overset{\text{def}}{=} P$.*

2 $H_{k+1} \overset{\text{def}}{=}$ The set of maximal elements of $[\downarrow (H_k + P)] \cap \mathcal{E}_q$.

3 Continue until $H_n = H_r$ for all $r \geq n$.

It is shown (Wang et al., 1998) that the SUMUP algorithm finds the E-set in any context P such that $D \subseteq P \subseteq V$. In fact it calculates $\epsilon(P)$. Since D is given in the decision system $\mathcal{D}$, we can straightforwardly calculate the least E-set.

Although the above algorithm can provably find the least E-set, it is computationally expensive. Note that our objective is to reduce the data and then to select instances based on the reduced dataset. So the least E-set is a means here, not a goal. We are interested in the least E-set for the following reasons: (1) it is consistent with the original dataset, due to the nature of equilabeledness; (2) it is unique and has a *comprehensive* coverage of simple tuples in V in the sense that not only the simple tuples in D are covered, but also the equilabeled sums (least upper bounds) of all combinations of these simple tuples; and finally (3) it is directly related to the well-known inductive bias (specific boundary) hence it is justifiable. The complexity comes from the combinatorial operation: for each class of tuples in the dataset, the SUMUP algorithm examines all possible partitions of it to see if it leads to a set of equilabeled hypertuples. If we modify the algorithm to examine part, not all, of the partitions (combinations), we will get a more efficient algorithm. This is what the EXTRACT algorithm (Wang et al., 1999) does.

Given a dataset D, the following algorithm finds a set of elements in $\omega(D)$ which has a disjoint coverage of D.

Algorithm 14.3.2 (Extract algorithm). *Given D_q and $\mathcal{E}_q$ as defined above.*

- *Initialization: let $X = D_q, H = \emptyset$.*

- *Repeat until X is empty:*

  1 Let $h \in X$ and $X = X \setminus \{h\}$.
  2 For $g \in X$, let $X = X \setminus g$. If $h + g$ is equilabeled then $h = h + g$.
  3 Let $H = H \cup \{h\}$.

14.4. MERGING HYPERTUPLES TO MAXIMIZE DENSITY

In the previous section we discussed how to merge hypertuples to preserve classification structure. In this section we discuss how to merge hypertuples to maximize density.

14.4.1. Density of Hypertuples

Here we introduce and justify a new measure of density which applies to both numerical and categorical data.

Let D be a dataset and [D] be the sublattice generated from D.

Definition 4. Let $h \in [D]$ be a hypertuple, and $x \in \Omega$ be an attribute. The *magnitude* of $h(x)$ [5] is defined as

$$\text{mag}(h(x)) = \begin{cases} \max(h(x)) - \min(h(x)), \text{ if } x \text{ is numerical} \\ |h(x)|, \text{ if } x \text{ is categorical} \end{cases} \quad (14.6)$$

The *volume* of h is defined as $\text{vol}(h) = \prod_{x \in \Omega} \text{mag}(h(x))$. The *coverage* of h is $\text{cov}(h) \stackrel{\text{def}}{=} \{d \in D : d \leq h\}$. The *density* of h is defined as $\text{den}(h) = |\text{cov}(h)|/\text{vol}(h)$. The *density* of hyperrelation H, $\text{den}(H)$, is then the average density of the hypertuples in H.

The above definition of density can not be directly applied to compare different hypertuples since different hypertuples may differ at different attributes, and different attributes may have different scales. Therefore we need to normalize the attributes up to a same *uniform scale*. The normalization can be achieved as follows. Let λ be the expected uniform scale. For an attribute $x \in \Omega$, the *normalization coefficient* is $s(x) \stackrel{\text{def}}{=} \lambda / \text{mag}(V_x)$. Note that V_x is the domain of attribute x. Then the volume of a hypertuple h after normalization is $\mathrm{vol}(h) = \prod_{x \in \Omega} s(x) \times \mathrm{mag}(h(x))$. The density definition can be normalized similarly.

This normalized notion of density is fine for hypertuples. But there is a problem for simple tuples. Consider Table 14.2. Suppose the uniform scale is 2. Then $s(A_1) = 2/6$ and $s(A_2) = 2/9$. Following the above definition, the volume for all simple tuples is 0 and so the density is ∞ since the projection of each simple tuple to (numerical) attribute A_2 contains only one value. This is not desirable, the reason for which will be seen in the next section. Therefore we need a method to allocate density values to simple tuples in such a way that the values can be compared with the density values of hypertuples, i.e., both simple tuples and hypertuples can be treated uniformly. Our solution is through *quantization of attributes*. For an attribute $x \in \Omega$ the measurement of a unit after normalization is $\mathrm{mag}(V_x)/\lambda = 1/s(x)$. For a hypertuple t, if $t(x)$ is less than the value of a unit, $1/s(x)$, the x-dimension of t should be treated as a unit. If t is a simple tuple then every dimension of t is treated as a unit and hence $\mathrm{vol}(t) = 1$ by definition. Since a simple tuple covers only itself, i.e., $\mathrm{cov}(t) = \{t\}$, we have $\mathrm{den}(t) = 1$. Consequently $\mathrm{den}(H) = 1$ if H is a *simple* relation. If t is a hypertuple, then $\mathrm{den}(t)$ may be greater or less than 1.

The parameter λ is the number of partitions of the domain of each dimension. It specifies the magnitude or granularity of a unit in each dimension. Note that the magnitudes of a unit in different dimensions may be different.

In the rest of this chapter whenever we talk of density we refer to the normalized and quantized density. This measure of density has a major advantage: *numerical and categorical attributes can be treated uniformly*. Tuples (simple or hyper) and relations (simple or hyper), either numerical or categorical or a mixture of the two, can thus be *uniformly* measured for their density. After normalization and quantization, it can be used to compare among hypertuples and among hyperrelations. This is different from most, if not all, existing measures for density (see, for example, Duda and Hart (1973), Ester et al. (1996)): they are mostly defined for numerical attributes only and they are not quantized.

Table 14.2   One class (class 1) of tuples in a dataset. Here attribute A_1 is categorical and A_2 is numerical.

	A_1	A_2	d		A_1	A_2	d
t_0	a	2	1	t_4	c	3	1
t_1	f	10	1	t_5	e	7	1
t_2	c	4	1	t_6	b	1	1
t_3	f	9	1	t_7	d	6	1

14.4.2.   Merging Hypertuples to Increase Density

Having a notion of density as defined above we now present our density-based approach to data reduction. Our philosophy is *merging tuples to increase the density of hyperrelations*: for any set of tuples in one class, if their sum has higher density then we are inclined to merge them and use their sum to replace this set of tuples.

More formally, let D_q be the set of tuples in class q, $P = \{D_q^1, \cdots, D_q^n\}$ be an arbitrary partition of D_q, and $H = \{\text{lub}(D_q^1), \cdots, \text{lub}(D_q^n)\}$. Here D_q^i is called a *cluster* of the partition, and $\text{lub}(D_i)$ is the least upper bound of all elements in D_i which is a hypertuple. H is called a *hyper-partition* of D_q since H is in fact a hyperrelation (Note P is not a hyperrelation). Again let $\mathcal{H} = \{H : H$ is a hyper-partition of $D_q\}$, the set of all hyper-partitions of D_q. Our objective is to find $H_0 \in \mathcal{H}$ such that $\text{den}(H_0) = \max\{\text{den}(H) : H \in \mathcal{H}\}$. In other words our expected hyper-partition should have the highest possible density. We call H_0 the *optimal hyper-partition* of D_q. With some abuse of language, we also call it optimal partition for simplicity.

The optimal partition of the data in Table 14.2 is shown in Table 14.3. Readers can check for themselves that any other hyperrelations obtained by merging simple tuples in the dataset using the lattice sum operation has lower density. For example, merging $\{t_0, \cdots, t_3\}$ and $\{t_4, \cdots, t_7\}$ results in a hyperrelation in Table 14.4, which has lower density.

Table 14.3   The optimal partition of the relation in Table 14.2 obtained by our method. The uniform scale used is 4, so the normalization coefficients are $s(A_1) = 2/3$ and $s(A_2) = 4/9$. The density of this hyperrelation is then 1.313. Note that the density values are normalized, and that the density for the original dataset is 1.

	A_1	A_2	Coverage	Density
t_0'	$\{a, b, c\}$	$\{1, 2, 3, 4\}$	$\{t_0, t_2, t_4, t_6\}$	1.500
t_1'	$\{d, e, f\}$	$\{6, 7, 9, 10\}$	$\{t_1, t_3, t_5, t_7\}$	1.125

Table 14.4   An arbitrary hyperrelation obtained by merging simple tuples in Table 14.2. The uniform scale used is the same as in Table 14.3, so are the normalization coefficients. The density of this hyperrelation is 0.5625.

	A_1	A_2	Coverage	Density
t_0''	$\{a, c, f\}$	$\{2, 4, 9, 10\}$	$\{t_0, t_1, t_2, t_3\}$	0.5625
t_1''	$\{b, c, d, e\}$	$\{1, 3, 6, 7\}$	$\{t_4, t_5, t_6, t_7\}$	0.5625

Now we discuss algorithmic issues relating to the optimal partition. A straightforward method is to evaluate the density function over all possible partitions, but this is clearly impractical since the number of partitions is astronomical. To avoid this combinatorial explosion, the density function can be evaluated for only a small set of partitions. There are some well-known techniques for identifying a small subset of partitions that has a good chance of containing the optimal partition. One approach is to optimize the criterion function using an iterative, hill-climbing technique; another is to use dynamic programming to eliminate many partitions and is still able to achieve an optimal solution (Jensen, 1969).

We take an iterative, hill-climbing approach. The general idea of this approach is as follows. Starting with an initial partition of the dataset (which corresponds to a hyperrelation), tuples are moved from one cluster to another in an effort to increase the density of the partition (hyperrelation). Thus each successive partition is a perturbation of the previous one and, therefore, only a small number of partitions is examined.

In our approach we don't need the number of clusters given in advance, so we can take the original data as an initial partition. To improve the density of partition, we can update an existing partition by merging each pair of tuples into a new tuple if the density of it increases. Each partition can be updated in many ways; that is, there may be many pairs of tuples such that merging them can improve density. In the spirit of hill-climbing approach, we choose the update such that the improvement is the greatest.

Note that although the number of clusters does not need to be given in advance, the uniform scale λ must be given. Based on the above discussion we designed an iterative, hill-climbing algorithm to find the optimal partition, DENSMER. The following is an outline of the algorithm.

Algorithm 14.4.1 (Densmer algorithm). *Input: a class D_q in dataset D and a uniform scale λ as defined as above.*

- *Step 1: initialization, $Q_0 = D_q$; $i = 0$;*

- *Step 2: If there are $x, y \in Q_i$ such that $\operatorname{den}(x + y) > \operatorname{den}(\{x, y\})$ and $\operatorname{den}(x + y)/\operatorname{den}(\{x, y\}) \geq \operatorname{den}(x' + y')/\operatorname{den}(\{x', y'\})$ for any $x', y' \in Q_i$, then*

    1 $Q_{i+1} = Q_i \cup \{x + y\}$;

    2 Remove all $z \in Q_{i+1}$ such that $z < x + y$;

    3 $i = i + 1$;

4 Repeat step 2.

- *Step 3: Merge the hypertuples in Q_i which cover common elements in D_q. Denote the result of this merge by H.*

As with most hill-climbing algorithms, the DENSMER algorithm may get stuck with local minima. Therefore the hyper-partition found by the algorithm may not be optimal – it is *quasi-optimal*.

14.5. SELECTION OF REPRENTATIVE INSTANCES

In the previous two sections we have discussed two different approaches to data reduction, both resulting in hyperrelations (sets of hypertuples) as data models. Since hypertuples need much more storage space than simple tuples and most learning algorithms deal only with simple tuples[6], we need to select instances (simple tuples) based on the hyperrelations obtained. In this section we discuss how to select representative instances from hyperrelations.

Representative instances are very useful since they tend to be much smaller in size and they are expected to result in similar learning performance. In our case, we already have a set of hypertuples (by either EXTRACT or DENSMER), so selecting representative instances can be done by calculating the center (a simple tuple) of each hypertuple as the representative instance for the hypertuple. The question is, what is the center of a hypertuple?

If all attributes are numerical, the center of a hypertuple can be justifiably taken to be the centroid of the set of simple tuples covered by the hypertuple. The difficulty arises when some attributes are categorical. Our solution is as follows. For continuous attributes, we use mean (average of all elements); for ordinal attributes we use median (middle element); and for categorical attributes, we use mode (most frequent element). Then the center of a hypertuple is a vector, each element of which is the center of the projection of the hypertuple onto the respective attribute.

Definition 5. Let h be a hypertuple, x be an attribute. The *center* of $h(x)$ is defined as

$$\text{cen}(h(x)) = \begin{cases} \text{mean of } \{t(x) : t \in \text{cov}(h)\}, & \text{if } x \text{ is continuous;} \\ \text{median of } \{t(x) : t \in \text{cov}(h)\}, & \text{if } x \text{ is ordinal;} \\ \text{mode of } \{t(x) : t \in \text{cov}(h)\} & \text{if } x \text{ is categorical.} \end{cases}$$

$$(14.7)$$

The *center* of h is defined as $\text{cen}(h) \overset{\text{def}}{=} \langle \text{cen}(h(x)) \rangle_{x \in \Omega}$

Note that $h(x)$ and $t(x)$ are the projections of h and t onto attribute x respectively, $\text{cov}(h)$ is the coverage of h.

With such a measure, we can calculate the centers (simple tuples) of all hypertuples and take them as the representative instances of the respective hypertuples. In this way there is only one representative instance for each hypertuple. Alternatively we can select a (fixed) proportion of instances (including the centers) from each hypertuple.

We have designed and implemented an algorithm, REPSEL, which is a straightforward implementation of the cen() function. It selects (generates) only one instance for each hypertuple.

14.6.   NN-BASED CLASSIFICATION USING REPRESENTATIVE INSTANCES

When representative instances are selected, they can be used to build classifiers. Algorithms like C5 or Cart can all be used for this purpose, since the selected instances are in the same format as the original data. But to better exploit the representative nature of these instances, we believe the NN approach is better suited.

The NN algorithm is one of the most venerable algorithms in machine learning. To classify a new instance, the Euclidean distance (possibly weighted) is computed between this instance and each training instance, and the new instance is assigned the class of the nearest neighboring instance. More generally, the k nearest neighbors are computed, and the new instance is assigned the class that is most frequent among these k neighbors.

Since datasets may contain both numerical and categorical attributes, we need to modify the standard Euclidean distance measure so that categorical attributes can be dealt with. We propose a modification based on the operations of normalization and quantization, as discussed in Section 14.4.

Let $t_1, t_2 \in V$ be two instances (simple tuples). The distance between t_1 and t_2 is defined as follows:

$$D(t_1, t_2) = \sqrt{\sum_{x \in \Omega} k(t_1(x), t_2(x))^2}$$

$$k(t_1(x), t_2(x)) = \text{mag}(\{t_1(x), t_2(x)\}) \times s(x)$$

where $t_i(x)$ is the projection of tuple t_i onto attribute x, $k(t_1(x), t_2(x))$ is the distance between the two instances with respect to attribute x, and the mag() and s() functions are defined in Section 14.4.

The training instances are the selected representative instances so there is no missing values. But there may be missing values in the new instances to be classified. In the context of NN algorithms, Aha (1990) evaluated three methods for distance computation with missing values. We adopt his *Ignore* method. This is one of the simplest methods for dealing with missing values: If an instance has a missing value for an attribute, then the distance for that attribute is 0.

14.7. EXPERIMENT

In this section we present experimental results of our study on instance selection. The objective of the experiment is to examine the quality of the selected representative instances in terms of their prediction capability compared with the prediction capability of the original data, when used with a decision tree classifier (C5) and our NN classifier.

The datasets we used are public datasets in UCI Machine Learning Repository, therefore the experiments can be repeated and compared. A brief description of the datasets is shown in Table 14.5.

Table 14.5 Description of the datasets.

Datasets	#Features	#Examples	#Class
Auto	25	205	6
Diabetes	8	768	2
Glass	9	214	6
Heart	13	270	2
Iris	4	150	3
Sonar	60	208	2
Vote	18	232	2

Note that some datasets have missing values. Missing values usually mean that the actual values are either not important, or not available. In either case the missing values do not contribute to the building of classifiers. The usual solution is to replace missing values by either the mean value in the case of continuous attributes, or the most frequent value in the case of discrete attributes. Since our work is based on hypertuples, we have a different solution. If we understand missing values as part of the data which do not contribute to building a classifier, a natural solution is to replace them by empty set since each hypertuple is a vector of sets and the empty set does not contribute to the model building operation.

To achieve our objective, we conducted two sets of experiments: (1) We split each dataset into training set (75%) and testing set (25%). The training sets were used for instance selection. We used the EXTRACT (DENSMER) algorithm to reduce a (training) dataset resulting in an initial model of data (hyperrelations). Then we fed the initial model to the REPSEL algorithm to select representative instances, which are then regarded as the final model of data. Finally we fed the model (a set of representative instances) to the C5 decision tree induction algorithm and our NN algorithm to examine how well the representative instances perform. The results are shown in Table 14.7. (2) We take EXTRACT-REPSEL-NN and DENSMER-REPSEL-NN as two integral classification algorithms and cross validate (5-fold) them using the datasets. The results are shown in Table 14.6, along with cross validation results of C5.

From the results we conclude the following. (1) The selected instances did not predict well with C5, but they performed quite well with NN. This provides evidence that instance selection should be tied to learning algorithms. A possible explanation is, the selected instances are evenly spread out although they covers different numbers of instances in the original datasets. NN takes advantage of the even spread of these instances, but C5 does not use this type of information. Note that our REPSEL algorithm selects (or generates) only one instance from each hypertuple. This suggests a possible solution: we can select a fixed proportion of instances from all hypertuples. This will be investigated in future work. (2) The number of representative instances is a small fraction of the size of the original data. This makes it possible to use computation demanding algorithms on large datasets. (3) There is no marked difference between our two algorithms (EXTRACT and DENSMER) with respect to the quality of the selected instances, though EXTRACT tends to generate slightly less number of instances.

14.8.   SUMMARY AND CONCLUSION

This chapter proposed a novel procedure for instance selection. Revolving around the notions of hypertuples and hyperrelations, the procedure sets out to select instances in two processes: building a model (a set of hypertuples) of data, and calculating the centers of hypertuples as representative instances. Two approaches were proposed for model building: one is based on the so-called *Lattice Machine* (Wang et al., 1998; Wang et al., 1999) and it aims to merge hypertuples while preserving classification structure; another is based on density estimation and it aims to merge hypertuples to maximize density. In the first approach we presented the EXTRACT algorithm, which was adapted from

Table 14.6 Prediction accuracy with 1/4 of the data by C5 algorithm and NN algorithm, along with the size of the set of selected representative instances. The training data were (1) the remaining 3/4 of the data; (2) the representative instances selected by EXTRACT from the remaining 3/4 of the data; and (3) the representative instances selected by DENSMER from the remaining 3/4 of the data. In this experiment the parameter λ, needed in the DENSMER algorithm, was set to 5.

Datasets	C5		REPSEL/EXTRACT			REPSEL/DENSMER		
	No.	Acc.	No.	Acc. C5	Acc. NN	No.	Acc. C5	Acc. NN
Auto	154	76.5	26	49.9	80.0	32	51.0	83.0
Diabetes	567	65.1	35	78.5	77.0	55	78.5	77.0
Glass	161	83.0	14	77.4	81.0	20	77.4	81.0
Heart	203	74.6	22	55.8	85.0	29	58.2	78.0
Iris	113	89.2	5	84.2	97	11	86.5	97.0
Sonar	156	69.2	39	54.5	93.0	33	46.2	93.0
Vote	174	98.3	51	90.0	87	62	90.1	87.0

Table 14.7 5-fold cross validation results by C5 and NN. In this experiment the parameter λ was also set to 5.

Datasets	C5: CV5	NN: CV5	
		REPSEL/EXTRACT	REPSEL/DENSMER
Auto	70.7	74.0	76.3
Diabetes	76.0	71.5	73.5
Glass	80.4	82.7	83.5
Heart	74.4	77.5	78.3
Iris	94.7	94.0	96.0
Sonar	71.6	88.0	86.4
Vote	97.0	93.2	88.4

the CASEEXTRACT algorithm by Wang et al. (Wang et al., 1999). In the second approach we presented the DENSMER algorithm. The key element in this approach is our new measure of density, which is a normalized and quantized measure, and which deals with numerical and categorical data uniformly. For the second process, we suggested calculating the center of a hypertuple and using the center to represent this hypertuple.

To use the selected instances for classification we suggested using an NN approach. For this we proposed a measure of distance, which takes advantage of our normalization and quantization operations and it can deal with both numerical and categorical data.

The proposed procedure was evaluated using 7 public datasets. When used with the NN classifier, the representative instances, which are just a small fraction of the original dataset in size, performed quite well in terms of their prediction capability. However, these instances performed poorly when used with C5. A possible reason is that only one instance was selected from each hypertuple. This suggests a direction in future work.

Notes

1. Here *model building* is taken as a generic term for classification, regression and clustering.

2. Formally, [A] is the smallest set which contains A and is closed with respect to +.

3. Note V_{x_i} can be finite or infinite. For the latter case the domain lattice is infinite. However we are only interested in the *finite* sublattice generated from a finite dataset.

4. Note that if $t \in \mathcal{L}$ and $x \in \Omega$, then $t(x)$ is the projection of t to its x-th component.

5. Note that $h(x)$ is the projection of h onto attribute x, $\min(h(x))$ is the minimal value in $h(x)$ while $\max(h(x))$ is the maximal value.

6. The CASERETRIEVAL algorithm (Wang et al., 1999) is designed to deal with hypertuples.

References

Aha, D. W. (1990). A study of instance-based algorithms for supervised learning tasks. Technical report, University of California, Irvine.

Duda, R. O. and Hart, P. E. (1973). *Pattern classification and scene analysis*. John Wiley & Sons.

Ester, M., Kriegel, H. P., Sander, J., and Xu, X. (1996). A density-based algorithm for discovering clusters in large spatial databases with noise. In *Proc. 2nd Int. Conf. on Knowledge Discovery and Data Mining*, pages 226–231. AAAI Press.

Grätzer, G. (1978). *General Lattice Theory*. Birkhäuser, Basel.

Jensen, R. E. (1969). A dynamic programming algorithm for cluster analysis. *Operations Research*, 17:1034–1057.

Mitchell, T. M. (1997). *Machine Learning*. The McGraw-Hill Companies, Inc.

Wang, H., Dubitzky, W., Düntsch, I., and Bell, D. (1999). A lattice machine approach to automated casebase design: Marrying lazy and eager learning. In *Proc. IJCAI99*, pages 254–259, Stockholm, Sweden.

Wang, H., Düntsch, I., and Bell, D. (1998). Data reduction based on hyper relations. In *Proceedings of KDD98, New York*, pages 349–353.

Chapter 15

KBIS: USING DOMAIN KNOWLEDGE TO GUIDE INSTANCE SELECTION

Peggy Wright
U.S. Army Corps of Engineers
Engineer Research and Development Center
Vicksburg, MS
wrightp@wes.army.mil

Julia Hodges
Department of Computer Science
Mississippi State University
Mississippi State, MS
hodges@cs.msstate.edu

Abstract   Prior to mining data for knowledge, selecting a potentially useful set of target data is necessary. Mining with missing attribute values increases uncertainty and decreases discovery accuracy. We present an instance selection method that determines the mining usability of an instance based on knowledge about which attributes are missing and the relative significance of the various attributes as defined by a domain expert. Knowledge-based instance selection (KbIS) is an instance utility metric that incorporates domain knowledge into a multi-criteria decision-making technique for instance selection.

Keywords: Domain knowledge, instance selection metric, ordered weighted aggregation, multi-criteria aggregation, missing values, record utility, feature presence, feature importance, feature selection.

15.1.   INTRODUCTION

Knowledge discovery is a means of extending limited human capabilities by using computer capabilities to analyze large, often complex datasets in order to extract more information than possible using conventional means. More time is spent in data preprocessing than in any other phase of the knowledge discovery process (Weiss and Indurkhya, 1997). Data selection is an important and time-consuming task. The data selection phase is made more complicated by missing data values. Missing values affect the potential usefulness of individual records and can lead to misleading or erroneous mining results. Missing values must be dealt with in any predictive model. Discarding records with missing values is a simple default solution, but it "destroys potentially valuable information" (Cheeseman et al., 1980).

Where data are scarce and particularly valuable, data selection should be applied in a considered, effective manner. KbIS (knowledge-based instance selection) attempts to improve on the default solution by measuring a record's information potential based on feature importance and presence. As previously stated, data preprocessing is a time-consuming task and data selection is probably the most important part of that task. Data selection typically includes selecting the target data and appropriate features or attributes. It can also include instance selection. Anything from selecting a random stratified sample to selecting records with all values present can be considered instance selection.

We are concerned with enhancing the data selection process by incorporating domain knowledge to use instance selection after feature selection has taken place. The development of an instance selection methodology is a means to further automate the data preprocessing phase and make it more effective and efficient. Early knowledge discovery models neither specifically included nor precluded instance selection (Fayyad et al., 1996; Brachman and Anand, 1996).

Figure 15.1 presents the "traditional" data selection method, which discards instances with missing values. In Figure 15.2, we demonstrate how instance selection can fit in the data selection process. Both assume that data selection is conducted after data cleaning (which may include elimination of obviously redundant or irrelevant features.)

The instance selection model we propose retains valuable records, where *valuable* is defined in terms of the discovery goal. We suggest that if the feature values available in an instance "outweigh" the value of those missing, the instance should be kept for mining.

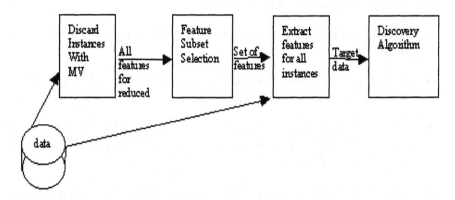

Figure 15.1 Traditional data selection model using the default method of discarding instances with missing values.

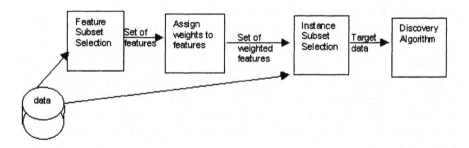

Figure 15.2 Extended data selection model using the KbIS framework to integrate domain knowledge.

15.2.　　MOTIVATION

A number of researchers have indicated the need for capabilities that can be provided by a methodology such as the one described in this paper. Some examples of such needs are:

- Automated/facilitated data selection and cleansing (Famili et al., 1997).

- Missing data problem (Famili et al., 1997; Fayyad et al., 1996; Holsheimer and Siebes, 1994).

- Including domain knowledge in the discovery process (Brachman and Anand, 1996; Lue and Hus, 1996; Grinstein, 1996; Owrang and Grupe, 1996).

As a result of researching this area, we concluded that using domain knowledge in a facilitated data selection framework could help address the missing value problem. To determine how this framework could be best utilized, we studied the shortcomings of the current data selection techniques that are used when missing values occur.

Current techniques for dealing with missing values include discarding records with missing values, ignoring missing values, and inferring replacement values. Deleting records with missing values is the easiest method (Holsheimer and Siebes, 1994) and may be appropriate when there are enough records present to build a complete predictive model. When there are not enough records available, or when records with missing values represent a particular portion of the population, this technique is neither effective nor appropriate.

Other techniques simply ignore missing values or assign the value *unknown*, which can have the effect of dissimilar records being grouped together. Another method uses one of many inferencing techniques to infer the missing values. When there are numerous missing values in a record, the resulting record with inferred values may not resemble the original record very much, making processing records with important missing attribute values more of a detriment than a benefit to the data mining process.

Famili (Famili, 1997) reported a naïve heuristic instance selection method that bases record retention on the percentage of missing values in a record. However, it does not consider the inherent value of an attribute. When a discovery goal is known prior to data mining, it is possible to determine the relative value an attribute contributes towards predicting the discovery goal. We theorize that this form of domain knowledge can be used to form the basis of a more intelligent instance selection method that considers the inherent value of the attributes present in an instance.

Deleting records with missing values may be considered a primitive instance selection technique as it "selects" instances that have all their attribute values present. The percentage heuristic is an extension of this primitive as it relates the number of attributes missing to the total number of attributes in the record. These techniques select records or instances based on the number of attribute values that are present. The other methods of dealing with missing values typically do not rely on instance selection but employ some scheme to infer the missing values or alter the discovery algorithm to deal with missing values in some other manner.

The heuristic percentage method reported by Famili (Famili, 1997) does not consider the inherent value of an attribute. When a discovery goal is known prior to data mining, it is possible to determine the relative value an attribute contributes towards predicting the discovery goal. We theorized that this form of domain knowledge could be used to form the basis of a more intelligent instance selection method that would consider the inherent value of the attributes present in an instance. The method we developed, KbIS, is based on knowledge about the significance or importance an attribute has relative to the discovery goal. KbIS considers attribute importance when the attribute value is present.

Essentially, KbIS measures information content in a manner different from information gain, which is based on Shannon's entropy (Mitchell, 1997). Both of these methods measure information based on a specific discovery goal. Information gain measures the specific value of each attribute in terms of partitioning the current dataset. KbIS measures the potential value of an instance based on the significance an attribute has to the goal.

15.3. METHODOLOGY

The following axioms serve to define desirable properties for knowledge-based instance selection. In the following axioms, U is the utility score for a record r, a is an attribute, s is the significance or importance of an attribute, and $n(r)$ is the number of attribute values present in a record.

1 $\forall r, r \equiv (a_1, \neg a_2, \ldots, \neg a_{n-1}, a_n)$. A record is a fixed vector of attributes whose values are either present or missing.

2 $\forall r, U(r) \equiv U(a_1 s_1, a_2 s_2, \ldots a_n s_n)$. The utility of a record is defined as some combination of each attribute's importance and its presence for all the attributes present.

3 $\forall r \ 0 \leq U(r) \leq 1$. The utility of a record is measured on the unit interval (non-negative).

4 $\forall r, \ U(a_1s_1, \neg a_2s_2) < U(a_1s_1, a_2s_2)$. The utility of a record increases as attribute values are added to the record (monotonic, increasing).

5 $\forall r, \ U(a_1s_1, a_2s_2) \geq U(a_1s_1) + U(a_2s_2)$. The utility of a record containing more than one attribute value is greater than or equal to the sum of the utilities of records containing the same individual attribute values.

6 $\forall r, \ r = (\neg a_1, \neg a_2, \ldots, \neg a_n), \ U(r) = 0.0$. The hypothetical record containing none of its attribute values has a utility score of 0.0.

7 $\forall r, \ r = (a_1, a_2, \ldots, a_n), \ U(r) = 1.0$. A record containing all its attribute values has a utility score of 1.0.

The utility of a record can be viewed as a multi-criteria aggregation based on the importance values of the features present in the record. This information metric is user-defined and is not probabilistic in nature, nor does the utility of a record depend on the content of other records. Each record or instance can be measured (assigned a utility score) based solely on its own merits relative to a selected knowledge discovery goal.

Once the desired system characteristics were formalized, we conducted an investigation to determine if a candidate methodology existed that would meet these requirements. Among the methodologies investigated, ordered weighted aggregation (OWA) seemed to best meet the KbIS requirements. OWA is a fuzzy-based multi-criteria aggregation that can be extended to include criteria importances (Yager, 1997b).

OWA operators are a special class of transform operators that have the following properties (Yager, 1998):

- Commutative: $T(a,b) = T(b,a)$

- Monotonic, non-decreasing: $T(a,b) \geq T(c,d)$, if $a \geq c$ and $b \geq d$ monotonic

- Idempotent: $T(a,a,a,a) = a$

OWA has its basis in fuzzy set theory and may be applied to such problems as classification, information retrieval, and decision analysis. The general form of an OWA operator produces an effective score F (Yager, 1993):

$$F = f(a_1w_1, a_2w_2, \ldots, a_nw_n)$$

where a represents an argument score, w is a weight applied to each attribute score, and f represents an aggregation function. The aggregation function f and the weights should be based on the nature of the aggregation task. The W values are the relative weights for each position and can be adjusted according to the problem. If we want to emphasize the larger scores, we would put most of the weight at the front of the vector. If we want to emphasize the smaller scores, we would put most of the weight at the end of the vector. The constraints placed on the A and W vectors are (Yager, 1997b):

$$a \in [0,1]$$

$$w \in [0,1]$$

$$\sum w_i = 1 \text{ (for all i)}$$

$$F(a_1, \ldots, a_n) = \sum b_j w_j \text{ (for all j) where } b_j \text{ is the } j^{th} \text{ largest } a$$
value.

The applicability of OWA was explored for each of the defining axioms. In OWA, each criterion has a score; the attribute presence in (1) can be viewed as an attribute or criterion score. (23) The OWA operator is a transform based on the unit interval such that there are two input vectors and one output which are all expressed in the unit interval $T:[0,1] \times [0,1] \Rightarrow [0,1]$. Axioms seven (7) and six (6) deal with special boundary cases. Since an OWA operator is idempotent, i.e., $F(a,a,a,a) = a$, these conditions are met iff the OWA operator is directly applied to the attribute presence vector A.

The same defining characteristic of an OWA operator that satisfied (3) also satisfies (2). By definition an OWA operator is monotonic nondecreasing and (4) requires monotonic increasing. Theoretically, the only case not covered is the equivalent case; with the way attribute presence is modeled in KbIS, this is not a problem. The OWA operator is a nonlinear function whose power lies in ordering the input vectors, and therefore (5) is met.

We investigated several options for implementing OWA operators. Given our problem of how to model and measure the value of each record when there are missing attribute values, we needed some method to represent feature presence. OWA starts with the a scores, which in this problem we relate to attribute scores. We used 1 to indicate that an attribute value was present and 0 to indicate it was missing. The a scores for each record are recorded as a vector A, where each score is either 0 or 1. Using the assumed w scores $\{0.4, 0.3, 0.2, 0.1\}$, we have the example:

1	1	0	1	A (feature presence)
0.4	0.3	0.2	0.1	W (position weights)
1	1	1	0	B (ordered A)
0.4	0.3	0.1	0.2	V (ordered W)

record value $= (1*0.4) + (1*0.3) + (1*0.1) + (0*0.2) = 0.80$

Determining effective position weights is key to using the OWA aggregation. These weights can be assigned or learned, determined by using a specified class of OWA operators, or modified by a linguistic quantifier (Yager, 1997a). Subjectively assigning the weights requires an in-depth knowledge about the application and knowledge about how the weights will be used. For our research, since the weights vary for each discovery goal, we felt that subjective weight assignments would not be effective. Each of the other methods used to automate learning the position weights was implemented and evaluated.

We first investigated a method that involved automated learning of the position weights. The weights were learned using an alpha value (α) based on the concept of *orness*, which is inversely proportional to *andness* (Yager, 1993; Yager, 1997a). Using a lower *orness* value results in a higher degree of *andness*. The general form of this function is shown below along with examples. Different alpha values were tested to observe the effect they had on the decision score for two records, one missing the least important feature and another missing the most important feature. The first example record is detailed below using an alpha value of 0.20. The results from the two test cases and the postulate cases are reported in Table 15.1

$w_1 = \alpha; w_2..w_{n-1} = (w_{i-1})(1 - \alpha); w_n = 1 - \sum w_{i,i=1 ton-1}$
$b_i = (1 - \alpha)(1 - u_i) + (u_i)(a_i)$
$D(r_1) = \sum b_i w_i, i = 1 ton$

1	1	0	1	(feature presence)
0.9	0.6	0.2	0.7	(feature importance)

$w_1 = 0.20$
$w_2 = (0.2)(0.8) = 0.16$
$w_3 = (0.16)(0.8) = 0.13$
$w_4 = (1.0 - 0.49) = 0.51$
$b_1 = (0.8)(0.1) + (0.9)(1) = 0.98$
$b_2 = (0.8)(0.4) + (0.6)(1) = 0.92$
$b_3 = (0.8)(0.8) + (0.2)(0) = 0.64$

$b_4 = (0.8)(0.3) + (0.7)(1) = 0.94$
ordered $B = \{0.98, 0.94, 0.92, 0.64\}$
$D(r1) = (0.98)(0.20) + (0.94)(0.16) + (0.92)(0.13) + (0.64)(0.51) = 0.79$

Table 15.1 Utility scores with learned position weights

Record	Learning Constant (α)						
Score	0.00	0.10	0.20	0.50	0.80	0.90	1.00
1111	1.00	0.93	0.89	0.86	0.88	0.89	0.90
1101	0.80	0.79	0.79	0.84	0.88	0.89	0.90
0111	0.10	0.32	0.48	0.71	0.73	0.72	0.70
0000	0.10	0.19	0.25	0.28	0.14	0.08	0.00

This method was designed to allow modification of the *min* and *max* operations. Since we were interested in having *almost all* of the features, a lower alpha value had the effect of easing the *all* constraint (*min* operator). The boundary case with all values present (7) is met only when the alpha value is at 0.0; the other boundary case (6) is met only when alpha is 1.0. Using the extreme values of alpha does not utilize this method properly, and there is no case where both postulates are met.

The second method investigated uses specially defined operators to describe the aggregation. *Min* and *max* are two such aggregation functions that are suited to special cases. The *min* function models the term *all* ($\forall$), while the max function models the term *any* (7). The *min* function is related to the connective *and*, and the *max* function is related to the connective *or*. Since we were concerned that most of the important features were present, we decided to use the *min* aggregation function in the following form (Yager, 1997a). The case of a record missing the least important feature is reviewed in detail and the scores for our test and postulate cases are summarized in Table 15.2.

$F(a\text{-}1, \ldots, a_n) = \text{Min}_j[a_j]$ for all j, $b_j = S(1\text{-}u_j, a_j)$, and S is a t-conorm such as Max

| 1 | 1 | 0 | 1 | (feature presence) |
| 0.9 | 0.6 | 0.2 | 0.7 | (feature importance) |

$F = \text{Min}[\ \text{Max}(0.1,1), \text{Max}(0.4,1), \text{Max}(0.8,0), \text{Max}(0.3,1)]$
$F = \text{Min}(1,1,0.8,1) = 0.8$

Table 15.2 Utility Scores with special operator

Record Score	Utility Score
1111	1.0
1101	0.8
0111	0.1
0000	0.1

Using this method, boundary condition (6) is not met. Additionally, the score for the third record, which is missing the most important value, is low. Having the other three attribute values present did not weigh significantly using this algorithm. In general, this operator selects the maximum score (1) when all features are present. When there are missing features, the record is worth the inverse of the weight $(1-u_j)$, of the most important attribute that is missing.

KbIS calculates record utility using feature weights. These weights or importance values may also be included in an OWA formula (Yager, 1998). The importance vector contains the domain knowledge about each attribute and the importance values are used to bias the final effective score based on the attribute's score and position weight. Using this algorithm, the importance values are used to calculate the position weights.

The importance vector values U are also contained in the interval $[0,1]$. The significance ratings assigned by the domain expert were translated into importance values and used to calculate the OWA weight vector W. The importance scores have the effect of quantifying attributes according to their relative significance to a particular goal. When combined with the attribute scores, this has the overall effect of measuring record usefulness based on an aggregation of feature importance and presence. The OWA method requires the A vector to be sorted based on the individual a scores, producing (ordered) vector B. The corresponding reordered importance value vector is denoted V.

Since the goal of KbIS is to select instances based on having *fewer than all* of the important attributes, we needed to further constrain or modify the OWA operator. Using linguistic quantifiers to modify or quantify an observation is a natural extension since OWA is based on fuzzy logic (Yager, 1993). Bosc and Lietard (Bosc and Lietard, 1997) propose using linguistic quantifiers with OWA operators for database querying. They can also be used in group decision-making (Kacprzyk et al., 1997). The following example uses the fuzzy-based OWA (ordered weighted aggregation) with a linguistic quantifier (Q) used to calculate the weights.

$$D(r_i) = \Sigma b_i w_i, \text{ where}$$
$$w_i = Q(S_i/T) - Q(S_{i-1}/T) \text{ and}$$
$$T = \Sigma v_i \text{ for } i=1ton$$
$$S = \Sigma v_j \text{ for } j=1toi$$
$$Q \text{ is the quantifier}$$

For KbIS, the decision whether to keep a record, $D(r_i)$, is calculated by aggregating the attribute weights of the features present. The D value determines if the instance is kept or discarded relative to a specified utility threshold. The optimal utility threshold appeared to vary from dataset to dataset and will be discussed further under analysis and conclusions. The initial quantifier used, $Q(r) = r^2$, is a fuzzy hedge for *most*. The record missing the least important feature is shown in detail and the other examples are summarized in Table 15.3.

1	1	0	1	A (feature presence)
0.9	0.6	0.2	0.7	U (feature importance)
1	1	1	0	B (sorted A)
0.9	0.6	0.7	0.2	V (sorted U)

$$T = 0.9 + 0.6 + 0.7 + 0.2 = 2.4$$
$$S_1 = 0.9, S_2 = 1.5, S_3 = 2.2, S_4 = 2.4$$
$$w_1 = (0.9/2.4)^2 - (0/2.4)^2 = (0.14 - 0) = 0.14$$
$$w_2 = (1.5/2.4)^2 - (0.9/2.4)^2 = (0.39 - 0.14) = 0.25$$
$$w_3 = (2.2/2.4)^2 - (1.5/2.4)^2 = (0.84 - 0.39) = 0.45$$
$$w_4 = (2.4/2.4)^2 - (2.2/2.4)^2 = (1.0 - 0.84) = 0.16$$
$$D(r) = (0.14)(1) + (0.25)(1) + (0.45)(1) + (0.16)(0) = 0.84$$

This quantifier, $Q(r) = r^2$, seemed to fit the problem because we wanted to include *most* of the important attributes in a record. It modeled the difference between the example records well. The possibility of using other linguistic quantifiers was investigated and this quantifier appeared to be better suited to the task at hand.

Table 15.3 Utility scores with linguistic quantifier *most*

Record Score	Utility Score
1111	1.00
1101	0.84
0111	0.39
0000	0.00

15.4.   EXPERIMENTAL SETUP

The goal of this experiment was to show the validity of knowledge-based instance selection (KbIS). As implemented, instance selection takes place after feature selection. Feature selection was done using domain expertise; the same feature subset was used for all comparison techniques. C4.5 was used as the learning algorithm. Since our approach is instance-based and takes place after feature subset selection, no valid feature subset selection technique is precluded by KbIS. The results reported here are based on one of the experiments we conducted using the National Bridge Inventory System (NBIS) data provided by the U.S. Department of Transportation, National Highway Administration. Datasets from two states were originally selected from the NBIS; random stratified selection was used to select the target data. The original datasets were preprocessed three separate times to randomly delete attribute values. In the first pass, the resulting datasets contained approximately 71% of the data, i.e., 29% of the overall attribute values were missing. The percentage of data values missing from each instance varied and 20% of the original data (every fifth record) was left unchanged. These datasets are named CR29% and MR29%.

The second pass of the data also randomly selected attribute values for deletion, but the probability of deletion was inversely proportional to the attribute importance. The resulting dataset was missing about 54% of the overall attribute values. Again, the percentage of missing data values varied from record to record and 20% of the data was left unchanged. This pass resulted in a dataset that was missing a high proportion of the data values with the most values missing from the least important features. These datasets are denoted CRP54% and MRP54%.

Since the first dataset contained purely random deletions, we decided to consider a dataset with approximately the same number of missing values, but with the deletions occurring among the least important attribute values. To produce the third datasets we used the same weighted random deletion algorithm as the second pass, but reduced the probability of deletion such that the resulting datasets contained approximately 29% missing values. These datasets are CRP29% and MRP29%.

Knowledge discovery was performed using the 20% missing value heuristic, i.e., all records having more than 20% missing values were deleted (Famili, 1997). The 20% figure was based on knowledge of the data characteristics and previous data mining experience. Experimentation using different percentages indicated that 20% was not necessarily the best percentage for our data. Therefore, the dataset with randomly deleted values was also processed such that data subsets representing different percentages of missing values were produced. All records with no missing values were selected into a file designated 0%. Likewise, all records missing more than 50% of the attribute values were written to a file designated 50%. This was done at 5% intervals from 0-50, creating 11 files.

Relevant features were selected by the domain expert and used on the previously built datasets. Feature selection was performed on the original dataset, and KbIS was used to measure the utility of each instance in this dataset. We varied the utility score to produce numerous datasets. Data subsets were created for utility values varying from 0.0 to 1.0 at intervals of .05, creating 21 separate files.

Knowledge discovery was performed on each of the 32 data subsets. Ten-fold cross validation was used to assure verifiable discovery results. Although different discovery parameters such as grouping and pruning were explored, there was no one set of parameters that consistently performed best on all 32 data subsets. Therefore, the standard cross validation provided with C4.5 (Quinlan, 1993) was used with blocksize (k)=10. Undoubtedly, the discovery parameters could be fine-tuned to each dataset to increase the learning performance.

15.5.   ANALYSIS AND EVALUATION

This analysis summarizes the experimental results obtained from using two datasets from different states consisting of 500 instances each. A baseline error rate on unseen cases was established on the original datasets and then the error rates were recorded for each of the three methods compared. These methods are the 20% missing value heuristic (20MVH), the best missing value heuristic (*MVH), and the best KbIS

(*KbIS). Both the *MVH results and the *KbIS results were chosen by selecting the lowest error rates observed on rules across all data subsets produced using the selection method.

Table 15.4 presents the discovery results for all datasets. The baseline error rates are in the rows labeled MORIG and CORIG. They vary from 10.8% for decision trees to 17.4% for rules. These prediction rates are considered acceptable based on data mining experience in this domain and military bridge assessment expertise. No instance selection methods were run on the original datasets because they had few naturally occurring missing values.

The results from those datasets that have on average 29% random missing attribute values are reported in the rows labeled MR29% and CR29%, with the best missing value heuristic performing best. *MVH outperformed *KbIS significantly on decision trees; on average it performed slightly better on rules.

The datasets that had weighted random deletions of approximately 29% are shown in the MPR29% and CPR29% rows. Both *MVH and *KbIS outperform 20MVH, and, for the randomly deleted values, *KbIS shows significant performance improvements on rules over *MVH.

*KbIS clearly outperformed the other two instance selection methods on the final datasets, MRP54% and CRP54%. The original error rates (particularly on the C dataset) are so high as to be of dubious worth, and *MVH did not improve the discovery results for these datasets. However, the error rates for rules obtained on the datasets selected using *KbIS were, by contrast, useful.

Table 15.4   C4.5 discovery results on all datasets

| | ERROR RATE ON UNSEEN CASES | | | | | | | | | | | |
| | NO SELECTION | | | 20MVH | | | *MVH | | | *KBIS | | |
Dataset	Recs	Trees	Rules	Recs	Trees	Rules	Recs	Trees	Rules	Recs	Trees	Rules
MORIG	500	10.8%	12.0%									
MR29%	500	20.2%	22.4%	102	19.6%	17.6%	88	7.9%	11.2%	153	13.0%	12.4%
MRP29%	500	23.0%	24.6%	42	29.5%	27.5%	141	21.3%	20.5%	57	22.7%	15.3%
MRP54%	500	35.4%	28.0%				109	40.4%	40.3%	304	31.2%	25.3%
CORIG%	500	14.4%	17.4%									
CR29%	500	29.0%	34.0%	101	24.8%	26.6%	86	23.5%	22.5%	219	26.6%	22.4%
CRP29%	500	25.2%	35.8%	34	33.3%	37.5%	263	25.4%	31.5%	163	22.7%	25.8%
CRP54%	500	50.2%	45.2%				98	53.1%	40.8%	74	36.4%	20.7%

Table 15.5 contains the percentage of improvement for each data subset produced from the first state. The first row shows the improvement

that *MVH had over the 20MVH; the second row, the improvement that *KbIS had over the 20MVH; and the third row, the improvement that *KbIS had over the *MVH. Table 15.6 contains the same information for the data subset produced from the second state.

Table 15.5 Instance selection comparison for the first state

	M500 IMPROVEMENT					
	MR29%		MRP29%		MRP54%	
	Trees	Rules	Trees	Rules	Trees	Rules
*MVH/20MVH	59.7%	36.4%	27.8%	25.5%		
*KBIS/20MVH	33.7%	29.5%	23.1%	44.4%		
*KBIS/*MVH	-64.6%	-10.7%	-6.6%	25.4%	22.8%	37.2%

The mean improvements in all categories for both states are shown in Table 15.7. On average, the instance selection method introduced here performs better than the other methods on those datasets that contain a high percentage of missing values. In particular, KbIS performs well on datasets that are missing attribute values but that still contain most of the important attribute values.

Table 15.6 Instance selection comparison for the second state

	C500 IMPROVEMENT					
	CR29%		CRP29%		CRP54%	
	Trees	Rules	Trees	Rules	Trees	Rules
*MVH/20MVH	5.2%	15.4%	23.7%	16.0%		
*KBIS/20MVH	-7.3%	15.8%	31.8%	31.2%		
*KBIS/*MVH	-13.2%	0.4%	10.6%	18.1%	31.5%	49.3%

15.6. CONCLUSIONS

We have introduced KbIS, an instance utility metric that incorporates domain knowledge about each feature in a multi-criteria aggregation. Results show that using KbIS on datasets with missing values leads to better learning accuracy over the "traditional" percentage heuristic reported in (Famili, 1997).

KbIS examines the inherent value of each record as it relates to the discovery goal. From our experimental results (including those reported

Table 15.7 Mean improvement for instance selection comparison

| | MEAN IMPROVEMENT | | | | | |
| | R29% | | RP29% | | RP54% | |
	Trees	Rules	Trees	Rules	Trees	Rules
*MVH/20MVH	32.5%	25.9%	25.8%	20.7%		
*KBIS/20MVH	13.2%	22.7%	27.4%	37.8%		
*KBIS/*MVH	-38.9%	-5.1%	2.0%	21.7%	27.1%	43.2%

here), we conclude KbIS shows promise of being a more intelligent method of instance selection when there are missing values than default instance selection methods such as the percentage heuristic. Although, KbIS is based on domain knowledge, the KbIS framework has the advantage of being domain independent.

References

Bosc, P. and Lietard, M. 1997. Quantified statements and some interpretations for the OWA operator. In The Ordered Weighted Averaging Operators: Theory and Application, R. R. Yager and J. Kacprzyk (Eds.). Kluwer, pp. 241-257.

Brachman, R. J. and T. Anand. 1996. The process of knowledge discovery in databases: A human-centered approach. In Advances in Knowledge Discovery and Data Mining, U. M. Fayyad, G. Piatetsky-Shapiro, P. Smyth, and R. Uthurusamy (Eds.). Menlo Park, CA: AAAI Press, pp. 37-57.

Cheeseman,P., J. Kelly, M. Self, J. Stutz, W. Taylor, and D. Freeman. 1988. AutoClass: A Bayesian classification system. Proc. 5th Int. Conf. on Machine Learning, pp. 54-64.

Famili, A. <Fazel.Famili@iit.nrc.ca>. Personal communication about 20% heuristic method. (3 November 1997).

Famili, A., W. M. Shen, R. Weber, and E. Simoudis. 1997. Data preprocessing and intelligent data analysis. Intelligent Data Analysis 1(1): <http://www.elsevier.com/locate/da> (21 January 1997).

Fayyad, U. M., G. Piatetsky-Shapiro, and P. Smyth. 1996. From data mining to knowledge discovery: An overview. In Advances in Knowledge Discovery and Data Mining, U. M. Fayyad, G. Piatetsky-Shapiro, P. Smyth, and R. Uthurusamy (Eds.). Menlo Park, CA: AAAI Press, pp. 1-34.

Grinstein, G. G. 1996. Harnessing the human in knowledge discovery. In KDD-96, Proc. 2nd Int. Conf. on Knowledge Discovery and Data Mining, Portland, Oregon, pp. 384-385.

Holsheimer, M. and A. Siebes. 1994. Data mining: The search for knowledge in databases. Amsterdam, The Netherlands: CWI, Report No. CS-R9406.

Huber, P. J. 1997. From large to huge: A statistician's reactions to KDD and DM. KDD-97, Proc. 3rd Int. Conf. on Knowledge Discovery and Data Mining, Newport Beach, California., pp. 304-308.

Kacprzyk, J., M. Fedrizzi, and H. Nurmi. 1997. OWA operators in group decision making and consensus reaching under fuzzy preferences and fuzzy majority. In The Ordered Weighted Averaging Operators: Theory and Application, R. R. Yager and J. Kacprzyk (Eds.). Kluwer, pp.193-206.

Liu, B. and W. Hsu. 1996. Post-Analysis of learned rules. Proc. 13th Natl. Conf. on AI and 8th Innovative Appl. of AI, pp. 164-168.

Mitchell, T. M. 1997. Machine Learning. Boston, MA: McGraw-Hill.

Owrang, M. M. and F. H. Grupe. 1996. Using domain knowledge to guide database knowledge discovery. Exp. Syst. with Appl. 10(2):173-180.

Quinlan, J. R. 1993. C4.5: Programs for Machine Learning. San Mateo, CA: Morgan Kaufmann.

Weiss, S. M. and N. Indurkhya. 1997. Predictive Data Mining: A Practical Guide. San Francisco, CA: Morgan Kaufmann.

Yager, R. R. 1998. On ordered weighted averaging aggregation operators in multi-criteria decision making. IEEE Trans. on Syst., Man, and Cybern. 18:183-190.

Yager, R. R. 1993. On ordered weighted averaging aggregation operators. In Readings in Fuzzy Sets for Intelligent Systems, D. Dubois, H. Prade, and R. R. Yager (Eds.). San Mateo, CA: Morgan Kaufmann, pp. 80-87.

Yager, R. R. 1997a. Criteria importances in OWA aggregation. Proc. 6th Int. Conf. on Fuzzy Syst., Barcelona Spain, pp. 1677-1682.

Yager, R. R. 1997b. On the inclusion of importances in OWA aggregations. In The Ordered Weighted Averaging Operators: Theory and Application, R. R. Yager and J. Kacprzyk (Eds.). Kluwer, pp. 41-59.

V

INSTANCE SELECTION IN MODEL COMBINATION

Chapter 16

INSTANCE SAMPLING FOR BOOSTED AND STANDALONE NEAREST NEIGHBOR CLASSIFIERS

David B. Skalak
IBM Data Mining and Analytics Group
Cornell Theory Center
Frank H.T. Rhodes Hall
Cornell University
Ithaca, NY 148
skalak@us.ibm.com

Abstract   Several previous research efforts have questioned the utility of combining nearest neighbor classifiers. We introduce an algorithm that combines a nearest neighbor classifier with a "small," *coarse-hypothesis* nearest neighbor classifier that stores only one prototype per class. We show that this simple *paired boosting* scheme yields increased accuracy on some data sets.

The research presented in this article also extends previous work on prototype selection for a standalone nearest neighbor classifier. We show that in some domains, storing a very small number of prototypes can provide classification accuracy greater than or equal to that of a nearest neighbor classifier that stores all training instances. We extend previous work by demonstrating that algorithms that rely primarily on random sampling can effectively choose a small number of prototypes.

Finally, we present a taxonomy of instance types that arises from the statistics collected on the performance of a set of sampled nearest neighbor classifiers as they are applied to each individual instance. This taxonomy generalizes the idea of an outlier.

Keywords: Boosting, nearest-neighbor, instance-based learning, classifier combination, model stability, sampling, prototype selection, instance selection.

16.1.   THE INTRODUCTION

Nearest neighbor classifiers predict the class of a previously unseen instance by computing its similarity to a set of stored instances called *prototypes*. *Prototype selection* — storing a well-selected, proper subset of available training instances — has been shown to increase classifier accuracy in many domains (Dasarathy, 1991). At the same time, using prototypes dramatically decreases storage and classification-time costs.

A large number of approaches have been tried to identify these salient instances that are stored by a nearest neighbor classifier; some are surveyed in Section 16.2 The goal of this article is to show that mere random sampling can select useful prototypes for two varieties of nearest neighbor classifiers: for standalone nearest neighbor classifiers and for nearest neighbor classifiers whose predictions are to be combined.

For a standalone nearest neighbor classifier, we show that a small number of samples of a minimal set of prototypes can often yield a classifier that is as accurate as one that retains all instances as prototypes. With this result as support, we then approach the problem of whether a small, sampled nearest neighbor classifier can be incorporated in what might be called a *paired boosting* architecture. In this classifier combination framework, the user is presented with a fully-fledged classifier, i.e., a strong learner (Harries, 1999), and the goal is to find a second classifier such that combining the predictions of the first classifier with the second yields more accurate predictions. Here, we pair a standard nearest neighbor classifier with a small, sampled nearest neighbor classifier, and combine their predictions with a decision tree.

Several previous attempts to combine nearest neighbor classifiers have shown that nearest neighbor classifiers resist accuracy-enhancing combination techniques that tend to work for other model classes (Breiman, 1992; Drucker et al., 1994), or at best yield an increase in application efficiency but no increase in accuracy through combination (Freund and Schapire, 1996). We discuss these projects in Section 16.2.2.

We pursue our investigation of paired boosting within a more general approach to combining classifiers that has been termed *stacked generalization* (Wolpert, 1992). In stacked generalization's most basic form, a layered architecture consists of a set of *component classifiers* that form the first layer, and of a single *combining classifier* that forms the second (somewhat degenerate) layer (Figure 16.1). Wolpert calls the component classifiers the *level-0 classifiers* and the combining classifier, the *level-1 classifier*. The composite classifier is called a *stacked generalizer*.

The crux of this article is that nearest neighbor classifiers can in fact be combined effectively, if one takes the extreme approach of radically

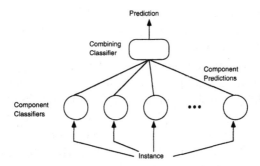

Figure 16.1 Composite classifier architecture

destabilizing a component classifier by storing only a very small number of prototypes. While stability in the face of small changes to a training set is generally considered a desirable characteristic of a standalone classifier (Turney, 1995). we observe that destabilization can be useful to combine classifiers that are inherently stable. The point is that these small classifiers can be sufficiently accurate to be fruitfully combined with other component classifiers to yield still greater accuracy.

Our instance-sampling approach leads also to a side result. We can hold each instance fixed, observe how sampled classifier classifies that instance, and gather statistics as to the frequency with which that instance is correctly classified by the sampled classifiers. When these statistics are used in tandem with the accuracy of each classifier on the entire training set, we observe that subsets of instances behave differently as classifiers are applied to them. The resulting heuristic taxonomy of instances broadens the notion of an outlier (Cleveland, 1993; Ripley, 1996).

Stepping back a bit, we see four reasons why *coarse-hypothesis* classifiers that incorporate simple classification hypotheses are worthy of additional investigation for standalone or combined classifiers. (1) Simpler solutions to classification tasks are less likely to be overlooked (Holte, 1993). (2) The problem of overfitting may be assuaged by simple hypotheses. Due to the flexibility present in typical combined classifiers, it is especially important to guard against overfitting the training set (Brodley, 1994). (3) The utility of simple, shallow models, such as *decision stumps* (Holte, 1993), has been demonstrated. (4) Simpler hypotheses are usually more comprehensible, an important consideration in data-mining applications, for example (Cherkauer and Shavlik, 1996). Quinlan has called for additional research that finds an appropriate tradeoff between accuracy and complexity, particularly with respect to boosted classifiers, which can be "totally opaque" (Quinlan, 1999, p.524). Quinlan proposes one line of research: start with an inscrutable boosted classifier and try to simplify it (Quinlan, 1999). In this article, we try the other direction: we start with simplicity and see how far we can progress towards accuracy.

The article is organized as follows. Previous approaches to prototype selection for a standalone nearest neighbor classifier and to nearest neighbor classifier combination are first reviewed (Section 16.2). We then propose an instance sampling approach to severely limit the number of prototypes stored in a single, standalone nearest neighbor classifier (Section 16.3). Moving to the paired boosting classifier combination problem, we investigate the utility of instance-sampling to create a classifier that can be combined with a standard nearest neighbor classifier that stores all instances as prototypes (Section 16.4). Finally, we present a taxonomy of instance types that results from looking more deeply into how a single instance behaves when classified by a collection of sampled classifiers (Section 16.5).

16.2.   **RELATED RESEARCH**

16.2.1.   **Creating a Standalone Nearest Neighbor Classifier**

Reducing the number of prototypes used for nearest neighbor retrieval has been a topic of research in pattern recognition and machine learning for over 30 years (Hart, 1968; Aha, 1990; Dasarathy, 1991). The problem is sometimes called the *reference selection problem* and the algorithms to perform this task have been called *editing algorithms* or *editing rules*.

Broadly speaking, there are three main approaches to reference selection: (1) instance-filtering, (2) stochastic search, and (3) instance-averaging. By far, the greatest bulk of work falls into the instance-filtering camp. A rule is used incrementally to determine which instances to store as prototypes and which to throw away. Examples of this approach have included storing misclassified instances (Hart, 1968; Gates, 1972; Aha, 1990); storing typical instances (Zhang, 1992); storing only training instances that have been correctly classified by other training instances (Wilson, 1972); and combining these techniques (Voisin and Devijver, 1987). Stochastic local search algorithms for prototype selection have been applied by Skalak (Skalak, 1994) and Cameron-Jones (Cameron-Jones, 1995). Still other systems deal with prototype selection by storing averages or abstractions of instances (Chang, 1974; de la Maza, 1991).

16.2.2.   Combining Nearest Neighbor Classifiers

Comparatively little research has been done on combining nearest neighbor classifiers. In fact, several research efforts can be viewed as demonstrating that nearest neighbor classifiers resist combination to increase accuracy.

Wolpert has applied the stacked generalization framework to the Net-Talk problem of phoneme prediction from letter sequences (Wolpert, 1992). The experimental architecture used a set of three 4-nearest-neighbor level-0 classifiers combined by a level-1 nearest neighbor classifier. The algorithm was not compared to other machine learning algorithms.

Breiman (Breiman, 1992) demonstrated that combining 100 nearest neighbor classifiers through the *bagging (bootstrap aggregating)* instance-resampling algorithm did not reduce the rate of misclassification below that of a single nearest neighbor classifier. He attributed this behavior to a characteristic of nearest neighbor classifiers: that the addition or removal of a small number of training instances does not change nearest neighbor classification boundaries very much. Decision trees and neural networks, he argued, are in this sense less *stable* than nearest neighbor classifiers.

In early experiments with boosting, another group of well-known researchers, Drucker, Cortes, Jackel, LeCun and Vapnik also observed informally that a classifier boosting algorithm was not effective for combining nearest neighbor classifiers. While specific results were not presented, experiments with other "non-neural based classifiers" (Drucker et al., 1994, p.61) were also found not to result in increased accuracy.

The Adaboost (Adaptive Boosting) boosting algorithm (Freund and Schapire, 1995) incorporates a large number of weak learners, changes the distribution of training examples as each weak learner is trained serially, and uses a weighted average to combine them. Freund and Schapire have used Adaboost to improve a nearest neighbor classifier on a handwriting recognition task (Freund and Schapire, 1996). The goal of this investigation was not to increase accuracy, but to increase efficiency through reducing the set of stored prototypes. The error for the boosted classifier was 2.7% versus 2.3% for a nearest neighbor classifier that used all instances as prototypes.

Ho (Ho, 1998) investigated an approach to a handwritten digit recognition problem that did provide an accuracy increase. Ho combined the decisions of a set of nearest neighbor classifiers, each of which incorporated a stochastically chosen proper subset of features into a Euclidean similarity metric. For a range of parameter selections, Ho demonstrated an increase in combined nearest neighbor accuracy over a single k-nearest neighbor classifier on this task.

Alpaydin (Alpaydin, 1997) also combined a set of nearest neighbor classifiers using voting, but relies on the order dependence of a classical prototype selection method, the Condensed Nearest Neighbor algorithm (Hart, 1968), to create nearest neighbor classifiers that make different predictions. Experiments showed that combining only three such nearest neighbor classifiers yielded an improvement in accuracy on three of six data sets, where each classifier stored from 14% to 67% of the original training instances.

The evidence provided by these efforts — both positive and negative — suggests that a somewhat different approach may be in order to the problem of nearest neighbor classifier combination. In the next sections, we investigate one such approach, which we apply first to the preliminary problem of creating a standalone nearest neighbor classifier and then to the problem of creating nearest neighbor classifiers whose predictions are combined.

16.3.   SAMPLING FOR A STANDALONE NEAREST NEIGHBOR CLASSIFIER

This section describes a generate-and-test algorithm that samples coarse nearest neighbor classifiers by sampling small sets of prototypes. The sampling plan we shall use is *stratified random sampling* (Hansen et al., 1953), where the strata are the instances of each class. In general, stratified sampling is used to assure that representatives of each stratum are sampled, and to improve the accuracy of a desired population estimate for a given sampling cost. We invoke prototype sampling in part because it provides a way to the search the collection of small sets of prototypes, a characteristic not found in editing algorithms that do not extensively search the space of sets of prototypes of a fixed cardinality, but instead increase or decrease the number of prototypes to improve the prototype set.

The nearest neighbor classifiers we construct are 1-nearest neighbor classifiers that use the Manhattan ("city block" or l_1) distance metric. The prototype with the smallest such distance to a test instance is its nearest neighbor.

Formally, if $x = (x_1, x_2, \ldots, x_n)$ and $y = (y_1, y_2, \ldots, y_n)$ are two instances, F_n is the set of numeric features, F_s is the set of symbolic features, then the distance between x and y, $d(x, y)$ is

$$d(x, y) = \sum_{i \in F_n} |x_i - y_i| + \sum_{j \in F_s} d_s(x_j, y_j)$$

where d_s is an exact-match metric for symbolic features:

$$d_s(x_j, y_j) = \begin{cases} 0 & \text{if } x_j = y_j \\ 1 & \text{otherwise} \end{cases}$$

16.3.1.   Algorithm

As input, the prototype sampling algorithm takes the training set T and four other parameters: φ (the fitness function to determine the usefulness of a set of prototypes), k (k-nearest neighbors), p (the number of prototypes sampled from each class), and m (the number of samples taken). These parameters are fixed in advance. Unless otherwise specified, for all the experiments presented, φ is the leave-one-out cross-validation accuracy of a nearest neighbor classifier on the training set, $k = 1$, $p = 1$ and $m = 100$. Thus, the total number of prototypes sampled is the number of classes (s) "exposed" in the training set T. The classes *exposed* in the training set are those classes for which instances of that class are present in the training set (Dasarathy, 1991). The classifiers are arguably minimal in the number of prototypes, since only one prototype is sampled from each class.

The sampling algorithm can be summarized as follows. We assume that a training set has been identified.

1 Select m random samples, each sample with replacement, of p instances from each class exposed in the training set.

2 Store each sample in a nearest neighbor classifier. To evaluate the utility of that classifier, compute the value of a *fitness function* on that sample. The *fitness* of a sampled classifier is its classification accuracy on the training set using a 1-nearest neighbor rule and leave-one-out cross-validation.

3 Return the classifier with the highest classification accuracy on the training set. In the event of an accuracy tie, return the tied classifier that was generated first.

16.3.2.   Evaluation

In Table 16.1 we give the classification accuracy of the baseline nearest neighbor algorithm and that of a sampling technique that stores only one prototype per class on a collection consisting primarily of benchmark data sets that have been previously used in classifier combination research, e.g., (Brodley, 1994; Ali and Pazzani, 1996). We sometimes call the baseline nearest neighbor classifier a *full* nearest neighbor classifier to indicate that it stores all training instances as prototypes. In this experiment, one hundred samples were taken. Two 10-fold cross-validation runs were made; mean test accuracies are given. A one-tailed t-test for matched pairs is used throughout to determine the statistical significance of differences in mean generalization accuracy.

Table 16.1 Classification accuracy (in percent correct, with standard deviation) of the nearest neighbor (NN) algorithm and the prototype sampling (PS) algorithm, one prototype per class.

Data	Prototypes	NN	PS	Sig.
Breast Cancer Ljubljana	2	67.5 ± 9.6	73.2 ± 8.2	0.019
Breast Cancer Wisconsin	2	96.2 ± 1.8	96.4 ± 2.4	0.422
Cleveland Heart Disease	2	77.0 ± 5.5	78.5 ± 8.0	0.298
Diabetes	2	68.7 ± 5.1	73.9 ± 4.1	0.007
Glass Recognition	6	73.3 ± 10.3	60.0 ± 8.5	1.000
Hepatitis	2	77.3 ± 11.4	82.7 ± 8.3	0.010
Iris Plants	3	91.3 ± 7.1	94.7 ± 5.5	0.053
LED-24 Digit	10	36.5 ± 9.4	32.2 ± 5.6	0.875
LED-7 Digit	10	71.2 ± 7.7	73.2 ± 7.8	0.095
Lymphography	4	80.7 ± 8.9	74.6 ± 5.2	0.968
Monks-2	2	48.8 ± 4.7	46.6 ± 5.0	0.802
Promoter	2	79.0 ± 9.9	64.5 ± 10.4	0.996
Soybean	19	91.0 ± 3.9	65.8 ± 7.3	1.000

Thus, on nine of thirteen benchmark data sets, a classifier that (i) is a minimal member of its model class (when measured by the number of prototypes) and (ii) is generated by a naive sampling algorithm that drew a small number of samples (when compared to the population size) has accuracy higher than or statistically comparable to a standard nearest neighbor classifier that stores all available instances as prototypes.

16.4. COARSE RECLASSIFICATION

The results of the investigation of sampled, coarse-hypothesis classifiers in the previous section suggest that they may be sufficiently accurate to be combined successfully. Our objective in this section is to find simple nearest neighbor classifiers whose predictions can be accurately combined with those of a nearest neighbor classifier that incorporates all training instances as prototypes. This simple classifier may be thought of as *complementary* to the strong-learner nearest neighbor classifier. We call this strong learner the *base* classifier (Figure 16.2). The complementary classifier we construct will be a minimal nearest neighbor classifier, which stores only one prototype per class. The combination method is a classical decision tree algorithm, ID3 (Quinlan, 1986); since the level-1 classification task is relatively simple, more sophisticated decision trees were not required (Quinlan, 1993).

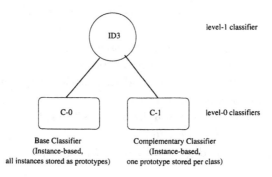

Figure 16.2 Paired boosting architecture.

In this section we show that an algorithm called Coarse Reclassification boosts nearest neighbor generalization accuracy on a variety of data sets by trying to maximize the training accuracy of the complementary component classifier. Coarse Reclassification uses the same idea as Breiman's bagging (Breiman, 1994): the utility of resampling a training set to create classifiers that are to be combined. However, bagging nearest neighbor classifiers reportedly did not work in Breiman's experiments because bagging's bootstrap sampling of the training set yielded stable nearest neighbor classifier components (Breiman, 1994). Breiman does not formally define stability, but uses the term to refer to the condition that leaving a small number training instances out of the training set (or putting a few more in) tends not to change the classifier concept hypothesis very much. For nearest neighbor classifiers, which store the training instances as prototypes, stability entails little change in the concept hypothesis when a few prototypes are left out.

To overcome the problem of nearest neighbor classifier stability, the Coarse Reclassification algorithm resamples the training set, but draws only a small number of prototypes in each sample. Storing few prototypes yields an unstable classifier, in the sense that substituting one prototype for another can make a large difference in the concept hypothesis of the classifier. The intuition behind Coarse Reclassification is that even though, in general, classifier instability is probably better avoided,

coarse-reclassification$(C_0, T, L_1, \varphi, k, p, m)$
  Key:
  C_0: level-0 base classifier
  C_1: level-0 complementary classifier
  L_1: level-1 classifier
  $\mathcal{G}$: stacked generalizer
  T: training set
  φ: fitness function, $\varphi(C_1, T)$ is the accuracy of C_1
    when applied to T
  p: number of prototypes from each class
  m: number of samples of prototype sets
  k: cardinality of neighborhood (k-nearest neighbor)
  ▷ *randomly sample prototypes*
  $C_1 = $ **sample-nearest-neighbor-classifiers**(T, φ, k, p, m)
  ▷ *train the level-1 classifier*
  $\mathcal{G} = $ **train-stacked-generalizer**(T, L_1, C_0, C_1)

Figure 16.3 Pseudocode for the Coarse Reclassification algorithm.

ironically, the deliberate introduction of instability may yield classifiers that may be combined more effectively.

To describe the algorithm, we examine in more detail the functions invoked in the pseudocode in Figure 16.3. First, we define notation that will be used in the rest of the article. A set of training instances will be denoted by $T = \{x_j | j = 1, \ldots, r\}$, where each instance x_j has been assigned a class target from a small set of class targets $S = \{S_i | i = 1, \ldots, s\}$. Let $C_{\mathcal{G}} = \{C_i | i = 0, \ldots, n\}$ be the set of n level-0 classifiers C_i for a stacked generalizer $\mathcal{G}$. In this chapter, we utilize only two level-0 components ($n = 1$), one of which is the given base classifier C_0, which we assume has been fully trained. Denote by $C_i(x_j)$, for $x_j \in T$ and $C_i(x_j) \in S$, the class predicted by the classifier C_i when applied to x_j.

`sample-nearest-neighbor-classifiers`. To find a complementary nearest neighbor classifier C_1, we randomly sample m sets of s instances (with replacement) from the training set T, where s is the number of classes exposed in T. This algorithm was described in Section 16.3.1. One instance is drawn from each class. A nearest neighbor classifier is constructed using each sample of s instances as its prototype set. Each of these m classifiers is then used to classify the instances in T. The classifier that displays highest fitness function value, classification accuracy on T, is taken as the complementary classifier C_1. As a default, we take $m = 100$ samples. Since C_1 stores only as many prototypes as

Table 16.2 Coarse Reclassification results. The second column gives the baseline nearest neighbor classifier accuracy (percent correct, with standard deviation). The third column gives the Coarse Reclassification accuracy. The fourth column gives the significance level at which the Coarse Reclassification algorithm accuracy is higher than the baseline nearest neighbor accuracy. The fifth column gives the average accuracy of the complementary component classifier added by the Coarse Reclassification algorithm. The sixth column gives the significance level at which the Coarse Reclassification accuracy is higher than that of the complementary component classifier.

Data	NN	CR	Sig. NN	Cmpnt	Sig. Cpmnt
Breast Cancer Ljubljana	67.5 ± 9.6	73.1 ± 6.9	0.020	71.4	0.002
Breast Cancer Wisconsin	96.2 ± 1.8	97.4 ± 1.9	0.075	96.5	0.005
Cleveland Heart Disease	77.0 ± 5.5	79.5 ± 7.1	0.147	79.9	0.848
Diabetes	68.7 ± 5.1	73.1 ± 5.1	0.019	72.9	0.254
Glass Recognition	73.3 ± 10.3	70.3 ± 9.9	0.992	53.3	0.000
Hepatitis	77.3 ± 11.4	80.2 ± 6.3	0.184	83.3	0.982
Iris Plants	91.3 ± 7.1	95.1 ± 3.8	0.178	95.8	0.938
LED-24 Digit	36.5 ± 9.4	27.2 ± 8.0	0.999	30.0	0.814
LED-7 Digit	71.2 ± 7.7	73.6 ± 7.8	0.008	73.3	0.381
Lymphography	80.7 ± 8.9	77.9 ± 10.7	0.860	74.0	0.062
Monks-2	48.8 ± 4.7	67.5 ± 5.8	0.000	48.9	0.000
Promoter	79.0 ± 9.9	79.6 ± 7.0	0.367	65.0	0.000
Soybean	91.0 ± 3.9	87.1 ± 4.4	0.999	64.4	0.000

there are classes, for most data sets there are a very small number of decision regions in the hypotheses created by the sampled classifiers.

train-stacked-generalizer. A stacked generalizer $\mathcal{G}$ is trained on a level-1 training set that is created by applying the level-0 classifiers C_0 and C_1 to the (level-0) training set. For each original training instance $x \in T$ with class S_i, a level-1 training instance is created: $(C_0(x), C_1(x), S_i)$. The set of these three-tuples of class labels is the training set used to train the level-1 learning algorithm.

Experiment. To evaluate this algorithm, we start with a full nearest neighbor classifier whose accuracy we want to boost by incorporating it in a stacked generalizer. We also fix the level-1 learning algorithm as ID3, with the gain ratio feature selection metric. We took $m = 100$ samples of prototypes, with one prototype per class, so $p = 1$. A 1-nearest neighbor algorithm was used, so that $k = 1$. The algorithm was run 50 times, consisting of five 10-fold cross-validation runs. The mean results are given in Table 16.2.

Analysis. Coarse Reclassification boosts nearest neighbor accuracy significantly at a significance level of $p \leq 0.05$ or better on four data sets

(Breast Cancer Ljubljana, Diabetes, LED-7 Digit and Monks-2). On two of these data sets (Breast Cancer Ljubljana and Monks-2), the composite classifier is significantly more accurate than both the full nearest neighbor component classifier and the complementary component classifier. The algorithm performed poorly on LED-24 Digit, Glass Recognition and Soybean data. On the remaining five data sets the accuracy of the two algorithms was comparable.

Coarse Reclassification demonstrates that a minimal classifier can be used to boost or maintain the accuracy of a full nearest neighbor classifier on a variety of these data sets. However, the poor performance of Coarse Reclassification on four data sets opens additional areas of investigation.

16.5.   A TAXONOMY OF INSTANCE TYPES

The Coarse Reclassification algorithm is simple; it merely tries to find a prototype set that maximizes training accuracy. One reason that the sampling algorithm may not work on some data sets is that training set accuracy may not always be a good indication of test set accuracy. In turn, one explanation of a training-testing mismatch is that there are instances in the training set that are not good indicators of test accuracy. In this situation, maximizing accuracy on the training set is not advisable. For example, if an instance has an erroneous class label, then probably it would not increase accuracy to classify correctly that mis-labeled instance, e.g., (Brodley and Friedl, 1996). We show that all of the data sets we have used contain instances that have the property that minimal nearest neighbor classifiers that classify those instances correctly will display lower, not higher, mean test accuracy. We suggest a taxonomy of instance types that arises pursuant to this notion. This taxonomy generalizes the idea of an *outlier*, an instance that has a strong influence on the calculation of some statistic, in particular a data instance that has been mis-entered or is not a member of the intended sample population.

To create the taxonomy, we once again sample classifiers. The taxonomy has a simple idea at its root. Understanding how each instance is differentially classified by a large set of classifiers may provide a clue to the structure of the space of instances. An instance may be a reliable predictor of a classifier's test accuracy if the classifiers that classify it correctly have *higher* mean classification accuracy on the training set than those that classify it incorrectly. An instance may be a suspect predictor of a classifier's test accuracy if the classifiers that classify it correctly have *lower* mean classification accuracy on the training set than those that classify it incorrectly. So for each instance, we create and compare two distributions of classifier accuracies: (i) the training set accuracies of classifiers that correctly classify that instance and (ii) the training set accuracies of classifiers that incorrectly classify that instance.

Table 16.3 suggests an instance taxonomy that is based on two criteria: the number of times that an instance is classified correctly by a sample of classifiers, and whether the classifiers that classify the instance correctly have higher training accuracy on average (%CorR, for "Percent correct if right") than those that misclassify the instance (%CorW, for "Percent Correct if wrong"). First, we characterize qualitatively the number of times an instance is classified correctly as *High, Medium* or *Low*. In the experiment below, we give a precise meaning to this informal characterization.

Table 16.3   Taxonomy of instances.

Type	Frequency Correct	%CorR ? %CorW
Core	High	>
Penumbral	Medium	>
Hard	Low	>
Misleading	High	<
Atypical	Medium	<
Outlying	Low	<

If Outlying instances are present, for example, it may not be advantageous to classify them correctly. Even if they are classified correctly during training, the average classification accuracy is likely to suffer. This effect may be a reflection of overfitting the training data, despite the application of classifiers that create coarse hypotheses only. (On the other hand, we may want to try affirmatively to get Outlying instances *wrong*. A similar notion is applied in the Deliberate Misclassification algorithm that we proposed in (Skalak, 1997) for component classifier construction.) Next, we give an example to demonstrate the presence of each type of instance in one of our data sets.

Example. To create the data in Table 16.4, 100 samples of nearest neighbor classifiers were selected by sampling sets of prototypes, one from each class, from the Soybean data set. For each instance, we computed the following statistics, which were derived over the entire sample of classifiers:

- the number of the sampled classifiers that correctly classify this instance (the *frequency correct*, in the "Freq.Cor." column)

- the mean accuracy over a training set of the classifiers that correctly classified this instance ("%CorR")

- the mean accuracy over a training set of the classifiers that incorrectly classified this instance ("%CorW")

- the statistical significance according to a t-test for a difference in the means of the previous two items ("Sig.")

We then culled an example from each class by hand to illustrate each taxonomic category.

Table 16.4　Examples of instances in the taxonomy from the Soybean data set.

Category	Case	Freq. Cor.	%CorR	%CorW	Sig.	Class
Core	260	89	53.95	46.90	0.000	frog-eye-leaf-spot
Penumbral	240	54	55.05	50.97	0.000	altern.-leaf-spot
Hard	163	15	55.80	52.71	0.032	anthracnose
Misleading	2	90	52.84	56.20	0.052	diap.-stem-canker
Atypical	50	62	52.25	54.68	0.023	phytophthora-rot
Outlying	283	14	48.84	53.88	0.000	frog-eye-leaf-spot

A next logical question is how many instances of each type appear in the Soybean and other data sets; answers are given in Table 16.5. In that table, "Undiff." means "undifferentiated," which is the leftover category for instances that do not fall into any of the other categories. An instance is placed into the Core, Penumbral or Hard categories if the mean training accuracy of classifiers that classify it correctly is significantly better (at the 95% confidence level) than the accuracy of those classifiers that classify it incorrectly. If this condition is satisfied, it was placed into one of those three categories according to the frequency correct, the number of times instance was correct classified by the classifiers in the sample. If the frequency correct for an instance was within (plus or minus) one standard deviation of the mean frequency, it was put in the Penumbral category. If the frequency correct was more than one standard deviation above the mean, it was placed into the Core category; less than one standard deviation, the Hard category. Analogous definitions were applied to the instances for which the classifiers that misclassified the instance had higher mean classification accuracy than those the correctly classified it, which were placed in the Misleading, Atypical and Outlying categories.

Misleading instances — instances that are often correctly classified but that lead to lower mean test accuracy — are rare. If a data set is

Table 16.5 Average number of instances of each type in the taxonomy.

Data	Undiff.	Core	Pen.	Hard	Mis.	Atyp.	Outly.
Breast Cancer Ljubljana	35.0	33.4	140.4	1.4	0.0	10.4	37.4
Breast Cancer Wisconsin	403.8	0.0	151.5	40.0	0.0	21.6	13.1
Cleveland Heart Disease	49.6	23.9	167.9	1.7	0.0	1.8	28.1
Diabetes	70.6	21.7	77.7	3.0	0.3	13.1	6.6
Glass Recognition	25.4	7.0	91.6	1.6	0.0	1.1	13.3
Hepatitis	52.0	0.0	63.0	3.9	5.1	6.2	4.8
Iris Plants	127.5	12.3	32.2	3.5	0.3	2.4	1.8
LED-24 Digit	144.0	78.0	198.8	3.4	0.0	3.5	22.3
LED-7 Digit	46.7	25.2	51.2	3.2	0.0	2.1	5.6
Lymphography	225.1	17.8	84.5	4.4	3.5	27.6	26.1
Monks-2	151.4	123.6	282.0	6.0	0.0	39.2	89.8
Promoter	56.1	8.1	29.2	2.1	0.0	0.1	0.4
Soybean	153.4	15.4	79.8	8.5	1.7	5.1	13.1

fairly easy, with mean accuracy for a minimal nearest neighbor classifier above 90% (e.g., Breast Cancer Wisconsin and Iris), there may be no Core instances. If nearly all the classifiers in a sample correctly classify an instance (say, 98 or 99, out of 100), there may be no classifier whose frequency correct is more than one standard deviation above the mean frequency correct.

This taxonomy currently is not used in our classifier construction algorithms. However, there appear to be opportunities to limit the set of training instances based on the taxonomic categories that each falls into, and we reserve this for future work.

16.6. CONCLUSIONS

The research presented in this article extends previous work on prototype selection for an independent nearest neighbor classifier. We show that in a variety of domains only a very small number of prototypes can provide classification accuracy equal to that of a nearest neighbor classifier that stores all training instances as prototypes. We extend previous work by demonstrating that algorithms that rely primarily on random sampling can effectively choose a small number of prototypes.

Although previous research results have questioned the utility of combining nearest neighbor classifiers, we introduce an algorithm that in most tested domains creates a paired boosting architecture that is no less accurate than a nearest neighbor classifier that uses all training instances as prototypes.

We also observe that a taxonomy of instances arises when we consider how each instance is differentially classified by a randomly sampled collection of minimal nearest neighbor classifiers. We demonstrate empirically that each of the data sets in this collection contains instances whose correct classification leads to lower accuracy on average.

Acknowledgments

Thanks to Claire Cardie and Edwina Rissland for insightful discussions.

References

Aha, D. W. (1990). *A Study of Instance-Based Algorithms for Supervised Learning Tasks: Mathematical, Empirical, and Psychological Evaluations.* PhD thesis, Dept. of Information and Computer Science, University of California, Irvine.

Ali, K. and Pazzani, M. (1996). Error reduction through learning multiple descriptions. *Machine Learning*, 24:173.

Alpaydin, E. (1997). Voting over Multiple Condensed Nearest Neighbors. *Artificial Intelligence Review*, 11:115–132.

Breiman, L. (1992). Stacked Regressions. Technical Report 367, Department of Statistics, University of California, Berkeley, CA.

Breiman, L. (1994). Bagging predictors. Technical Report 421, Department of Statistics, University of California, Berkeley, CA.

Brodley, C. (1994). *Recursive Automatic Algorithm Selection for Inductive Learning.* PhD thesis, Dept. of Computer Science, University of Massachusetts, Amherst, MA. (Available as Dept. of Computer Science Technical Report 96-61).

Brodley, C. and Friedl, M. (1996). Identifying and eliminating mislabeled training instances. In *Proceedings of the Thirteenth National Conference on Artificial Intelligence and the Eighth Innovative Applications of Artificial Intelligence Conference*, pages 799–805. AAAI Press/MIT Press, Menlo Park, CA.

Cameron-Jones, M. (1995). Instance Selection by Encoding Length Heuristic with Random Mutation Hill Climbing. In *Proceedings of the Eighth Australian Joint Conference on Artificial Intelligence*, pages 99–106. World Scientific.

Chang, C. L. (1974). Finding Prototypes for Nearest Neighbor Classifiers. *IEEE Transactions on Computers*, c-23:1179–1184.

Cherkauer, K. and Shavlik, J. (1996). Growing simpler decision trees to facilitate knowledge discovery. In *Proceedings of the Second International Conference on Knowledge Discovery and Data Mining*, pages 315–318. AAAI Press, San Mateo, CA.

Cleveland, W. (1993). *Visualizing Data.* Hobart Press, Summit, NJ.

Dasarathy, B. V. (1991). *Nearest Neighbor (NN) Norms: NN Pattern Classification Techniques.* IEEE Computer Society Press, Los Alamitos, CA.

de la Maza, M. (1991). A Prototype Based Symbolic Concept Learning System. In *Proceedings of the Eighth International Workshop on Machine Learning*, pages 41–45, San Mateo, CA. Morgan Kaufmann.

Drucker, H., Cortes, C., Jackel, L., LeCun, Y., and Vapnik, V. (1994). Boosting and other machine learning methods. In *Proceedings of the Eleventh International Conference on Machine Learning*, pages 53–61. Morgan Kaufmann, San Francisco, CA.

Freund, Y. and Schapire, R. (1995). A Decision-Theoretic Generalization of On-Line Learning and an Application to Boosting. In *Proceedings of the Second European Conference on Computational Learning Theory*, pages 23–37. Springer Verlag, Barcelona, Spain.

Freund, Y. and Schapire, R. (1996). Experiments with a new boosting algorithm. In *Proceedings of the Thirteenth International Conference on Machine Learning*, pages 148–156. Morgan Kaufmann, San Francisco, CA.

Gates, G. W. (1972). The Reduced Nearest Neighbor Rule. *IEEE Transactions on Information Theory*, IT-18, No. 3:431–433.

Hansen, M., Hurwitz, W., and Madow, W. (1953). *Sample Survey Methods and Theory*, volume I. John Wiley and Sons, New York, NY.

Harries, M. (1999). Boosting a strong learner: Evidence against the minimum margin. In *Proceedings of the Sixteenth International Conference on Machine Learning*, pages 171–180. Morgan Kaufmann, San Francisco, CA.

Hart, P. E. (1968). The Condensed Nearest Neighbor Rule. *IEEE Transactions on Information Theory (Corresp.)*, IT-14:515–516.

Ho, T. (1998). Nearest Neighbors in Random Subspaces. In *Proceedings of the Second International Workshop on Statistical Techniques in Pattern Recognition*, pages 640–648, Sydney, Australia. Springer.

Holte, R. C. (1993). Very Simple Classification Rules Perform Well on Most Commonly Used Datasets. *Machine Learning*, 11:63–90.

Quinlan, J. (1999). Some elements of machine learning (extended abstract). In *Proceedings of the Sixteenth International Conference on Machine Learning*, pages 523–525. Morgan Kaufmann, San Francisco, CA.

Quinlan, J. R. (1986). Induction of Decision Trees. *Machine Learning*, 1:81–106.

Quinlan, J. R. (1993). *C4.5: Programs for Machine Learning.* Morgan Kaufmann, San Mateo, CA.

Ripley, B. (1996). *Pattern Recognition and Neural Networks.* Cambridge, Cambridge, England.

Skalak, D. (1997). *Prototype Selection for Composite Nearest Neighbor Classifiers.* PhD thesis, Dept. of Computer Science, University of Massachusetts, Amherst, MA.

Skalak, D. B. (1994). Prototype and Feature Selection by Sampling and Random Mutation Hill Climbing Algorithms. In *Proceedings of the Eleventh International Conference on Machine Learning*, pages 293–301, New Brunswick, NJ. Morgan Kaufmann.

Turney, P. (1995). Technical Note: Bias and the Quantification of Stability. *Machine Learning*, 20:23–33.

Voisin, J. and Devijver, P. A. (1987). An application of the Multiedit-Condensing technique to the reference selection problem in a print recognition system. *Pattern Recognition*, 5:465–474.

Wilson, D. (1972). Asymptotic Properties of Nearest Neighbor Rules using Edited Data. *Institute of Electrical and Electronic Engineers Transactions on Systems, Man and Cybernetics*, 2:408–421.

Wolpert, D. (1992). Stacked Generalization. *Neural Networks*, 5:241–259.

Zhang, J. (1992). Selecting Typical Instances in Instance-Based Learning. In *Proceedings of the Ninth International Machine Learning Workshop*, pages 470–479, Aberdeen, Scotland. Morgan Kaufmann, San Mateo, CA.

Chapter 17

PROTOTYPE SELECTION USING BOOSTED NEAREST-NEIGHBORS

Richard Nock

Université Antilles-Guyane

Département Scientifique Interfacultés, Campus Universitaire de Schoelcher,

97233 Schoelcher, Martinique, France

rnock@univ-ag.fr

Marc Sebban

Université Antilles-Guyane

Département de Sciences Juridiques, Campus Universitaire de Fouillole,

97159 Pointe-à-Pitre, Guadeloupe, France

msebban@univ-ag.fr

Abstract   We present a new approach to Prototype Selection (PS), that is, the search for relevant subsets of instances. It is inspired by a recent classification technique known as Boosting, whose ideas were previously unused in that field. Three interesting properties emerge from our adaptation. First, the accuracy, which was the standard in PS since Hart and Gates, is no longer the reliability criterion. Second, PS interacts with a *prototype weighting scheme*, *i.e.*, each prototype receives periodically a real confidence, its significance, with respect to the currently selected set. Finally, Boosting as used in PS allows to obtain an algorithm whose time complexity compares favorably with classical PS algorithms.

Three types of experiments lead to the following conclusions. First, the optimally reachable prototype subset is almost always more accurate than the whole set. Second, more practically, the output of the algorithm on two types of experiments with fourteen and twenty benchmarks compares favorably to those of five recent or state-of-the-art PS algorithms. Third, visual investigations on a particular simulated dataset gives evidence in that case of the relevance of the selected prototypes.

Keywords: Boosting, prototype weighting, nearest neighbor.

17.1.   INTRODUCTION

With the development and the popularization of new data acquisition technologies such as the World Wide Web (WWW), computer scientists have to analyze potentially huge data sets (DS). Available technology to analyze data has been developed over the last decades, and covers a broad spectrum of techniques and algorithms. However, data collected are subject to make interpretation tasks hazardous, not only because of their eventually large quantities, but also when these are raw collections, of low quality. The development of the WWW participates to the increase of both tendencies. The reduction of their effects by a suitable preprocessing of data becomes then an important issue in the fields of Data Mining and Machine Learning. Two main types of algorithms can be used to facilitate knowledge processing. The first ones reduce the number of description variables of a DS, by selecting relevant attributes, and are commonly presented as *feature selection* algorithms (John et al., 1994). The second ones reduce the number of individuals of a DS, by selecting relevant instances, and are commonly presented as *prototype selection* (PS) algorithms (Gates, 1972; Hart, 1968; Sebban and Nock, 2000; Zhang, 1992). The principal effect of both types of algorithms is to improve indirectly the reliability and accuracy of post-processing stages, particularly for machine learning algorithms, traditionally known to be sensitive to noise (Blum and Langley, 1997). They have also an important side effect. By reducing the "useful" DS size, these strategies reduce both space and time complexities of subsequent processing phases. One can also hope to reduce the size of formulas obtained by a subsequent induction algorithm on the reduced and less noisy DS. This may facilitate interpretation tasks. PS raises the problem of defining relevance for a prototype subset. From the statistical viewpoint, relevance can be partly understood as the contribution to the overall accuracy, that would be *e.g.* obtained by a subsequent induction. We emphasize that removing prototypes does not necessarily lead to a degradation of the results: we have observed experimentally that a little number of prototypes can have performances comparable to those of the whole sample, and sometimes higher. Two reasons come to mind to explain such an observation. First, some noises or repetitions in data could be deleted by removing instances. Second, each prototype can be viewed as a supplementary degree of freedom. If we reduce the number of prototypes, we can sometimes avoid over-fitting situations. With this definition, there are two types of irrelevant instances which we should remove. The first ones are instances belonging to regions with very few elements: their vote is statistically a poor estimator, and a little noise might affect dramatically their vote. It is also common in statistical analyzes to search and remove such points, in regression, parametric estimations, etc. Removing them does not necessarily brings a great reduction in the size of the retained prototypes, but it may be helpful for future predictions tasks. The second ones are instances belonging to regions where votes can be assimilated as being randomized. Local densities are approximately evenly distributed with respect to the overall class distributions. These instances are not necessarily harmful for prediction, but a great reduction in size can be obtained after removal if they are numerous.

Historically, PS has been firstly aimed at improving the efficiency of the Nearest Neighbor (NN) classifier (Hart, 1968). Conceptually, the NN classifier (Cover and Hart, 1967) is probably the simplest classification rule. Its use was also spread and encouraged by early theoretical results linking its generalization error to Bayes. However, from a practical point of view, this algorithm is not suited to very large DS because of the storage requirements it imposes. Actually, this approach involves storing all the instances in memory. Pioneer work in PS firstly searched to reduce this storing size. Hart (Hart, 1968) proposes a *Condensed NN Rule* to find a *Consistent Subset, CS*, which correctly classifies all of the remaining points in the sample set. However, this algorithm will not find a *Minimal Consistent Subset, MCS*. The *Reduced NN Rule* proposed by Gates (Gates, 1972) tries to overcome this drawback, searching in Hart's *CS* the minimal subset which correctly classifies all the learning instances. However, this approach is efficient if and only if Hart's *CS* contains the *MCS* of the learning set, which is not always the case. It is worthwhile remarking that in these two approaches, the PS algorithm deduces only one prototype subset. It is impossible to save more or less instances, and to control the size of the subset. Moreover, we have no idea about the relevancy of each instance selected in the prototype subset.

More recently, Skalak (Skalak, 1994) proposes two algorithms to find sets of prototypes for NN classification. The first one is a Monte Carlo sampling algorithm, and the second applies random mutation hill climbing. In these two algorithms, the size of the prototype subset is fixed in advance. Skalak proposes to fix this parameter to the number of classes. Even if this strategy obtains good results for simple problems, the prototype subset is too simple for complex problems with overlaps, *i.e.* when various matching observations have different classes. Moreover, these algorithms require to fix another parameter which contributes to increasing the time complexity: the number of samples (in the Monte Carlo method) or the number of mutation in the other approach. This parameter also depends on the complexity of the domain, and the size of the data.

Brodley (Brodley, 1993) with her *MCS system*, and Brodley and Friedl (Brodley and Friedl, 1996) with Consensus Filters, deal with the PS problem. While the approach presented in (Brodley and Friedl, 1996) is rather an algorithm for eliminating mislabeled instances, than a real PS algorithm, we have observed in (Sebban and Nock, 2000) that it could constitute a good pre-process of the PS problem. This approach constructs a set of base-level detectors and uses them to identify mislabeled instances by consensus vote, *i.e.* all classifiers must agree to eliminate an instance. Finally, in the algorithm RT3 of Wilson and Martinez (Wilson and Martinez, 1997), an instance e is removed if its removal does not hurt the classification of the instances remaining in the sample set, notably instances that have e in their neighborhood. More precisely the algorithm is based on sequentially ordered procedures, each tailor-made to remove some particular kind of irrelevant instances. Two of them are the following ones: (1) it uses a noise-filtering pass, that removes instances misclassified by their k nearest neighbors, that helps to avoid "over-fitting" the data; (2) it removes instances in the center of clusters before border point, by sorting instances according to the distance to their nearest *enemy* (*i.e.* of a different class).

In this work, we propose a new PS algorithm which attempts to correct some of the defaults of the previous methods. Its main idea is to use recent results about Freund and Schapire's AdaBoost Boosting algorithm (Freund and Schapire, 1997; Schapire and Singer, 1998). This is a classification technique in which an induction algorithm is repetitively trained, over a set of examples whose distribution is periodically modified. The current distribution favors examples that were badly classified by the previous outputs of the induction algorithm, called *weak hypotheses*. This ensures that the induction algorithm is always trained on an especially hard set of instances (Schapire and Singer, 1998). The final output consists of a weighted majority vote of all outputs, where the weight of each weak hypothesis is a real confidence in its predictive abilities.

Our adaptation of ADABOOST to PS has the following original features:

- Each step of the algorithm consists in choosing a *prototype* instead of calling for a weak hypothesis. This removes the time spent for repetitive induction. In the PS framework, we avoid the principal criticism often made to Boosting (Quinlan, 1996): the prohibitive time complexity.

- Each selected prototype receives a weight, equivalent to that of Boosting for weak hypotheses. This weight can be explained in terms of relevance.

- The best prototype, having the highest coefficient, is kept at each stage of the algorithm. Equivalently, we minimize a criterion derived from Boosting which is not the accuracy, this latter criterion being the conventional criterion optimized even in PS.

- When a prototype is chosen, the distribution of all remaining prototypes is modified. This favors those that are not well explained by the currently selected set.

- The algorithm stops at the user's request. Therefore, one can fix the desired size of the subset. This is the only user-dependent parameter of the algorithm.

When comparing this approach to the previously cited ones, some differences appear, one of which is extremely important to us. The central mechanism for PS is a *dynamic weighting* scheme. Each selected prototype is given a real number which can be reliably interpreted in terms of relevance. Furthermore, whenever a prototype is selected, the distribution of the remaining ones is modified. This will influence and guide the choice of *all* future prototypes, toward those being reliable *and* completing accurately the previously selected prototypes.

The remaining of this paper is organized as follows. First, we introduce the notion of weak hypothesis, and its link with prototype weighting and selection. This is the central mechanism for adapting Boosting to PS. Then, we present the whole PS algorithm. Finally, we present thorough results on three types of experiments: the evolution of the selected subset's accuracy computed using NN, some comparisons with previous methods on readily available benchmarks, and finally some visual results on a particular problem, showing in that case which kind of prototypes are selected.

17.2. FROM INSTANCES TO PROTOTYPES AND WEAK HYPOTHESES

Consider a simple dataset, in which instances are represented by points on the plane. Each *instance* (or *example*) is a line of data given by a triple (x_1, x_2, y), where (x_1, x_2) is an *observation* , and y is its *class*. The objective of PS is to select a representative subset of the instances. In this subset, the instances selected become *prototypes*.

Let LS stands for the set of available instances, to which we usually

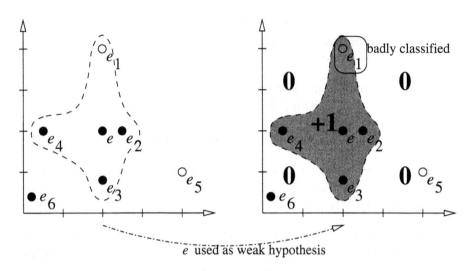

Figure 17.1 A simple case with $k = 1$: the points for which e is the nearest neighbor (left) and e used as a weak hypothesis (right).

refer as the *learning sample* in classical machine learning studies. For instance, as shown in Figure 17.1, fix $LS = \{e, e_1, e_2, e_3, e_4, e_5, e_6\}$. e_1 and e_5 belong to the *negative class*, abbreviated $-$, and the others belong to the *positive class*, $+$. This is a simple two-classes problem. In Figure 17.1 (left drawing), instances from class $-$ are symbolized by an empty circle, and the others by a filled circle. The dashed line symbolizes the set of points for which e is the nearest neighbor.

Given one new instance to be classified, the classical k-NN algorithm proceeds by taking the majority class among its k nearest neighbors, as shown by Figure 17.1 (left, $k = 1$). In that case, if e is an observation to be classified, and distances are computed using the Euclidean metric, then e is given e_2's class $+$, the right one.

k-NN is a simple class of local voting procedures where votes are basically unweighted. In a more general setting, if we replace it by an ordinary voting scheme where each voting instance becomes one complex formula, and weighted votes are allowed, then Boosting (Freund and Schapire, 1997; Schapire and Singer, 1998) gives an efficient way to make the whole construction.

Boosting is concerned with the stepwise construction of voting methods, by repeatedly asking for voting, *weak hypotheses* to a weak learner W, which, put altogether, form an accurate *strong hypothesis*. The original algorithm of Boosting, called ADABOOST (Freund and Schapire, 1997; Schapire and Singer, 1998), gives to each currently selected weak hypothesis h_t a real weight α_t, adjusting its vote into the whole final set of hypotheses. While reliable hypotheses receive a weight whose magnitude is large, unreliable hypotheses receive a weight whose magnitude tends to zero. ADABOOST is presented below in its most general form, described in the two classes framework: the sign of the output formula H gives the class of an observation. When there are more than two classes, say c, the algorithm builds c voting procedures, for discriminating each class against all others.

ADABOOST (LS, W, T):
  Initialize distribution $D_1(e) = 1/|LS|$ for any $e \in LS$;
  for $t = 1, 2, ..., T$
    Train weak learner W on LS and D_t, get weak hypothesis h_t;
    Compute the confidence α_t;
    Update for all $e \in LS$: $D_{t+1}(e) = \dfrac{D_t(e)e^{-\alpha_t y(e)h_t(e)}}{Z_t}$;
    /*Z_t is a normalization coefficient*/
  endfor
  return the classifier

$$H(e) = sign(\sum_{t=1}^{T} \alpha_t h_t(e))$$

The key step of ADABOOST is certainly the *distribution update*. In the initial set of instances, each element can be viewed as having an appearance probability equal to $1/|LS|$ multiplied by its number of occurrences. At run time, ADABOOST modifies this distribution so as to re-weight higher all instances previously badly classified. Suppose that the current weak hypothesis h_t receives a large positive α_t . In ADABOOST's most general setting, each weak hypothesis h_t is allowed to vote into the set $[-1; +1]$, but this is not a restriction, since the role of the α_t is precisely to extend the vote to $I\!R$ itself. A negative observation e badly classified by h_t has, before renormalization, its weight multiplied by $e^{-\alpha_t y(e)h_t(e)}$. Since $h_t(e) > 0$, $\alpha_t > 0$ and $y(e) = -1 < 0$, the multiplicative factor is > 1, which tends indeed to re-weight higher the example. This would be the same case for badly classified, positive observations.

The adaptation of ADABOOST to NN and then to Prototype Selection (PS) is almost immediate. Suppose that we have access for each instance e to its *reciprocal neighborhood* $R(e) = \{e' \in LS : e \in N(e')\}$, where $N(.)$ returns the neighborhood. In Figure 17.1, $R(e)$ is symbolized by the dashed curve. Suppose we want to weight all instances in LS. If we consider e as a weak hypothesis, $R(e)$ gives all points in LS for which e

will give a vote. The output of this weak hypothesis as such, takes two possible values (see also Figure 17.1):

- $y(e)$ $(\in \{-1; 1\})$ for any instance in $R(e)$ (+1 in Figure 17.1),

- 0 for any instance not in $R(e)$,

For multiclass problems, with $c > 2$ classes, we make the union of c biclass problems, each of which discriminates one class, called +, against all others, falling into the same class, −. The overall selected subset of prototypes is the union of each biclass output. This allows to save the notation $+1/-1$ for $y(e)$. The presentation of the algorithm ADABOOST does not give the way to choose weak hypotheses. Suppose that we have access to more than one h_t. Which one do we choose ? In our specific framework, the whole set of weak hypotheses is the set of remaining instances, and this question is even more crucial. Freund, Schapire and Singer (Freund and Schapire, 1997; Schapire and Singer, 1998) have proposed a nice way to solve the question. Name W_e^+ (resp. W_e^-) as the fraction of instances in $R(e)$ having the same class as e (resp. a different class from e), and W_e^0 is the fraction of instances to which e gives a null vote (those not in $R(e)$). Formally,

$$W_e^+ = \sum_{e' \in R(e):y(e')=y(e)} D_t(e')$$

$$W_e^- = \sum_{e' \in R(e):y(e')\neq y(e)} D_t(e')$$

$$W_e^0 = \sum_{e' \in LS\backslash R(e):y(e')} D_t(e')$$

Then, following Freund, Schapire and Singer (Freund and Schapire, 1997; Schapire and Singer, 1998), the example e we choose at time t should be the one minimizing the following coefficient:

$$Z_e = 2\sqrt{\left(W_e^+ + \frac{1}{2}W_e^0\right) \times \left(W_e^- + \frac{1}{2}W_e^0\right)}$$

and the confidence α_e can be calculated as

$$\alpha_e = \frac{1}{2}\log\left(\frac{W_e^+ + \frac{1}{2}W_e^0}{W_e^- + \frac{1}{2}W_e^0}\right)$$

In these formulae, the subscript e replaces the t subscript of ADABOOST without loss of generality, since we choose at each time t an example

$e \in LS$. In the framework of PS, examples with negative α_e are likely to represent either noise or exceptions. Though exceptions can be interesting for some Data Mining purposes, they are rather risky when relevance supposes accuracy, and they also prevent reaching small subsets of prototypes. We have therefore chosen not to allow the choice of prototypes with negative values of α_e. The algorithm PSBoost, shown below, presents our adaptation of ADABOOST to PS. Remark that whenever the best remaining instance e has an $\alpha_e < 0$, the algorithm stops and return the current subset of prototypes. It must be noted that this situation was never encountered experimentally.

It is not the purpose of this paper to detail formal reasons for calculating α_e as such, as well as the choice of Z_e. Informally, however, we can present a few crucial points of the proofs in the general ADABOOST's settings. First, minimizing the accuracy of the voting procedure can actually be done rapidly by minimizing the normalization coefficient Z_t of ADABOOST. Coefficients α_t can be calculated so as to minimize Z_t, which gives their preceding formula. Then, putting these α_t in the normalization coefficient Z_t gives the formula above. For more information, we refer the reader to the work of Freund, Schapire and Singer (Freund and Schapire, 1997; Schapire and Singer, 1998). Our adaptation of ADABOOST to PS is mainly heuristic, since we do not aim at producing a classifier, but rather at selecting an *unweighted* set of prototypes.

Some useful observations can be done about the signification of Z_e in the light of what is relevance. First, as argued in the introduction, Z_e penalizes instances belonging to regions with very few elements. Indeed, if an instance e belongs to a region with very few prototypes, it is unlikely to vote for many other instances, and W_e^0 will be large, preventing to reach small Z_e. Second, Z_e also penalizes instances coming from regions with approximately evenly distributed classes. Indeed, if a prototype belongs to a region with evenly distributed instances, W_e^+ and W_e^- tend to be balanced, and this, again, prevents to reach small Z_e. To compare with approaches such as RT3 (Wilson and Martinez, 1997), note that the Z_e criterion allows to cope with various kind of irrelevance, which necessitated in (Wilson and Martinez, 1997) a specific tailor-made mechanism for each of them to be addressed. Finally, the distribution update allows to make relevance quite *adaptive*, which appears to be very important when selecting prototypes one by one. Indeed, the distribution is modified whenever a new prototype is selected, to favor future instances that are not well explained by the currently selected prototypes. Finally, note that, as Z_e does, α_e also takes into account relevance, in the same way. The higher α_e, the more relevant prototype e.

PSBOOST (LS, N_p):
  Initialize distribution $D_1(e) = 1/|LS|$ for any $e \in LS$;
  Initialize candidates set $LS_* = LS$;
  Initialize $LS' = \emptyset$;
  for $t = 1, 2, ..., N_p$
    pick $e = argmax_{e' \in LS_*} \alpha_{e'}$;
    if $\alpha_e < 0$ *then break*;
    $LS' = LS' \cup e$;
    $LS_* = LS_* \backslash \{e\}$;
    Update for any $e' \in R(e)$:

$$D_{t+1}(e') = \frac{D_t(e')e^{-\alpha_{e'}y(e')y(e)}}{Z_e};$$

    Update for any $e' \in LS_* \backslash R(e)$:

$$D_{t+1}(e') = \frac{D_t(e')}{Z_e};$$

/*Z_e is a normalization coefficient*/

  endfor
  return LS'

17.3.   EXPERIMENTAL RESULTS

In this section, we apply algorithm PSBOOST. We present some experimental results on several datasets, most of which come from the UCI database repository[1]. Dataset LED is the classical LED recognition problem (Breiman et al., 1984), but to which the original ten classes are reduced to two: even and odd. $LED24$ is LED to which seventeen irrelevant attributes are added. $H2$ is a hard problem consisting of two classes and ten features per instance. There are five features irrelevant in the strongest sense (John et al., 1994). The class is given by the XOR of the five relevant features. Finally, each feature has 10% noise. The $XD6$ problem was previously used by Buntine and Niblett (Buntine and Niblett, 1992): it is composed of ten attributes, one of which is irrelevant. The target concept is a disjunctive normal form formula over the nine other attributes. There is also classification noise. Other problems were used as they appeared in the UCI repository in the 1999 distribution.

Unless otherwise stated, each original set is divided into a learning sample LS (2/3 of the instances) and a "validation" set VS (the remaining third). This section has a triple objective. First, we assess the ability of our algorithm to correctly determine the relevant prototypes, without damaging accuracy computed on LS. Second, we compare performances of our method with Hart's, Gates's, Skalak's, Brodley and Friedl's, Wilson and Martinez's algorithms. Third, we analyze on a visual example what are the selected prototypes.

17.3.1.   Optimal Reachable Subsets

In order to bring to the fore the interest of PSBOOST, we have applied for each dataset the following experimental set-up.

Table 17.1 Synthetic results on compression rates and optimal classification accuracies (Datasets sorted by Comp.%)

Dataset	$|LS|$	$|LS^*|$	Comp. %	Acc_{LS}	Acc_{LS^*}
German	500	50	90.0	67.4	**70.1**
LED	300	75	75.0	83.0	**84.5**
Pima	468	125	73.3	69.7	**74.3**
Echocardio	70	20	71.4	59.0	**62.3**
Breast-W	400	175	56.3	97.7	**98.0**
Xd6	400	175	56.3	73.5	73.5
Vehicle	400	175	56.3	68.2	**69.7**
Horse Colic	200	125	37.5	70.8	**74.4**
H2	300	200	33.3	56.6	**57.1**
LED24	300	250	16.7	66.5	**70.5**
Hepatitis	100	90	10.0	70.9	70.9
White House	235	225	4.2	92.5	92.5

The accuracy Acc_{LS} is pre-computed on VS, using a 1-NN classifier consisting of the whole set LS. Acc_{LS} is represented for each benchmark by an horizontal dotted line (Figure 17.2). The first two columns of the Table 17.1 summarize the datasets, and presents the LS cardinalities. PSBOOST is run on LS for different values of N_p. For each deduced learning subset $LS' \subset LS$, the accuracy $Acc_{LS'}$ is computed on VS using a 1-NN classifier. The different values of $Acc_{LS'}$ are plotted for each benchmark on Figure 17.2, and then interpolated. The selected prototype sample LS^* is the smallest LS' set for which $Acc_{LS^*} = \max_{LS' \subseteq LS} Acc_{LS'}$. Finally, we compare LS^* with strict randomization: 10 subsets of LS having size $|LS^*|$ are independently sampled. Their average accuracy on testing is plotted (with a star) for each benchmark, on Figure 17.2. Though we use the word "validation" for VS, it must be pointed out that this validation does not strictly refer to usual Machine Learning studies. In particular, even if LS' is tested on VS, subset VS should not be taken for a *test* subset of LS. Indeed, LS^*

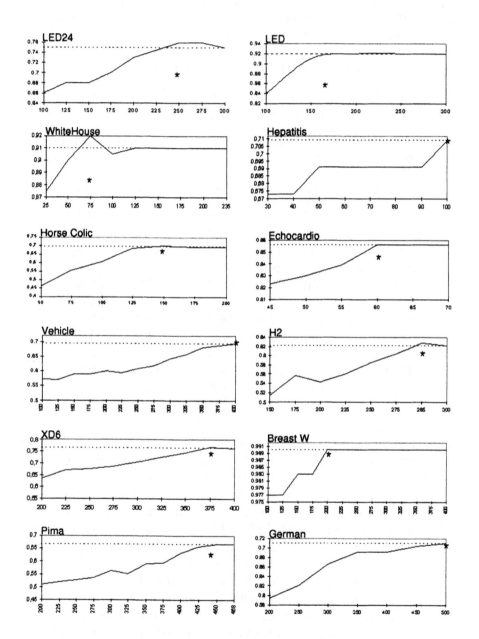

Figure 17.2 Accuracy curves on 12 datasets with different prototype subsets. The dotted line represents the accuracy with the whole original learning set.

is *optimized* by tests on *VS*. Our main objective in this section is to prove that PSBoost, when ran with an adequate value of N_p, can find reduced subsets of prototypes highly competitive with respect to *LS*, and randomly chosen subsets of the same size. In particular, it does not give experimental value to PSBoost with respect to other algorithms (this is the subject of the next subsection). Rather, it is intended to show the "compression" abilities of PSBoost. This can all the more be done reliably using comparisons with the star plot, as it was measured under the same experimental conditions, and the results obtained are statistically highly significant. In addition, this does not require, like traditional validation, any further splitting of a dataset to create the test set. Therefore, this represents an economic way to compare results without using too restricted subsets. As such, it may be viewed as another adequate "validation" measurement. That is the reason why we keep this usual terminology.

Table 17.1 presents $|LS^*|$, the Compression rate (Comp. %), the Acc_{LS^*} (%) and the Acc_{LS} (%) value for each dataset. The compression rate is the ratio $(|LS| - |LS^*|)/|LS|$. For 9 datasets among 12, LS^* allows to compute a better accuracy than *LS*, that confirms the positive effects of our strategy to delete irrelevant instances. Figure 17.2 presents in a more detailed way the twelve curves. We can draw the following general remarks. First, the datasets can be divided into two groups: (i) those which are evenly learned by a smaller number of instances (*White House, Xd6, Hepatitis, H2*), (ii) and those which are better learned by a smaller number of instances (*LED, LED24, Horse Colic, EchoCardio, Breast-W, Pima, Vehicle, German*). For this latter case, the PSBoost's strategy is always the same: the better instances are progressively selected, that leads to a regular improvement of the accuracy. Once the optimum is reached, it remains only irrelevant instances which decrease the global accuracy. Once again, it confirms the positive effect of our algorithm. Moreover, we observe from Figure 17.2 that the randomized accuracy symbolized by the star is always under that of LS^*. This is finally a clear advocacy that the results of PSBoost are not due to randomization itself: a sign test gives indeed a threshold probability $p < 0.05\%$.

17.3.2. Comparisons

In order to compare performances of PSBoost with the approaches of Brodley and Friedl, Hart, Gates, Skalak, Wilson and Martinez (and another one of ours), we ran two main experiments. In a first experiment, we decided to fix in advance $k = 1$. The experimental set-up applied is the following. Hart's algorithm CNN is run. We determine

the *consistent subset CS* and use this one as a learning sample to test
VS with a simple 1-NN classifier. The number of prototype $|CS|$ and
the accuracy Acc_{cnn} on VS are presented in Table 17.2. Then, Gates's
algorithm RNN is run. We determine the *minimal consistent subset*
MCS and use this one as a learning sample to test VS with a sim-
ple 1-NN classifier. The number of prototype $|MCS|$ and the accuracy
Acc_{rnn} on VS are presented in Table 17.2. Then, Skalak's algorithm by
a Monte Carlo (M.C.) sampling is applied, with a number of samples
$n_s = 50$ and 100. We choose the number of prototypes $N_p = |CS|$, for
comparisons. Then, PSBOOST is run, fixing in advance $N_p = |CS|$. We
compute the accuracy on VS with a simple 1-NN classifier. Note that
$|CS|$ is not *a priori* the optimal size of the prototype subset (cf Figure
17.2 for $|LS^*|$). We fix $N_p = |CS|$ to allow comparisons. Finally, we
compute the accuracy with $|CS|$ prototypes, randomly (*Rand.*) chosen
into the learning sample LS. In the majority of cases (9 among 14), the

Table 17.2 PSBOOST vs 3 prototype selection algorithms on 14 benchmarks (the
best values among all PS algorithms are bold-faced underlined; PSBOOST is noted
PSB).

Dataset	LS Acc	CNN |CS|	CNN Acc	RNN |MCS|	RNN Acc	PSB Acc	M.C. (n_s=) 50	M.C. (n_s=) 100	Rand. Acc
LED24	66.5	151	64.0	141	62.5	**70.5**	70,0	66,0	59.0
LED	83.0	114	69.5	114	69.5	**83.5**	89,5	82,5	80.0
Wh. House	92.5	35	86.0	25	86.5	88,5	88,0	**92,0**	88.0
Hepatitis	70.9	32	**74.6**	29	72.7	71.8	65,5	74,5	61.8
Horse Colic	70.8	110	67.3	92	67.3	**73.8**	68,5	69,6	60.1
Echocardio	59.0	33	**63.9**	33	**63.9**	59.0	59,0	63,0	60.1
Vehicle	68.2	178	61.9	178	61.9	**69.7**	64,8	69,0	61.1
H2	56.6	177	**59.9**	177	**59.9**	56.1	59,4	58,0	54.7
Xd6	73.5	157	72,5	157	72,5	71.0	72,5	**74,0**	69.0
Breast W	97.7	57	96.3	50	96.3	**97.7**	97,3	**97.7**	94.0
Pima	69.7	245	59.0	245	59.0	**73.7**	69,3	69,0	66.0
German	67.4	241	60.8	241	60.8	**68.8**	65,8	65,6	65.6
Ionosphere	91.4	52	81.9	40	80.2	**87.1**	81,3	84,3	80.0
Tic-tac-toe	78.2	225	76.0	218	76.3	**78.5**	77,7	74,6	74.4
Average	75.1		71.0		70.7	**75.0**	73,5	74,4	69.6

$|CS|$ prototypes selected by PSBOOST allow to obtain the best accuracies among all four algorithms. Moreover, if we compare an average
accuracy on all the datasets (that has a sense only for the present comparison), we can conclude that PSBOOST seems to be the best approach,
not only in accuracy terms, but also as a good compromise between performances and complexity of the algorithm. The Monte Carlo method
presents goods results only for simple problems, easy to learn or with

a small learning set, for which a little number of samples is necessary (*Vote, Hepatitis, or Breast Cancer*). On the other hand, hard problems with overlaps would require a higher number of samples, which would increase a lot the complexity of the algorithm. That explains the poor performances on this kind of dataset (*German, Horse colic, etc.*).

To address more in depth the performances of PSBOOST, we have chosen to compare it with a recent approach (of Sebban and Nock (Sebban and Nock, 2000)) and that of Wilson and Martinez (Wilson and Martinez, 1997). The reason to use the RT3 approach of Wilson and Martinez (Wilson and Martinez, 1997) is that this algorithm is well-known in PS for obtaining very small selected subsets. Both algorithms do not necessitate to fix the desired number of prototypes. The experimental setup was closer to conventional Machine Learning studies, since the results for each dataset were the average of the results over ten-fold cross-validations. We have chosen to compare PSBOOST with both algorithms by running it with the PS complexity of the other algorithms. This helps both to find accurate values for N_p, and to compare the algorithms with each other. Table 17.3 shows the results on 20 benchmarks; for each result, we give the average accuracy "Acc" (computed using a 3-NN rule), and the standard dev. σ over all folds. The last line of Table 17.3 shows the average percentage of LS which is retained after the selection. From the results exposed, it comes that PSBOOST outperforms the other algorithms with their complexities. Two paired t-tests for "PSrcg vs PSBoost$_1$" and "RT3 vs PSBoost$_2$" show in both cases the superiority of PSBOOST up to respective threshold risks ≈ 0.035 and ≈ 0.0003. The algorithm also outperforms the CONSENSUS FILTER (CF) approach ((Brodley and Friedl, 1996), not shown due to space constraints), not significantly on the accuracy (they realize 74.37%), but on the size of the selected subset (CF select 81.30% of the data on average). Another interesting part of the results is that PSBOOST also beats significantly the 3-NN rule while keeping less than 2/3 of the original set LS, when ran with the complexity of the PSRCG algorithm of Sebban and Nock (Sebban and Nock, 2000). Therefore, PSBOOST seems to be the best approach to cope with relevance, among all the algorithms which were tested.

17.3.3. Visualization

After these quantitative assessments of the PSBOOST performances, we go further in analyzing what kind of prototypes are selected by PSBOOST. In order to address this problem, we simulated a two-classes problem in the Euclidean plane (Figure 17.3, left). Each class contains

Table 17.3 Further comparisons between PSBOOST and two prototype selection algorithms on 20 benchmarks (the accuracy of the 3-NN is 73.86% on average); the best values over each of the two %|LS| are bold-faced.

Dataset	PSRCG		PSBOOST$_1$		RT3		PSBOOST$_2$	
	Acc	σ	Acc	σ	Acc	σ	Acc	σ
Audiology	**57.43**	12.87	53.52	10.18	59.96	13.59	**61.00**	10.79
Austral	78.41	5.06	**80.27**	5.16	70.67	3.68	**71.68**	4.60
Bigpole	**59.92**	8.16	58.91	7.70	**58.89**	7.25	57.29	8.34
Breast-W	**97.04**	1.33	96.19	1.54	**87.72**	18.42	81.24	24.31
Brighton	**94.60**	2.61	94.46	5.66	**90.56**	3.02	90.27	5.65
Bupa	66.76	7.81	**68.77**	6.54	**63.35**	11.05	62.45	8.12
Echocardio	**62.58**	9.99	61.87	11.95	58.30	13.00	**61.87**	9.04
German	70.87	3.98	**72.65**	3.29	69.87	3.67	**72.65**	4.15
Glass2	72.10	9.54	**74.49**	6.00	55.07	14.33	**60.29**	7.81
H2	44.91	6.28	**47.85**	9.00	48.27	6.47	**50.55**	11.22
Heart	74.56	8.02	**79.92**	5.83	**74.91**	5.77	72.72	6.42
Hepatitis	**81.20**	6.81	76.63	9.56	68.25	7.88	**76.56**	10.26
Horse Colic	69.48	6.33	**72.95**	7.31	**70.30**	8.84	67.89	4.39
Ionosphere	71.46	12.21	**76.73**	12.03	74.06	10.66	**76.20**	8.86
LED17(2C)	69.93	12.70	**76.12**	11.01	64.62	10.39	**70.43**	14.50
LED2	87.63	3.81	**89.20**	3.18	66.79	8.72	**74.68**	6.88
Pima	66.79	5.07	**67.18**	3.63	62.65	6.97	**65.87**	3.54
Vehicle	**70.17**	2.96	69.95	5.04	58.82	3.42	**68.51**	5.45
White House	**91.47**	5.31	90.57	4.25	81.52	12.53	**90.11**	6.59
XD6	80.29	3.26	**80.63**	3.93	72.26	3.84	**74.70**	5.48
Avg.	73.38	6.71	**74.44**	6.64	67.84	8.68	**70.34**	8.32
Avg. %|LS|	62.36		62.36		10.06		10.06	

500 instances normally distributed, with equal standard deviation and different means; thus, $|LS| = 1000$. The experiments were carried out choosing in advance $N_p \ll |LS|$ (here $N_p = 100$). The 100 selected prototypes are presented in Figure 17.3 (center).

The chosen prototypes are located in regions where the density of their class is greater than elsewhere (this reduces W_e^0); furthermore, they are located where their class tends to be majority (this reduces $W_e^+ W_e^-$). This way to proceed is then different to the strategies of Hart and Gates, which tends to select instances belonging to the class boundary, *i.e.* randomly classified, why not badly classified (see Figure 17.3, right).

17.4. CONCLUSION

We have proposed in this paper an adaptation of Boosting to Prototype Selection. As far as we know, this is the first attempt to use Boosting in that field. Even more, Boosting was previously sparsely

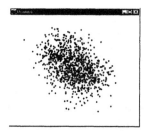  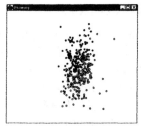

Figure 17.3 1000 points normally distributed (left), 100 selected prototypes by PS-BOOST (center), and the prototypes selected by the CNN rule (right) (GRAF Plotting Interface).

used in Machine Learning to Boost Nearest Neighbor classifiers (Freund and Schapire, 1997). In this work, the use of Boosting was also much different from ours, not only in the motivations (Classification, to recognize handwritten digits), but also in the way to grow the strong hypotheses. In particular, weak hypotheses consisted in whole Nearest Neighbor classifiers, and not in simple instances. In the field of PS, the novelty of the Boosting approach stems from the following crucial aspect. PS is achieved by an *adaptive weighting scheme*. Each prototype receives a weight which quantifies its relevance. Equivalently, the prototype to look for can be found at each stage of the algorithm by optimizing a criterion being not the accuracy, which was in turn a common aspect to previous state-of-the art PS algorithms. After a prototype is selected, each remaining instance updates a distributional weight, so as to be weighted higher if it is badly explained by the currently selected prototypes. This is the adaptive part of the selection mechanism.

While we do not use, in this paper, the weight in any classification rule, we can note that this strategy is similar to build a weighted classifier. It deserves then future investigations and comparisons in the field of weighting instances. Actually, first experiments using all the weighted learning instances in a weighted classification rule (not exclusively a NN rule), seem to show the efficiency of such a method, and will be the subject of future thorough studies.

The GRAF software package is available upon request to Marc Sebban. It runs on PC-Win95.

Notes

1. http://www.ics.uci.edu/~mlearn/MLRepository.html

References

Blum, A. and Langley, P. (1997). Selection of relevant features and examples in Machine Learning. *Artificial Intelligence*, pages 245–272.

Breiman, L., Freidman, J. H., Olshen, R. A., and Stone, C. J. (1984). *Classification and Regression Trees*. Wadsworth.

Brodley, C. (1993). Adressing the selective superiority problem: automatic algorithm/model class selection. In *Proc. of the 10 [th] International Conference on Machine Learning*, pages 17–24.

Brodley, C. and Friedl, M. A. (1996). Identifying and eliminating mislabeled training instances. In *Proc. of AAAI'96*, pages 799–805.

Buntine, W. and Niblett, T. (1992). A further comparison of splitting rules for Decision-Tree induction. *Machine Learning*, pages 75–85.

Cover, T. M. and Hart, P. E. (1967). Nearest Neighbor pattern classification. *IEEE Transactions on Information Theory*, pages 21–27.

Freund, Y. and Schapire, R. E. (1997). A Decision-Theoretic generalization of on-line learning and an application to Boosting. *Journal of Computer and System Sciences*, 55:119–139.

Gates, G. W. (1972). The Reduced Nearest Neighbor rule. *IEEE Transactions on Information Theory*, pages 431–433.

Hart, P. E. (1968). The Condensed Nearest Neighbor rule. *IEEE Transactions on Information Theory*, pages 515–516.

John, G. H., Kohavi, R., and Pfleger, K. (1994). Irrelevant Features and the subset selection problem. In *Proc. of the 11 [th] International Conference on Machine Learning*, pages 121–129.

Quinlan, J. R. (1996). Bagging, Boosting and C4.5. In *Proc. of AAAI'96*, pages 725–730.

Schapire, R. E. and Singer, Y. (1998). Improved boosting algorithms using confidence-rated predictions. In *Proc. of the 11 [th] Annual ACM Conference on Computational Learning Theory*, pages 80–91.

Sebban, M. and Nock, R. (2000). Prototype selection as an information-preserving problem. In *Proc. of the 17 [th] International Conference on Machine Learning*. to appear.

Skalak, D. B. (1994). Prototype and feature selection by sampling and random mutation hill-climbing algorithms. In *Proc. of the 11 [th] International Conference on Machine Learning*, pages 293–301.

Wilson, D. and Martinez, T. (1997). Instance pruning techniques. In *Proc. of the 14 [th] International Conference on Machine Learning*, pages 404–411.

Zhang, J. (1992). Selecting typical instances in instance-based learning. In *Proc. of the 9 [th] International Conference on Machine Learning*, pages 470–479.

Chapter 18

DAGGER: INSTANCE SELECTION FOR COMBINING MULTIPLE MODELS LEARNT FROM DISJOINT SUBSETS

Winton Davies and Pete Edwards

{wdavies, pedwards}@csd.abdn.ac.uk

Department of Computing Science,

Kings College,

University of Aberdeen

Scotland, UK.

Abstract   We introduce a novel instance selection method for combining multiple learned models. This technique results in a single comprehensible model. This is to be contrasted with current methods that typically combine models by voting. The core of the technique, the DAGGER (Disjoint Aggregation using Example Reduction) algorithm selects training example instances which provide evidence for each decision region within each local model. A single model is then learned from the union of these selected examples.

We describe experiments on models learned from disjoint training sets which show that DAGGER performs as well as weighted voting on this task and that it extracts examples which are more informative than those that can be selected at random. The experiments were conducted on models learned from disjoint subsets generated with a uniform random distribution. DAGGER is actually designed for use on naturally distributed tasks, with non-random distribution. We discuss how one view of the experimental results suggests that DAGGER should work well on this type of problem.

Keywords: Combining Multiple Models, Selective Sampling.

18.1. INTRODUCTION

This paper describes a novel instance selection algorithm used to speed up the supervised concept formation task. However it differs from other instance selection methods in that is applied after the initial application of a standard inductive algorithm, in this case C4.5 (Quinlan, 1993). The same inductive learning algorithm is then applied to the resulting much reduced training set. Why might this be useful ? Consider a national supermarket chain: each local store's learning agent learns a model from its own database. A small set of examples is selected by each store, which are then transmitted to headquarters. Finally, these sets of examples can be used to form a national model. A second scenario involves a task where training examples arrive in batches. Each time a new set of examples arrives, a model can be learned using the current set of new examples, together with the examples selected in the previous time period. Once the model has been learned, a new set of examples are selected, and then held until the next set of examples arrives.

This overall technique is termed combining multiple models or integrating multiple learned models. We will describe the current approaches to combining multiple models in Section 18.2 With one recent exception (Domingos, 1997b; Domingos, 1998), existing approaches to combining multiple models do not result in a single model of the same type as the initial set of models. In addition, the normal motivation for combining multiple models is to increase accuracy. To this end, the work done is actually increased, as each model is learned using a significant proportion of the training set multiple times. For example, boosting uses approximately 63% of total training examples each time (Bauer and Kohavi, 1999). Our goal is to learn from naturally distributed, disjoint datasets, or at least from datasets that can be easily partitioned, and the work farmed out to multiple processors. This introduces another key difference with previous work on combining multiple models: They are based on a uniformly random sample. Our instance selection algorithm is specifically designed with non-random distributions in mind.

The DAGGER (Disjoint Aggregation with Example Reduction) algorithm is described in Section 18.3 The basic concept is to reduce the size of each local subset of training examples by removing all those examples that are irrelevant to the local model. These reduced subsets of training examples are then collected, and a new model learned from this new combined training set. An alternative view is to consider the approach as selecting the most informative training examples used in creating the model. The algorithm analyses each decision region defined by each of the models. For each decision region, the algorithm selects a near minimum set of examples which together provide a first order approximation of the attribute-value distribution within that decision region. In Section 18.4 we provide an informal proof that DAGGER, when applied to a single model with all possible, will result in the selection of a set of examples which allow the exact reproduction of the original tree. The general intuition is that it is possible to provide evidence of how justified a given decision region is, with respect to the examples contained within it. This leads to the obvious question "Why not just take a random sample of a given size?"

The experiments described in Section 18.5 are designed to explore this question and to compare our approach to the most common alternative, that of weighted voting. For the purposes of this paper, we only study a random distribution of the examples. We focus on confirming two hypotheses about the accuracy of the DAGGER algorithm. These are:

1 That it is approximately the same as the weighted vote approach, given disjoint subsets.

2 That it is better than that of a model learned from a random sample of the same size as the training set selected by DAGGER.

The results, in Section 18.6 do indeed appear to confirm both these hypotheses. Furthermore, they seem to confirm the suitability of the DAGGER approach in learning from non-random distributions. The results for the degenerate case: a single model, selecting the examples using DAGGER, then re-learning from these, suggest that the DAGGER approach will work well on non-random distributions of disjoint subsets. This is discussed in Section 18.7 where we also suggest some future directions for experimentation.

We are aware that this paper does not address related work in the field of instance selection. However we hope that it is complementary to other papers in this collection, and that it serves to introduce a new paradigm to which other instance selection methods may be applied.

18.2. RELATED WORK

There has been a lot of attention in recent years in learning and combining multiple models. *Bagging* (Breiman, 1996) and *Boosting* (Freund, 1990) have become accepted as good approaches to increase accuracy, although it should be noted, not reducing processing load. *Bagging* combines multiple models learned from bootstrap samples (or sampling with replacement). Each sample typically comprises two thirds of the original data set. *Simple voting* is used to combine the models during classification. A new example is classified into a particular decision region by each model. The class of the decision region selected for each model are grouped by class, and then summed. The majority class is the classification predicted by the combined model. Learning each model may be distributed, as may the voting process. *Boosting* is an iterative process, which learns a series of models, which are then combined by a vote whose value is determined by each of the classifiers' accuracy. At each step, weights are assigned to each training example, which reflect its importance. These weights are modified so that erroneously classified examples are "boosted", causing the classifier to pay more attention to these. For a good overview (and an extensive evaluation) of *Bagging* and *Boosting* and related techniques we refer to (Bauer and Kohavi, 1999).

Neither *Bagging* nor *Boosting* are directly relevant to our work, because although they are methods for combining multiple models, they are applied specifically to highly overlapping subsets of data, and the final models are computed by voting. However they form a direct link with what we are trying to achieve. Domingos (1997a) noted in his investigation of *Bagging*: "We would like to find the simplest decision tree extensionally representing the same model as a bagged ensemble". In two later papers (Domingos, 1997b; Domingos, 1998), he describes in detail his method for combining multiple models. We regard Domingos' goal as being very similar to ours. His algorithm, CMM, generates synthetic training examples which are then classified by each local model. These are then combined with all the original training examples, and a final decision tree is learned from this new training set. The approach does indeed produce single simple trees which, on average, retain 60% of the accuracy gains of Bagging. The first difference between DAGGER and CMM is that we are dealing with disjoint subsets, and second, is that we actively select a subset of the initial training examples, rather than take all of them together with the new ones.

Ting and Low (1997) also describe an approach using weighted voting. They find an estimate of the predictive accuracy of each decision region using cross-validation. This predictive accuracy value is then used in to weight the final vote. Their aim is similar to ours in that they wish to combine models learned on separate datasets. Ali and Pazzani (1996) investigated combining decision trees learned from disjoint datasets using several different voting mechanisms. Their main conclusion was that voting increases accuracy by eliminating uncorrelated errors amongst the models. This is relevant to our findings comparing the behavior of DAGGER with the weighted vote approach. It is also the sole paper we have found which evaluated voting on models learned from disjoint subsets.

Chan and Stolfo (1995) suggest using meta-learning rather than voting to combine multiple models. The two meta-learning approaches (called *arbiter* and *combiner*) work by learning a secondary model, which integrate the multiple models together. In the *arbiter* approach, an additional model is learned, using as a training set those examples (from a hold-out set) on which the initial models cannot agree upon. During the classification task, the *arbiter* is used to resolve any ties in the vote between the models. The *combiner* approach learns a second level model: i.e. a model that predicts which of the base models will be correct on a given example. The *combiner* approach comes much closer to our aims, as it suggests it is possible to learn a description of the "expertise" of each local model. However, it does not result in a single model, although there may be a way to rebuild the multiple trees, using the *combiner* as a guide.

Although we have not evaluated our approach with respect to distributing the computational load, we would like to draw attention to (Provost and Hennessy, 1995) work on combining rules learned over disjoint samples using distributed processing. Their approach relies on using a syntactically restricted version space for the rules, coupled with communication between the processes. They use an evaluation metric together with their "invariant partitioning property" to ensure that at least one agent will find each one of the globally consistent rules (i.e. all those rules that would have been found if run on all the data). This is a very attractive approach, but potentially suffers from restricting the hypothesis space too much. We would also like to draw attention to (Provost and Kolluri, 1999) who provide an extensive survey of the state of the art in scaling up inductive learning algorithms.

18.3. THE DAGGER ALGORITHM

As described in the introduction, our goal is to create a single model, from a set of models each learned from a disjoint set of data. We explore the idea of finding examples which show the extent to which the decision region is justified. The idea is to find diametrically opposed examples which pin down a decision region, in the same way that two opposite corners can define a rectangle.

We select a *minimal spanning example set* from every decision region of each model. The DAGGER algorithm (see Figure 18.1 below) is used to select the examples. Once the examples have been selected from each model, they are collected together, and a new single model learned from them. The *minimal spanning example set*, is a subset of examples that demonstrates all the values of the attributes of the set of examples within the region. For example, consider the following:

1 Our training examples have 4 attributes A, B, C, D , each with two possible values (e.g. $A = a1, a2$). There are two classes $+, -$.

2 Consider a decision region which corresponds to "+ *if (A = a1)*".

3 The following four examples are found in this region. They are represented by the following set of vectors: $\{< +, a1, b1, c1, d1 >, < +, a1, b2, c1, d1 >, < +, a1, b1, c2, d1 >, < +, a1, b2, c2, d1 >\}$.

4 One of the two possible minimal spanning sets for this region is $\{< +, a1, b1, c1, d1 >, < +, a1, b2, c2, d1 >\}$ Note that this shows the values: $\{a1, b1, b2, c1, c2, d1\}$ Of course, we would only expect to find *A=a1* , as this defines the decision region. However, the decision region suggests that *D* can be either *d1* or *d2* , yet only examples with *d1* , are found within this decision region.

We can also describe the *minimal spanning example set* as providing a first order approximation of the degree of support of the decision region. It turns out that finding a *minimal spanning example set* is NP-Complete (by reduction to k-Set Cover) (Halldórsson, 1996). However, this paper also gives an approximation algorithm for this problem. This is a greedy hill-climbing approach combined with random perturbation.

The algorithm was initially designed to work only with nominal attribute values, but it was later modified to handle continuous attributes. The DAGGER approach only works with axis orthogonal decision regions (i.e. Hypercubes). For example, the *"default rule"* approach, found in C4.5Rules (Quinlan, 1993), is not axis orthogonal. The *default rule* is the difference of the whole decision space with the union of the axis orthogonal decision regions of represented by the rules. It also relies on being able to capture which training examples fall into which decision region. We used C4.5 revision 8 (Quinlan, 1993), as the base learning algorithm.

Initially, C4.5 is applied to each (disjoint) training set to learn a decision tree. The examples used to learn each tree are classified into the appropriate leaf node (defining a decision region), and are collected into an array. We collect the set of distinct attribute values that appear for each attribute in the decision region. Next, an example is chosen at random, to seed the minimal spanning example set. Then the algorithm tries to cover the attribute values, using a hill climbing procedure.

This simply iterates through all the examples, testing to see how many attribute values would be covered by each example. The one that covers the most attribute values is added to the minimal spanning set, the attribute values it covers removed, and the process repeated until all attribute values are covered.

The minimal spanning set is now checked against the best so far. If it smaller in size, then it replaces the best set. Next, a chosen proportion of examples are removed from the best set, and the remaining examples

```
DAGGER(array training_set, int n) {
   for i from 1 to n /* number of models/agents */
      { model[i] = C4.5(training_set[i])}
   finalSet = {}
   for i from 1 to n
   {
      for j from 1 to |decision_regions(model[i])|
      {
         dr[i,j] = classify(training_set[i],
                            decision_region(model[i]),j)
         for k from 1 to |attributes(training_set[i])|
            { av[i,j,k] = distinct_attribute_values(k,dr[i,j])}
         /* Find minimal spanning example set (MSES) */
         |MSES_best| = ∞
         MSES_best = undefined
         MSES_new = random_example(dr[i,j])
         for m from 1 to 20  /* annealing iterations */
         {
            MSES_new = hill_climbing_cover(MSES_new,
                                           av[i,j,k],
                                           dr[i,j])

            if |MSES_new| < |MSES_best|
               then {MSES_best= MSES_new }
            MSES_new = delete_random(MSES_best,30%)
         }
         for example in |MSES_best|
            {weight(example) = |dr[i,j]| / |MSES_best|}
         finalSet = finalSet ∪ MSES_best
      } /* for j */
   } /* for i */
   C4.5(finalSet)
}
```

Figure 18.1 The DAGGER Algorithm

become the seed minimal spanning set. The process is repeated for a number of iterations, and finally leaving us with an approximately minimal spanning set at the end.

Finally, one small addition is made to the *minimal spanning example set* : A weight equal to the total number of examples in the decision region divided by the size of the minimal spanning sample set, is assigned to each example chosen. This is critical for TDIDT algorithms, which rely on example counts to guide hill-climbing search.

18.3.1. Continuous Attributes

C4.5 handles continuous attributes by a cut point. Every time a continuous attribute is considered as a candidate for a split, a cut point is imagined between each the point formed by the examples on the axis of that attribute. Information Gain is then computed for a pseudo attribute based on whether the value of each example's attribute x is: $x < cut$ or $x \geq cut$.

To demonstrate the range of the decision region, it is necessary to select two examples, that satisfies the predicate x = min value of dimension axis in decision region and x = max value of dimension axis in decision region

A slight modification of this approach is required in practice, to avoid selecting too many examples. Due to the relatively random distribution of examples, it might typically take 4 examples to define a 2 dimensional decision region. However, by relaxing the constraint for the exact minimum and maximum values of the region for each axis so that it is merely close to the boundary of the region, we are able to typically reduce the number of examples to two per region, as opposed to two examples per continuous attribute.

Thus we have two predicates per continuous attribute. The predicates are: x < (min value + n% width of dimension axis in region) and x ≥ (max value - n% width of dimension axis in region)). A value of n = 10 seems to be appropriate.

18.3.2. Noise

C4.5 handles noise in a number of ways. Firstly, it does not require a decision region to contain examples of only one attribute. Secondly it allows the user to specify the smallest number of examples that should be in a decision regions. Finally it has a pruning mechanism that "rolls back" a split if further splitting does not yield an increase in information.

The result of this noise tolerance is that regions may not contain examples of only one class. This would potentially cause a problem with the basic DAGGER algorithm. There are several possibilities approaches to dealing with this issue.

1 Ignore the class when selecting examples.

2 Select only from the majority class examples.

3 Select examples that justify the extent to which this region belongs to each class.

DAGGER was modified to allow each of these approaches to be used. In general, the third approach seemed to be the most promising. As the selected examples are weighted according to the density of all the examples in the region, insignificant noisy examples do not dominate. However, if there is a justifiable sub-region, then it will show up. We have not conducted systematic experiments to compare the different approaches. This is an avenue for future work.

18.3.2.1 Computational Complexity. Space does not permit a discussion of the computational complexity of the DAGGER algorithm. However, an analysis show it to be:

$$Order(IV(A - D)^2 E)$$

Where:

I	=	Annealing Iterations
V	=	Average Number of Values Per Attribute
A	=	Average Number of Attributes
D	=	Average Depth of Decision Tree
E	=	Total Number of Examples in Training Set

18.4. A PROOF

The DAGGER algorithm is highly heuristic and its behaviour is governed by the distribution of the examples available to the agent. Thus it is extremely difficult to make any claims regarding its universality. Thus the following interesting proof only concerns the extreme degenerate case of all the training examples being available to a single agent.

To Prove: Given a complete set of all possible training examples, and a decision tree learned from these, that the examples selected by the DAGGER algorithm can be fed back to the same base learning algorithm and will produce the same set of decision regions as if all training examples were present.

Given: A base learning algorithm (L) that recursively selects a single binary/propositional attribute (A) to split on based on some function (F) of the weighted counts of positive (P) and negative (N) examples that are still not separated. We assume a divide and conquer algorithm.

Proof: All possible examples are initially present. The learning algorithm (L) proceeds by checking to see if all the examples are positive or negative, else it chooses an attribute (A), based on function (F). The examples are partitioned into two new sets, each containing half of the examples. L is then applied to each of these sets. This is recursive, terminating when the examples are either all positive or negative at the leaf node.

At the end of this process, the leaf nodes entirely partition the space into various sized hypercubes. These hypercubes have the property that they are disjoint, and that the union of them covers the entire space. In addition they are each entirely positive or negative examples.

Dagger chooses two examples in each hypercube, such that all attributes which the hypercube have as don't care, are represented in the two examples. These are weighted as a function of half the number of examples contained in the cube.

For example: a, b, c, d are the attributes. One of the hypercubes might be described as $(a \wedge b)$, The examples chosen might be $(a \wedge b \wedge c \wedge d)$ and $(a \wedge b \wedge \neg c \wedge \neg d)$. There are 8 examples within this cube. Thus the weight for each selected example is 4. This process is repeated for each leaf node.

Now, we have to show that applying L to this set of reduced examples produces the same tree. Note: The choice of attribute by F can only be made by a function of the weighted examples.

Let us note that the initial space is exactly tiled by the leaf hypercubes. This is important to the proof.

For each attribute being considered, then this will either pass to the side of each cube, or will divide the cube exactly in half. If it passes one of the hypercubes, then that hypercube will contribute its 2 examples, whose joint weight is the same as all the original examples in that hypercube.

If the hypercube is divided by the attribute split, then the two sides of the attribute split will have one of the examples, with a weight equal to half the examples in that cube. This is also the number of examples that would have been attributed by this split cube. Note that it is the very property of DAGGER's choice of the two examples (that they must differ on every attribute not in the hypercube description), that means this will always be true.

For example, consider the above hyper-cube $(a \wedge b)$ and the two examples given. Now, attributes a and b, will cause the entire hyper-cube to be considered as falling on the a or b side of the partition. If we choose c, then the first example will go to the c side, and the other to the $\neg c$ side and likewise for d.

Now given the property of the set of hypercubes for the initial region being considered, (i.e. they exactly tile the whole region), then for the region being considered, then the sum of the Ps and N's for each side of the split will be the same as if we were considering the original number of the individual examples. Thus the function F will generate the same value as before for each attribute, and L will choose the same "best" one.

Thus starting with the initial region, and descending recursively, will result in exactly the same decision on which attribute to split on. The exact tiling condition holds with each recursive call because the same attribute as originally chosen is picked again. Thus the tree produced on the weighted selected examples will be identical to the original tree. QED.

18.5. THE EXPERIMENTAL METHOD

The experiments were designed to test the following two hypotheses:

1 That the accuracy of the DAGGER approach is approximately the same as the weighted vote approach, given disjoint subsets.

2 That the accuracy of the DAGGER approach is better than that of a model learned from a random sample of the same size as the training set selected by DAGGER.

The first hypothesis will confirm that one can create a single, comprehensible, model in place of the weighted vote model without sacrificing accuracy. We did not expect our approach to do as well as *bagging*, where each model is learned from much larger, overlapping training sets, therefore we did not test against this. The second hypothesis will confirm that the selective sampling approach of DAGGER algorithm actually out performs a random sample of the same size. It is not intuitive that when learning decision trees that selective sampling could indeed do better than a random sample. We have not yet undertaken experiments with non-random distributions, which is the ultimate target of the DAGGER algorithm.

To conduct these experiments, we were restricted to datasets with nominal attributes. We used well known datasets from the Machine Learning Repository (Merz and Murphy, 1998). We are aware of the criticism leveled against using these datasets as the basis for comparing learning algorithms. In our defense, we wish to point out that for the first hypothesis we use them as an equality benchmark, rather than using them to prove that we do better than the weighted vote approach. For the second hypothesis, our algorithm does not actually perform the learning task: it selects the examples that will be given the learning algorithm. We believe that it is therefore legitimate to compare C4.5 decision trees learned from examples selected by DAGGER against C4.5 decision trees learned from a random selection of the same size.

We were restricted to using larger datasets, so that they could be divided into disjoint training sets of a reasonable size. The smallest dataset we used consisted of 440 examples (the 1984 House Voting Records). This requirement left us with nine datasets, some of which were too large and hence were reduced to sizes mentioned in Table 18.1 in Section 18.6 below. The reduction was done by random selection, and repeated at each iteration. For testing, we used a 30% hold out test set on each iteration. The same test sets and training sets were used for each set of comparisons we made, allowing us to use paired test confidence intervals. The LED-24 artificial dataset was generated with 10% noise as standard. The Adult Income dataset was discretized, assigning one of three values to both the age and hours worked attributes, and ignoring all other continuous attributes.

We decided to limit the number of models to 32, and to use powers of 2 as the increment factor. Thus we used 1, 2, 4, 8, 16, and 32 models. Each model is learned from a randomly selected, disjoint subset of the training data. Every model was learned using C4.5 in decision tree mode, with pruning turned off. For the weighted voting tests, we propagated the training examples to the leaves of each local model (a tree), and used the size of the examples in the majority class as the weight for each classification. Please note that the single model case is obviously degenerate, as we already have a single model, but we included it in our tests, and it has proved to show some interesting results.

For each of the nine dataset we carried out the procedure outlined in Figure 18.2, and recorded the results. We assume the existence of a weighted_vote algorithm, which simply runs C4.5 on each of the partitions, and then computes the final class of an unseen example by simple weighted majority vote by each tree formed from the partition.

18.6.   RESULTS

Table 18.1   Sizes of Training Sets and Average No. of Examples Selected by DAGGER

Datasets/Models	Total Examples	1	2	4	8	16	32
Splice	2233	435	534	703	935	1208	1512
Votes	304	33	47	66	104	151	208
Nursery	7258	412	639	921	1247	1583	1952
LED	6300	1643	1666	1759	1916	1870	1923
Connect-4	4729	1687	1817	1982	2128	2271	2371
TicTacToe	670	140	206	269	303	317	348
Chess	2237	110	178	297	450	634	899
Adult	9950	2306	2775	3137	3657	4056	4500
Mushroom	5686	50	95	169	299	524	907

```
TEST(array dataset) {
    for iterations from 1 to 30 { /* for statistical accuracy */
        select from dataset: training_set(70%), test_set(30%)
        for m in (1, 2, 4 ,8 ,16, 32) {
            divide training_set into m partitions[]
            record_accuracy(dagger(partitions[],m))
            record_accuracy(weighted_vote(partitions[],m))
            for P in (1, 2, 4, 8)
                {record_accuracy(C4.5(random(training_set,
                                            P×|finalSet|))) }
            record_accuracy(C4.5(initial_training_set))
        } /* end model (m) loop */
    } /* end of iterations loop */
}
```

Figure 18.2 The Test Procedure

Table 18.2 Relative Accuracy of DAGGER vs. Weighted Vote

Set	1 model	2 models	4 models	8 models	16 models	32 models
Splice	-2.14±1.11	-2.06±0.72	-1.47±0.41	-0.48±0.47	0.33±0.55	2.68±0.68
Votes	1.31±0.60	-1.35±0.67	-1.18±0.54	-1.48±0.56	-1.00±0.61	1.37±0.75
Nurs.	-2.83±0.55	-0.89±0.29	0.32±0.31	2.36±0.30	5.27±0.33	7.05±0.26
LED	5.85±0.32	-0.09±0.28	-2.87±0.26	-3.90±0.33	-3.73±0.34	-3.16±0.27
Conn4	-0.12±0.44	-2.07±0.40	-2.75±0.39	-2.19±0.48	-0.20±0.47	2.19±0.54
TTT	-13.56±2.48	-4.87±1.82	3.30±1.57	11.10±1.65	15.69±1.78	18.48±1.57
Chess	-2.64±0.83	-0.51±0.52	0.48±0.39	1.99±0.51	3.49±0.40	4.93±0.66
Adult	5.31±0.33	0.47±0.27	-1.97±0.35	-3.38±0.33	-4.51±0.23	-5.30±0.30
Mush.	-0.03±0.02	-0.27±0.47	0.15±0.05	0.48±0.13	1.29±0.10	1.13±0.10

Dataset/Models	1	2	4	8	16	32
Splice	-	-	-	-	=	+
Votes	+	-	-	-	-	+
Nursery	-	-	+	+	+	+
LED	+	=	-	-	-	-
Connect-4	=	-	-	-	=	+
TicTacToe	-	-	+	+	+	+
Chess	-	=	+	+	+	+
Adult	+	+	-	-	-	-
Mushroom	-	=	+	+	+	+

The second column of Table 18.1 shows the total number of training examples available at each test point, before being divided into the disjoint subsets used for each model. The remainder of the columns give the total number of examples selected by DAGGER as a function of the

Table 18.3   Relative Accuracy of DAGGER vs. Random Selection of Same Size

Set	1 model	2 models	4 models	8 models	16 models	32 models
Splice	0.61±1.32	1.86±0.87	1.30±0.48	1.07±0.60	0.80±0.40	0.63±0.34
Votes	3.12±1.66	0.46±1.26	1.24±0.90	1.22±1.10	1.22±0.97	-0.31±0.80
Nurs.	8.35±0.94	6.53±0.61	4.50±0.50	3.42±0.30	2.05±0.28	1.26±0.33
LED	6.09±0.51	6.39±0.48	6.22±0.50	6.04±0.45	6.27±0.48	7.19±0.45
Conn4	4.03±0.63	3.58±0.70	3.35 ±0.67	2.62±0.66	2.61±0.62	2.67±0.45
TTT	2.53±3.07	3.99±1.93	3.32±1.88	2.09±2.33	0.89±1.52	1.23±1.52
Chess	3.64±1.05	3.77±0.69	2.43±0.76	1.04±0.47	-0.20±0.39	-1.01±0.36
Adult	±0.43	3.67±0.33	3.37±0.52	2.58 ±0.53	2.08±0.44	1.51±0.47
Mush.	3.52±1.20	2.14±0.87	1.65±0.33	1.07±0.25	0.27±0.08	-0.09±0.11

Datasets/Models	1	2	4	8	16	32
Splice	=	+	+	+	+	+
Votes	+	=	+	+	+	=
Nursery	+	+	+	+	+	+
LED	+	+	+	+	+	+
Connect-4	+	+	+	+	+	+
TicTacToe	=	+	+	=	=	=
Chess	+	+	+	+	=	-
Adult	+	+	+	+	+	+
Mushroom	+	+	+	+	+	=

Table 18.4   Relative Accuracy of DAGGER vs. 2x Random Selection

Set	1 model	2 models	4 models	8 models	16 models	32 models
Splice	-1.09±1.25	0.23±0.72	-0.19±0.49	0.32±0.47	NA	NA
Votes	2.01±0.67	1.31±0.98	1.18±0.68	0.24±1.00	1.37±2.24	NA
Nurs.	4.15±0.77	3.15±0.47	1.26±0.33	0.45±0.24	-0.30±0.22	-0.90±0.24
LED	5.95±0.44	6.08±0.41	6.11±0.39	5.91±0.43	6.09±0.43	6.89±0.39
Conn4	1.27±0.60	1.44±0.40	0.73±0.48	0.18±0.45	0.32±0.42	0.51±0.88
TTT	-6.62±2.70	-3.13±2.14	-3.67±1.30	-3.83±1.60	-4.35±1.45	-5.09±2.97
Chess	1.50±1.08	1.82±0.67	0.32±0.36	-0.41±0.26	-1.07±0.36	-1.58±0.35
Adult	4.09±0.44	4.10±0.32	3.62±0.39	3.57±0.43	2.81±0.33	2.35±0.35
Mush.	1.89±0.42	1.30±0.49	1.03±0.26	0.28±0.09	0.05±0.07	-0.22±0.08

Datasets/Models	1	2	4	8	16	32
Splice	=	=	=	=	NA	NA
Votes	+	+	+	=	=	NA
Nursery	+	+	+	+	-	-
LED	+	+	+	+	+	+
Connect-4	+	+	+	=	=	=
TicTacToe	-	-	-	-	-	-
Chess	+	+	=	-	-	-
Adult	+	+	+	+	+	+
Mushroom	+	+	+	+	=	-

number of models. These are the total number of examples selected, not the per-model number.

Table 18.5 Relative Accuracy of DAGGER applied to a single model vs. Random Selections of Varying Sizes

Sset	1x	2x	4x	8x	ALL
Splice	0.61±1.32	-1.09±1.25	-1.96±1.10	NA	-2.14±1.11
Votes	3.12±1.66	2.01±0.67	2.16±0.82	1.24±0.92	1.31±0.60
Nursery	8.35±0.94	4.15±0.77	0.92±0.63	-1.41±0.62	-2.83±0.55
LED	6.09±0.51	5.95±0.44	5.95±0.58	NA	5.85±0.32
Conn4	4.03±0.63	1.27±0.60	NA	NA	-0.12 ±0.44
TTT	2.53±3.07	-6.62±2.70	-14.13±2.63	NA	-13.56±2.48
Chess	3.64±1.05	1.50±1.08	-0.39±0.99	-1.89±0.80	-2.64±0.83
Adult	4.08±0.43	4.09±0.44	5.14±0.24	NA	5.32±0.31
Mush.	3.52±1.20	1.89±0.42	1.63±0.36	0.64±0.19	-0.03±0.02

Dataset/Size	1x	2x	4x	8x	ALL
Splice	=	=	-	NA	-
Votes	+	+	+	+	+
Nursery	+	+	+	-	-
LED	+	+	+	NA	+
Connect-4	+	+	NA	NA	=
TicTacToe	=	-	-	NA	-
Chess	+	+	=	-	-
Adult	+	+	+	NA	+
Mushroom	+	+	+	+	-

It should also be noted, that while the total number of examples selected increases with the number of models they are selected from, that the per model number decreases. For example, the worse case is the adult income dataset, which with 32 models selects approximately 1 in 2 of the examples in the training sets. However, this is just over 150 examples per model. Also note that when the concept to be learned is easy, only few examples need to be selected. For example, in the degenerate single model case, DAGGER selects only 50 examples to represent the decision regions found in the mushroom dataset.

We have chosen an unusual format for displaying our results. The results tables are divided into two sets of columns. Each column indicates the number of models used. The left hand set of columns gives a sign test for the results. A "+" indicates that DAGGER was more accurate than a competitor approach, "=" indicates no significant difference, and a "-" means DAGGER did worse. The right hand set of columns gives the absolute difference in accuracy between DAGGER and the opposing approach. These tables are given with bounds calculated using two sided 95% confidence intervals for paired tests.

Table 18.2 shows the results for our first hypothesis, that DAGGER is approximately competitive with the weighted vote approach to combining multiple models. The signed tests show that 23 out of 54 cases, we do slightly better, 25 out of 54 we do slightly worse, and in 6 cases there is no discernible difference. We would argue that these results show that DAGGER performs approximately the same as weighted voting. We are not concerned with out performing voting, as we have the benefits of a single model. An interesting observation can be made: in none of the cases show in Table 18.2, does one approach completely dominate the other. There is a tendency for DAGGER to do better with more models.

Table 18.3 shows the results for DAGGER against a random selection of the same size (for the average size of the selection see Table 18.1). We believe that our second hypothesis is thus confirmed empirically. In 44 cases the model learned by DAGGER outperforms a model learned from a random selection of the same size. In 9 cases there is no discernible difference, and only in one case does it do worse.

In Table 18.4, we give the results comparing the accuracy of the model learned using the examples selected by DAGGER against the accuracy of a model learned from a random sample of twice the size. Here, we see DAGGER's performance decrease. In 28 cases DAGGER does better, in 11 cases no difference, and in 12 cases DAGGER does worse. Finally, note that in 3 cases (denoted by NA) it is not possible to select a random set of twice the size of the examples selected by DAGGER, due to the size of the underlying training set.

We do not show the results for 4 and 8 times the size of the DAGGER selection, because in a majority of cases it was not possible to select these sizes. However, in Table 18.5, we show all the results for the degenerate case, that of a single initial model. This is different from the previous results. The columns give a comparison of DAGGER applied to the degenerate single model approach against the accuracy of models learned using random selections of various sizes. These sizes are varied as multiples of the size DAGGER selected example set. We believe it is very interesting that in are four datasets, DAGGER can select examples that results in a new model whose accuracy is equivalent to the whole training set. It should also be noted that compared with models using a random selection of four times (4x) the size, DAGGER continues to do well. The fact that this is the degenerate single model case is discussed below.

18.7. DISCUSSION AND FUTURE WORK

We believe the results show that the DAGGER algorithm does indeed produce a single, comprehensible, model which performs as well as using weighted voting to combine multiple models. We have also shown that it achieves this by selecting examples which are more indicative of the underlying decision regions than can be achieved by simply randomly selecting examples.

We only have a limited informal proof of how this occurs. Our belief is that the examples selected by DAGGER express a local set of bounds for each decision region. It may be in fact that it simply results in a very uniform sample that better reflects example distribution.

The results shown in Table 18.5 are the most encouraging in some ways. If the disjoint training subsets were to be drawn according to some local distribution, then each training subset will completely reflect a particular portion of the total decision space. In contrast, with a random distribution, the examples are widely dispersed across the whole space. With non-random distribution, DAGGER should be able to select the best examples that provide evidence for decision regions within its unique view of the decision space. Table 18.5, the degenerate single model case seems to suggest that given the whole decision space, it is possible to actively select training examples which are much more informative than others. Although in this case we have the whole decision space in front of us, we would suggest that this is more akin to the problem where the decision space is broken up into pieces, rather than the random distribution examined in all the other (non-degenerate) multiple model cases.

There is a non-random distribution case that DAGGER would not handle well: Distribution based on the target class. In this case, each model would simply be "the entire decision space is true/false". However, if the process could be iterated, with the selected examples from each model added to all the other training subsets, then it is possible that the models might converge.

In conclusion, we believe there is still much to do to validate our approach. Firstly, we must develop a methodology for evaluating non-random distributions. Initially we plan to use artificial distributions, but ideally tests must be carried out on naturally occurring non-random distributions. The second task is to evaluate the effect of noise on the DAGGER algorithm, and modify the noise handling if necessary. Finally, we must extend our evaluation to include datasets with continuous valued attributes. Initial experiments (not reported here) show positive results.

References

K. M. Ali and M. J. Pazzani, (1996).*Error Reduction through Learning Multiple Descriptions* , Machine Learning, 24, pages 173-203, Kluwer Academic Publishers, Boston.

E. Bauer and R. Kohavi, (1999).*An Empirical Comparison of Voting Classification Algorithms: Bagging Boosting and Variants.* Machine Learning, 36 (1/2), pages 105-139. Kluwer Academic Publishers, Boston.

L. Breiman, (1996).*Bagging Predictors*, Machine Learning, 24, pages 123-240, Kluwer Academic Publishers, Boston.

P. K. Chan and S. J. Stolfo, (1995). A Comparative Evaluation of Voting and Meta-Learning on Partitioned Data, In*Proceedings of the Twelfth International Conference on Machine Learning (ML95)* , Morgan-Kaufmann, pages 90-98, Lake Tahoe, CA.

P. Domingos, (1997a). Why Does Bagging Work? A Bayesian Account and its Implications. In*Proceedings of the Third International Conference on Knowledge Discovery and Data Mining* , AAAI Press, pages 155-158. Newport Beach, CA.

P. Domingos, (1997b). Knowledge Acquisition from Examples Via Multiple Models. In*Proceedings of the Fourteenth International Conference on Machine Learning*, Morgan Kaufmann, pages 98-106, Nashville, TN.

P. Domingos, (1998).*Knowledge Discovery Via Multiple Models* . Intelligent Data Analysis, 2 (3), 1998.

Y. Freund, (1990).*Boosting a weak learning algorithm by majority*, Information and Computation, 121(2), pages 256-285.

M. M. Halldórsson, (1996), Approximating k-set cover and complementary graph coloring, In*Proceedings of 5th International Conference on Integer Programming and Combinatorial Optimization* , Lecture Notes in Computer Science 1084, Springer-Verlag, pages 118-131.

C. J. Merz and P. M. Murphy, (1998).*UCI Repository of Machine Learning Databases*, [http://www.ics.uci.edu/ mlearn/MLRepository.html], University of California, Irvine.

F. J. Provost and D. N. Hennessy, (1995). Distributed Machine Learning: Scaling up with Coarse Grained Parallelism, In*Proceedings of the Second International Conference on Intelligent Systems for Molecular Biology (ISMB94)*, AAAI Press, pages 340-348, Stanford, CA.

F. J. Provost and V. Kolluri, (1999).*A Survey of Methods for Scaling Up Inductive Algorithms.* Data Mining and Knowledge Discovery, 3(2), pages 131-169, Kluwer Academic Publishers, Boston.

J. R. Quinlan, (1993), C4.5: Programs for Machine Learning, Morgan Kaufmann, San Mateo, CA.

K. M. Ting and B. T. Low, (1997). Model Combination in the Multiple-Data-Batches Scenario. In*Proceedings of the Ninth European Conference on Machine Learning (ECML-97).* Springer-Verlag, pages 250-265, Prague, Czech Republic.

VI

APPLICATIONS OF INSTANCE SELECTION

Chapter 19

USING GENETIC ALGORITHMS FOR TRAINING DATA SELECTION IN RBF NETWORKS

Colin R Reeves and Daniel R Bush
School of Mathematical and Information Sciences
Coventry University
Priory Street, Coventry, CV1 5FB, UK.
C.Reeves@coventry.ac.uk

Abstract

The problem of generalization in the application of neural networks (NNs) to classification and regression problems has been addressed from many different viewpoints. The basic problem is well-known: minimization of an error function on a training set may lead to poor performance on data not included in the training set—a phenomenon sometimes called over-fitting.

In this paper we report on an approach that is inspired by data editing concepts in k-nearest neighbour methods, and by outlier detection in traditional statistics. The assumption is made that not all the data are equally useful in fitting the underlying (but unknown) function—in fact, some points may be positively misleading. We use a genetic algorithm (GA) to identify a 'good' training set for fitting radial basis function (RBF) networks, and test the methodology on two artificial classification problems, and on a real regression problem. Empirical results show that improved generalization can indeed be obtained using this approach.

Keywords: Genetic algorithms, radial basis functions, classification, regression, generalization, forecasting.

19.1.   INTRODUCTION

In many applications of artificial neural networks (ANNs) for data mining, it is customary to partition the available data into (at least) two sets, one of which is used to train the net, while the other is used as a 'test set' to measure the generalization capability of the trained net. It is not so commonly realized that this partition may influence the performance of the trained net quite significantly. In some earlier experiments (Reeves and Steele, 1993) we found that different partitions of a given data set gave substantial differences in error rates for a classification problem.

The 'ideal' training set in any problem would be a faithful reflection of the underlying probability distribution over the input space. However, most actual training sets are far from ideal—even in the 'simple' case of a random sample from a uniform distribution over a 2-dimensional input space, 'holes' and 'clumps' will readily appear unless the sample is large. Unfortunately, the need for a large training set often conflicts with the need for a computationally manageable task, and for a reasonably-sized 'test' set on which to base conclusions as to the capability of the final network. Of course, for real problems we have to accept whatever data we have to hand.

19.2.   TRAINING SET SELECTION: A BRIEF REVIEW

It is intuitively obvious that some points in a given set of data will have a greater effect on the fitted model than others. In stating this, we are in good company. Here, for example, is the famous mathematician Daniel Bernoulli writing in 1777 in the context of astronomical calculations:

> *...is it right to hold that the several observations are of the same weight or moment...? Such an assertion would be quite absurd.* (Bernoulli, 1777)

Since Bernoulli's time, many others have pondered the implications of this statement, and have answered it in many ways. In modern statistics it is now common to investigate the 'influence' of each data point on the coefficients obtained in a regression problem. Methods of detecting points with special influence are comprehensively surveyed in (Cook and Weisberg, 1982). In the case of linear least squares problems, the identification of influential points can be done with relatively little extra effort in terms of computational requirements.

The concept of influence is also related to the question of identifying *outliers*—points that appear not to fit the general pattern of the data, i.e. they are assumed to belong to a different population. Of course, we can never be sure that such points *do* come from a different population, but often such points are rejected in order to obtain what is hoped to be a better fit to the unknown 'true' model. The circumstances in which this can plausibly be done were precisely what exercised Bernoulli in his original enquiry. There is now a considerable statistical literature on the subject—a good review is that given in (Barnett and Lewis, 1984). A related area of interest is the growing use of 'robust' methods for regression, which often entail deleting or giving differing weights to particular data points.

The question of 'data editing' has also been extensively investigated in the application of k-nearest neighbour methods for classification. A good (if now slightly dated) source for research on this topic is the collection of seminal papers in (Dasarathy, 1991). In k-nearest neighbour classification, a 'new' point is given the classification of a majority of its k nearest neighbours in an existing 'training' set. (Of course, more sophisticated versions are possible, but this is not an issue here.) It is thus obvious that points near to the boundaries between classes will have more influence than those which are far away. Points with little influence can therefore be deleted, provided that a sufficient number of near-boundary points are retained as a set of exemplars.

Here the motivation has often been primarily computational—to avoid the need for calculating the k nearest neighbours of every point—but as with the case of robust regression, it has also been recognized that exclusion of some points can enhance the reliability of the estimates made from the data.

In some neural network applications, a similar situation also applies. For example, there may be situations where there are too many data points—as in (Reeves, 1995), for example—and in such cases a similar argument can be made for deleting points that are far from the boundary on the grounds of computational expense. Indeed, it can be argued that almost always (and not just in such cases) the data are not equally useful in training the net. It is intuitively clear that those data points which fall near the decision 'boundary' between two classes are likely to be more influential on the weight-changing algorithm than points which are well 'inside'. Similarly, if several points from the same class are very close to each other, the information they convey is virtually the same, so are they all necessary? (Similar, although not identical, points can be made concerning regression problems.)

Of course, as the boundaries are initially unknown, the status of any particular data point is also unknown. In (Reeves, 1995), an approach was described that used an initial crude estimate of the decision boundaries to select appropriate training data, followed by a phased addition of points to the training set. This procedure required the specification of a parameter θ and a test for the potential boundary status of a point, a definition of convergence, and a decision on the size of the initial training set. All of these proved somewhat sensitive and problem-specific.

There were other problems with this approach. Not only was the computational burden high, since multiple re-optimizations using back-propagation were necessary, but there was also the somewhat arbitrary nature and size of the resulting training set.

In this research we adopted a different approach. First, we used radial basis function (RBF) nets, since training is significantly quicker than for nets that use back-propagation. Secondly, instead of an *ad hoc* approach to training set selection, we used a genetic algorithm (GA) that attempted to select a training set with the objective of minimizing an error function. Some earlier work (Reeves and Taylor, 1998) had found that using a genetic algorithm was a promising approach to finding 'better' training subsets for classification problems. The work reported in this paper extends the earlier work to regression problems.

Some other recent pieces of related research should also be mentioned for completeness. There is a comprehensive study (Plutowski, 1994) into the general question of training set selection in the context of function approximation (i.e. regression problems). A similar approach has also been reported for regression problems in (Röbel, 1994), while Tambouratzis and Tambouratzis (1995) have described a technique based on Smolensky's harmony theory. However, these approaches are based on a selection criterion that relates to the training data only. In what follows, we propose a means of allowing the data points themselves to indicate whether or not they should be included on the basis of the *generalization* performance observed. The means whereby this is accomplished will be the genetic algorithm.

19.3.   GENETIC ALGORITHMS

Genetic algorithms (GAs) are becoming so familiar that a full discussion is probably not required. Several books (Goldberg, 1989; Reeves, 1993; Bäck, 1996) provide copious details concerning the general principles, but a brief introduction may still be in order here.

Genetic algorithms process a *population* of strings, often referred to in the GA literature as *chromosomes*. The recombination of strings is carried out using simple analogies of genetic *crossover* and *mutation*, and the search is guided by the results of evaluating some objective function f for each string in the population. Based on this evaluation, strings that have higher *fitness* (i.e., represent better solutions) can be identified, and these are given more opportunity to breed. The individual symbols in a particular string are often called *genes* and the symbols of the alphabet from which the genes are drawn are *alleles.* In many applications, a binary alphabet is appropriate, but this is by no means essential.

Crossover is simply a matter of replacing some of the genes in one parent by the corresponding genes of the other. One-point crossover, for example, is the following: Given parents P1 and P2, with crossover point X, the offspring will be the pair O1 and O2:

```
P1    1 0 1  0 0 1 0        O1    1 0 1  1 0 0 1
          X
P2    0 1 1  1 0 0 1        O2    0 1 1  0 0 1 0
```

The other common operator is mutation in which a gene (or subset of genes) is chosen randomly and the allele value of the chosen genes is changed. In the case of binary strings, this simply means complementing the chosen bit(s). For example, the string O1 above, with genes 3 and 5 mutated, would become 1 0 0 1 1 0 1. A simple template for the operation of a genetic algorithm is shown in Figure 19.1.

Choose an initial population of chromosomes;
while termination condition not satisfied **do**
    repeat
        if crossover condition satisfied **then**
        {select parent chromosomes;
        perform crossover}
        if mutation condition satisfied **then**
        perform mutation}
        evaluate fitness of offspring
    until sufficient offspring created;
    select new population;
endwhile

Figure 19.1 A genetic algorithm template—a fairly general formulation, accommodating many different forms of selection, crossover and mutation. It assumes user-specified conditions (typically, randomized rules) under which crossover and mutation are performed, a new population is created (typically, when a fixed number of offspring have been generated), and whereby the whole process is terminated (typically, by a limit on the total number of offspring generated).

19.3.1. Crossover for Set-Selection

The simple one-point crossover operator defined above is of little relevance to the type of problem considered here. The question is one of selecting a subset from a much larger set of points, and in (Radcliffe, 1993) the problematic aspects of the binary representation and simple crossover in such cases have been eloquently pointed out. An 'obvious' way of representing a subset is to have a string of length N, where there are N points altogether, with the presence or absence of each point indicated by 1 or 0. However, the application of simple crossover would make it difficult to maintain a subset of fixed size, quite apart from other problems that are discussed in (Radcliffe, 1993). We therefore adopted the representation and operator described by Radcliffe and George (Radcliffe and George, 1993) for creating a child from two parents chosen on the basis of their fitness.

Radcliffe's $RAR(w)$ operator works as follows: Suppose we have two 'parent' training sets, represented by subsets of the points $\{1, \ldots, N\}$. We place w copies of every element that appears in both parents into a bag, along with w copies of every 'barred' element—those that appear in neither parent. Finally, a single copy of every element that appears in one parent but not both is placed in the bag, together with its barred equivalent. Elements are drawn randomly without replacement and stored as elements of the 'child' until either the child is complete, or the bag has no unbarred elements left. In the latter case, the child is completed by randomly choosing from the currently barred elements (including any that had previously been included in the bag).

The parameter w is capable of tuning the effect of this crossover-like operator either in the direction of complete 'respect' (every element in both parents is always inherited by the child) as $w \to \infty$, or towards complete 'assortment' (all elements in the child come from one parent but not both) as $w \to 0$. In (Radcliffe and George, 1993), $w = 3$ was found to be a good compromise, and some preliminary experiments suggested this was a good choice for our problem also.

19.4.   EXPERIMENTS

All the experiments reported here used radial basis function (RBF) nets: mathematically, the distance between each input vector $\mathbf{x}$ and each of k *centres* $\mathbf{c}_i$ is calculated. These distances are then passed through a non-linearity $\phi(\cdot)$ (often a Gaussian function), and a weighted sum of these values used to predict the output y. In effect, it fits a non-linear model

$$\rho(\mathbf{x}) = \lambda_0 + \sum_{i=1}^{k} \lambda_i \phi\left(\frac{||\mathbf{x} - \mathbf{c}_i||}{\sigma_i^2}\right).$$

These nets have the advantage that non-linearities can be incorporated into the first-level question of how many RBF 'centres' there should be, where they should be located, and what scaling parameter(s) σ should be used. The choice of architecture and parameters for this problem was as follows: the k-means clustering algorithm was used once to find k centres, and the σ values estimated using a P-nearest neighbour method (with $P = 10$) as described in (Moody and Darken, 1990). The problem then reduces to a second-level one of solving a system of linear equations, which can be accomplished in a number of ways—for example, singular value decomposition (SVD), as was used in the research reported here.

It can be seen that the distance measure takes account of the square of the difference between each input variable and the corresponding centre. In cases where one of the input variables has a much smaller range than the other variables, it is clear that the activation of the unit will be highly insensitive to this variable. In such cases a simple transformation of the input data is usually recommended. Here we found the mean and variance of each input variable in the complete data set, and re-scaled the inputs to have zero mean and unit variance.

19.4.1.   Circle-In-Square Problem

In order to investigate the systematic selection route more fully, it was decided that initially a fairly simple artificial 2-dimensional problem would be experimentally studied. This was a 2-class problem (the 'circle in a square problem') with 2 inputs $\{(x_1, x_2) : -1 \leq x_i \leq 1; i = 1, 2\}$, defined as follows:

$$x_1^2 + x_2^2 < \frac{2}{\pi} \rightarrow C_1$$

$$x_1^2 + x_2^2 \geq \frac{2}{\pi} \rightarrow C_2.$$

It should be noted that the class boundary in this problem is sharp—that is, the problem in finding the correct boundaries is not due to any intrinsic errors or noise, but rather to the fact that the input data may not be spread 'evenly' in the input space.

In this case we created a training set of 200 data points, generated randomly in the square $[-1, 1]^2$, and an additional 'validation' set of a further 200 randomly generated points. Finally, we used a third randomly generated set as a test set. The approach would be to use a GA to select an 'exemplar set' of $n < 200$ from the first set; the fitness would be evaluated using the 2nd set, while the performance of the network finally chosen would be measured on the 3rd set. In using such a test problem the errors in the model generated could actually be evaluated exactly, without the need for the 3rd set. However, in real problems this would obviously not be possible, so an additional item of interest here is whether using the 3rd set creates any difficulties, since the final performance measure is itself merely an estimate.

The first step was to ascertain the appropriate network architecture. In this case it is intuitively clear that one RBF centre at the origin (0,0) would provide the best model. Some preliminary experiments were carried out to investigate the effect of different numbers of centres using the full training set. The mis-classification (or error) rates obtained on the test set confirmed our expectation that one centre was appropriate. It was also interesting to note that these error rates differed little from those obtained by an exact calculation.

We then applied the GA with a population of size $\min(2n, 100)$, a mutation rate of $1/n$, a crossover probability of 1 and an incremental reproduction strategy (i.e. one offspring is produced and immediately inserted into the population in place of one of the currently below average strings). To determine a stopping criterion, we ran the GA for an initial epoch of 200 offspring generations and recorded the best exemplar set found. For subsequent epochs of 50 offspring generations, the best solution is compared, and if no improvement has been made over the previous epoch the algorithm terminates. The GA was run for values of n running from 5 to 195 in steps of 5. For comparison, the effect of using a randomly selected subset of size n was also investigated at each value of n.

The whole procedure was run 30 times with different groups of training, validation and test sets. Qualitatively the results were surprisingly similar. Even with a subset as small as 5 data points, the performance in terms of error rate can be improved substantially over a random choice. Further investigation showed that the centre co-ordinates were always close to the origin, as expected. However, as expected, overall performance improves as the subset size increases. Figure 19.2 compares performance of the best subset (averaged over 30 replications) at different subset sizes with that of a randomly selected subset of the same size. The error rate using the entire training set is also shown for comparison.

As can be seen, the GA-selected subsets consistently out-perform the randomly selected training set, and produce better performance than the entire training set—even with as few as 5 exemplars. It also appears that the procedure is quite robust to the choice of subset size, although in general the lowest error rates were most consistently obtained somewhere between $n = 50$ and $n = 70$.

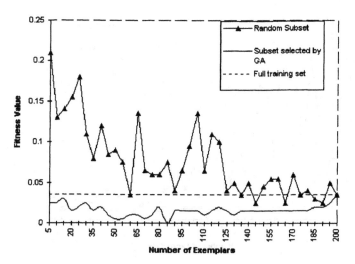

Figure 19.2 Error rates for different sizes of subset for the circle-in-square data set. Note that the error rates are those obtained on the unseen test set.

19.4.2. Continuous XOR Problem

The method was also tested on another classification problem—a continuous analogue of the well-known XOR problem. In this 2-class problem points in the opposite quadrants of a square come from the same class. Formally, for inputs $x_i \in [-1, 1]$,

$$x_1 x_2 < 0 \rightarrow C_1$$

$$x_1 x_2 > 0 \rightarrow C_2.$$

Similar experiments on this problem have been reported in (Reeves and Taylor, 1998). However, that work relied on a different implementation of the GA, so the experiments were repeated for the new (object-oriented) version. Although exact agreement was impossible because of implementation details, the results are qualitatively very similar, and are included here for completeness.

Again, we used training, validation and test sets of size 200 points, with the same GA parameters as in 19.4.1. For this problem, we would expect the ideal RBF net to have 4 centres, located at the centre of each of the quadrants. Again, experiments on the full training set confirmed this expectation. The GA selection procedure was applied to this problem for different subset sizes as before, with the results displayed in Figure 19.3.

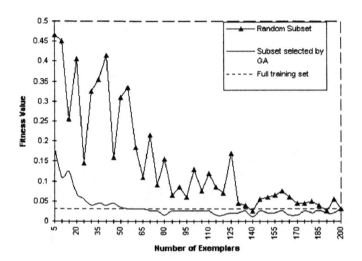

Figure 19.3 Error rates for different sizes of subset. Note that the error rates are those obtained on the unseen test set.

It is interesting that in this case the subset needed to be considerably larger ($n > 70$) before the GA-selected training sets improved on the results obtained from the full training set. An investigation of the centre co-ordinates also showed that they became closer to the quadrant centres as the subset size increased. For small subsets, at least one centre was usually quite substantially different from its expected position. This problem would seem to be somewhat harder to solve than the earlier one, and in practice whether it would be worth the extra computation would depend on the likely gain resulting from a relatively small improvement in performance. However, such methods can be useful for classification problems in the real world—in (Reeves and Taylor, 1998) we discuss an application to a large real world mortgage loan problem for which this approach produced better generalization than existing methods.

19.5. A REAL-WORLD REGRESSION PROBLEM

The results of the above tests were sufficiently encouraging for a more interesting application to be investigated, where a fairly small reduction in error could lead to substantial reductions in cost. This case was a regression problem, in which we used data provided by BG Technology plc, whose concern is to forecast daily demand for gas, based on weather forecast data for the day in question. Previous work had identified 12 important variables which are currently used in a NN model to forecast gas demand on a regional as well as on a national basis. This model uses the familiar multi-layer perceptron (MLP) network and although it has proved fairly reliable, it was felt that improvements should be possible, and that the method outlined above was a reasonable avenue to explore. However, training MLPs is extremely time-consuming, and it was agreed that the goal was not specifically to improve the performance of their current models, but rather to establish that the concept of training set selection was valid. Thus it clearly made sense to investigate the training set selection approach using a model with much speedier training performance, such as RBF nets like those used above.

BG currently uses different models for different seasons of the year, and it was decided to attempt to find a subset of exemplars from full training sets of 1000 days' demand data for the East Midlands local distribution zone (LDZ) for summer and 856 days' data for the winter months. The figures for 1996 were used to provide validation and test sets. These were created by arbitrarily assigning equal numbers of points from the summer or winter months respectively. Once the data sets had been partitioned they were normalized so that each variable had a mean of 0 and a standard deviation of 1.

19.5.1.   Simulation Results

Initially, the full training set was used and performance evaluated on the test set for different numbers of RBF centres. In each case the fitness of a net was evaluated by the mean absolute percentage error (MAPE) in the forecast for the relevant period of 1996. The training data were presented 5 times, each time in a different order. It was clear that, for both summer and winter, if more than about 7 centres are used the performance is not significantly enhanced, so on the prrinciple of parsimony, only 7 centres were used in all the subsequent experiments.

The GA was then used to find a suitable training set with subsets of $100, 200, \ldots, 900$ data points for the summer data, and $100, 200, \ldots, 800$ points for the winter data. A population of 100 strings was used; all other parameters were the same as those used in Section 19.4.1, and the GA was run with an initial epoch of 2000 offspring generations. Subsequently a termination test was made every 100 offspring generations— termination occurring if no improvement had been made during that time. The computational requirements were more demanding than in the earlier tests, so at each subset size only 5 trials were made and the results averaged as shown in Figures 19.4 and 19.5.

The graphs displayed compare the average performance of a 7-centre RBF net using the full 1000 data points and GA-selected subsets of

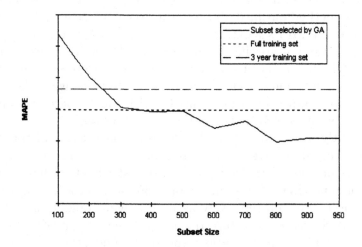

Figure 19.4 MAPE at different subset sizes for the summer data. For reasons of commercial confidentiality, the scale on the *y*-axes is omitted.

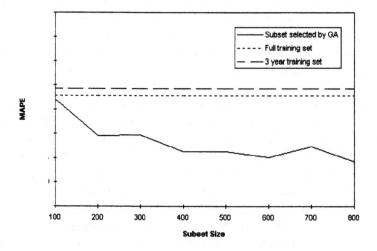

Figure 19.5 MAPE at different subset sizes for the winter data. For reasons of commercial confidentiality, the scale on the *y*-axes is omitted.

different sizes, along with the MAPE obtained using BG's current 3-year training data sets. It is interesting to note that in both cases the 1000-point data set actually does better than the current data set of about 530 points, which suggests that *ceteris paribus* a larger training set should improve performance. However, in both cases smaller data sets of selected points can do even better. For the winter data, even 100 selected points achieve a lower MAPE, but it is harder to do so for the summer data. In the latter case, selected subsets of at least 500 points were needed. Taking both cases together, it would seem that selecting subsets of about 800 points provide the best performance. Of course, the lines depict the average MAPE, but in every case the standard deviation was small; the reduction in MAPE at best was about 10%, which was statistically highly significant, and practically important.

19.5.2.   Further Analysis

From one standpoint, the use of 1996 information as validation and test data might seem unusual, but from the company's point of view, the aim was to try to find the most appropriate past data for predicting the most recent patterns of demand. Forecasting systems generally rely on past data to predict the future, but it is always recognized that not all the past is equally useful. That an improvement was obtained by using a selection process seems to bear this out.

The composition of the selected subsets was next analysed in order to look for an explanation for the improvement observed. In the first place, it was suggested that 'older' data might be less useful than those points obtained more recently. The data points in the selected subsets were therefore categorized by their year of origin, as shown in Tables 19.1 and 19.2.

In the case of the summer data, there is apparent evidence for this, in that the early years (1990 and 1991) tend to be under-represented. However, 1990 was one month short compared to the others, so the expected pattern is not uniform. If the figures are adjusted for this, it is clear that the 'recency' effect, although present, is hardly a major factor. In the case of the winter data, the pattern is rather more interesting. Here data from the winters of 1990 and 1995 are consistently under-represented, despite an almost uniform underlying distribution of the years in this case. This would suggest that gas demand in these years was substantially different from demand in the year for which the forecasts were made (1996), and points to the need for further research into the classification of benchmark information in such problems.

Table 19.1 Yearly breakdown of subsets selected by the GA for forecasting the summer months' demand

Subset	% of exemplars present in					
Size	1990	1991	1992	1993	1994	1995
100	13.00	17.00	18.00	18.00	16.00	18.00
200	13.50	18.50	17.00	16.00	17.00	18.00
300	13.33	14.67	17.67	20.00	16.67	17.67
400	14.25	16.25	17.50	17.00	17.25	17.75
500	13.80	15.60	17.80	18.40	16.80	17.60
600	13.83	16.00	17.50	18.67	16.67	17.33
700	13.14	15.43	16.71	18.86	17.14	18.71
800	13.75	15.50	17.25	18.25	17.25	18.00
900	13.33	16.00	17.56	18.33	17.22	17.56

Table 19.2 Yearly breakdown of subsets selected by the GA for forecasting the winter months' demand

Subset	% of exemplars present in					
Size	1990	1991	1992	1993	1994	1995
100	10.00	20.00	20.00	19.00	18.00	13.00
200	9.00	22.00	19.50	20.00	18.50	11.00
300	9.67	21.67	19.00	20.33	19.67	9.67
400	10.00	20.75	20.00	20.25	19.50	9.50
500	10.20	20.60	19.20	20.60	19.20	10.20
600	10.17	20.83	19.83	10.17	18.83	10.17
700	10.00	20.14	19.29	21.29	19.00	10.29
800	9.75	20.50	20.13	20.25	19.37	10.00

A final analysis was made which examined the nature of the points included in the selected subsets in a different way. As discussed earlier, not all points will be equally informative in their impact on a forecast. It would thus be interesting to know what it is that these discarded points have in common. Each point in this case is a 12-dimensional vector, where each dimension has been normalized as described above. Points were classified by counting on the number of variables for which they were outside the range $(-2, +12)$. The idea is simply to be able to characterize points that are 'unusual' in that they are outside the 12-dimensional sphere thus described. On the assumption of a Normal distribution, these correspond approximately to 95% confidence intervals along each dimension, but this interpretation is not necessary. In fact the distribution observed is not what would be expected for a 12-dimensional Normal distribution, having more points inside the sphere, and also more points that are unusual along 4 or more dimensions. Table 19.3 shows the detailed breakdown of the full data set.

Table 19.3 Number of points containing r unusual variables taken from the full summer and winter training sets.

No. of unusual dimensions r	0	1	2	3	4	5	6	7+	Total
No. of points (summer)	687	159	55	28	27	25	10	9	1000
No. of points (winter)	585	135	66	31	27	12	-	-	856

Next, the composition of the selected subsets was examined to see what type of points were included. If there were no bias towards points of a particular type, the composition of each subset should have the same frequencies in each category as the full data set in each case. Tables 19.4 and 19.5 express the ratios of the frequency actually observed to that expected.

Table 19.4 Ratios of observed to expected frequencies of each category for the selected subsets used to forecast the summer months

Subset Size	No. of unusual dimensions							
	0	1	2	3	4	5	6	7+
100	0.960	1.006	1.090	1.071	0.740	1.600	1.000	2.222
200	0.997	0.880	1.090	1.071	1.111	1.400	0.500	1.111
300	1.009	0.922	1.030	0.952	1.111	1.066	0.666	1.111
400	0.989	1.022	0.727	1.071	1.203	1.100	1.250	1.666
500	1.013	0.943	0.981	1.071	0.888	1.040	1.000	1.111
600	1.001	0.995	0.909	1.071	1.111	1.000	1.000	0.926
700	1.004	0.997	0.935	1.020	1.005	1.028	1.000	0.952
800	0.989	0.998	1.045	1.026	1.064	1.050	1.000	1.111
900	0.999	0.992	0.989	0.992	1.028	1.022	1.000	1.111
950	0.999	0.993	1.014	1.015	1.013	1.010	0.947	1.052

It can be seen from these data that for the most part the ratios are close to 1, which implies that points from each category are being selected in proportion to their frequency in the full data set. Admittedly, the interpretation of the data for small subsets is difficult, since the expected frequencies are small values anyway, but there is perhaps a slight tendency for rather more of the 'most unusual' points to be selected. This tendency soon disappears as the subset size increases. Moreover, if a χ^2 goodness-of-fit test is carried out on each subset, in no case does the hypothesis of proportionate selection come into any doubt whatsoever, even for the small subsets. It was in any case the subsets of size 500 or above where the improvement was most noticeable, and for these instances, while there is little overall deviation from proportionality, a few 'unusual' points are inevitably omitted.

Table 19.5 Ratios of observed to expected frequencies of each category for the selected subsets used to forecast the winter months

Subset	No of unusual dimensions					
Size	0	1	2	3	4	5+
100	0.980	0.824	1.037	1.380	1.585	1.426
200	0.965	1.014	1.167	0.966	1.109	1.426
300	0.975	1.014	0.994	1.104	1.268	1.188
400	0.998	0.982	1.037	0.897	1.109	1.070
500	0.992	1.001	1.037	0.994	1.077	0.998
600	0.999	1.025	0.972	0.966	0.951	1.070
700	1.003	0.996	1.000	0.986	0.951	1.019
800	1.000	0.982	1.005	1.035	1.030	0.980

Thus it would appear that improved performance is due, not so much to the omission of points that might be called 'outliers', as to the selection of points that allow the RBF centres to lie at their 'best' positions, just as we observed for the artificial problems in sections 19.4.1 and 19.4.2.

19.6.   CONCLUSIONS

This paper has described a novel approach to the question of improving the generalization capability of a neural net. Genetic algorithms have been used in conjunction with ANNs in many ways, but usually with the intent of finding an appropriate architecture, or of improving the training algorithm, or of selecting suitable features of the input space. In this research we have shown that GAs can also be used effectively to find a smaller subset of the training data for both classification and regression problems. Not only might this be useful in problems where there is a huge amount of data, but it also appears that generalization performance can also be improved.

Of course, it is undeniable that an approach such as this is computationally intensive. Many thousands of RBF nets are implicitly trained and discarded in the course of a single run. However, such training does not need to be on-line, and an improvement in generalization of even a small amount can be practically significant in many data-mining contexts.

Research is clearly needed to investigate training set selection methods more comprehensively, but this work has demonstrated their potential. In the forecasting problem, using MLPs would fit better with current practice at BG, but embedding back-propagation (or even enhanced training schemes) in the GA approach might be too heavy a computational burden. One avenue to explore would be to see whether RBF or other nets can sensibly used as surrogates for choosing a training set which can then be transferred to the MLP.

In the work described above, the network architecture and training parameters were assumed to be fixed before the training set was selected. For the artificial problems, we can be fairly confident that these architectures were correct, but for real-world data, we cannot be so sure. In practice, it might well turn out that the architecture influences the type of training set selected. The converse, of course, may equally well be true. It is hoped that future research will confront this problem by investigating the possibility of co-evolution of training sets and network architecture.

Acknowledgment   We wish to acknowledge the provision of data by BG Technology plc, and the helpful discussions with members of the Gas Research and Technology Centre.

References

Reeves, C.R. and Steele, N.C. (1993). Neural networks for multivariate analysis: results of some cross-validation studies. *Proc. of 6th International Symposium on Applied Stochastic Models and Data Analysis*, Vol II, 780–791. World Scientific Publishing, Singapore.

Bernoulli, D. (1777). The most probable choice between several discrepant observations and the formation therefrom of the most likely induction. *Biometrika*, 48: 3–13. Translated by Allen, C.G. (1961).

Cook, R.D. and Weisberg, S. (1982). *Residuals and Influence in Regression*. Chapman and Hall, New York.

Barnett, V. and Lewis, T. (1984). *Outliers in Statistical Data*. Wiley, Chichester.

Dasarathy, B.V. (1991). *Nearest Neighbor (NN) Norms: NN Pattern Classification Techniques.* IEEE Computer Society Press, Los Alamitos, CA.

Reeves, C.R. (1995). Training set selection in neural network applications. In Pearson, D.W., Albrecht, R.F. and Steele, N.C. (Eds.) (1995) *Proc. of 2nd International Conference on Artificial Neural Nets and Genetic Algorithms,* 476–478. Springer-Verlag, Vienna.

Reeves, C.R. and Taylor, S.J. (1998). Selection of training sets for neural networks by a genetic algorithm. In Eiben, A.E., Bäck, T., Schoenauer, M. and Schwefel, H-P. (Eds.) *Parallel Problem-Solving from Nature—PPSN V,* 633–642. Springer-Verlag, Berlin.

Plutowski, M. (1994). *Selecting Training Exemplars for Neural Network Learning.* PhD Dissertation, University of California, San Diego.

Röbel, A. (1994). *The Dynamic Pattern Selection Algorithm: Effective Training and Controlled Generalization of Backpropagation Neural Networks.* Technical Report, Technical University of Berlin.

Tambouratzis, T. and Tambouratzis, D.G. (1995) Optimal training pattern selection using a cluster-generating artificial neural network. In Pearson, D.W., Albrecht, R.F. and Steele, N.C. (Eds.) (1995) *Proc. of 2nd International Conference on Artificial Neural Nets and Genetic Algorithms,* 472–475. Springer-Verlag, Vienna.

Goldberg, D.E. (1989) *Genetic Algorithms in Search, Optimization, and Machine Learning.* Addison-Wesley, Reading, Massachusetts.

Reeves, C.R. (Ed.) (1993). *Modern Heuristic Techniques for Combinatorial Problems.* Blackwell Scientific Publications, Oxford, UK; re-issued by McGraw-Hill, London, UK (1995).

Bäck, T. (1996). *Evolutionary Algorithms in Theory and Practice: Evolution Strategies, Evolutionary Programming, Genetic Algorithms.* Oxford University Press, Oxford, UK.

Radcliffe, N.J. (1993). Genetic set recombination and its application to neural network topology optimisation. *Neural Computing and Applications,* 1: 67–90.

Radcliffe, N.J. and George, F.A.W. (1993). A study in set recombination. In Forrest, S. (Ed.) (1993) *Proceedings of 5th International Conference on Genetic Algorithms,* 23–30. Morgan Kaufmann, San Mateo, CA.

Moody, J. and Darken, C.J. (1990). Fast learning in networks of locally-tuned processing units. *Neural Computation,* 1: 281–294.

Chapter 20

AN ACTIVE LEARNING FORMULATION FOR INSTANCE SELECTION WITH APPLICATIONS TO OBJECT DETECTION

Kah-Kay Sung

Department of Computer Science, National University of Singapore

sungkk@comp.nus.edu.sg

Partha Niyogi

Lucent Technologies, Bell Laboratories, Murray Hill, NJ 07974, USA.

niyogi@research.bell-labs.com

Abstract In certain real-world learning scenarios where there are enormous amounts of training data, the training process can become computationally intractable. Researchers have attempted to address this problem by performing *instance selection*, which is to automatically identify and preserve a sufficiently small but highly informative data sample for training.

In most classical formulations of example-based learning, the learner *passively* receives randomly drawn training examples from which it recovers the unknown target function. *Active Learning* describes a different example-based learning paradigm where the learner explicitly seeks for new training examples of high utility, and can thus be viewed as a form of instance selection. This chapter presents a Bayesian formulation for active learning within a classical function approximation learning framework, and shows how one can derive precise example selection algorithms for learning some simple target function classes more accurately with less training data. We then present a real-world learning scenario on object (face) detection, and show how our active learning formulation leads to a useful instance selection heuristic for identifying and retaining high utility training data.

Much of this work was done when the authors were postgraduate students at the Center for Biological and Computational Learning, Massachusetts Institute of Technology, USA.

Keywords: Active learning, function approximation, instace selection, data sampling strategies, face detection.

20.1.   INTRODUCTION

Example-based learning has been an attractive framework for extracting knowledge from empirical data, especially in problem domains where prior human knowledge may be unreliable or incomplete. Typically, the system to be modeled is treated as an input-output functional mapping of relevant domain-dependent attributes. Learning involves recovering this unknown functional mapping from training data consisting of actual input-output examples, where an important goal is to recover a function that generalizes well on new input patterns.

In example-based learning, although more training data often improves generalization results, the training process can itself become computationally intractable with increasingly large data sets. This issue is becoming more evident today, because there are complex problems waiting to be solved in many domains, where large amounts of training data are available but cannot be fully exploited because of limited computation resource (Fayyad et al., 1996). To address this problem on learning with large data sets, researchers in data-mining (see for example (Aha et al., 1991; Blum and Langley, 1997; Musick et al., 1993; Syed et al., 1999)) have attempted, with varying degrees of success, to constrain computational requirements by performing *instance selection*, which is to automatically select small but highly informative data samples for training.

In most classical formulations of example-based learning, the learner *passively* receives training examples from which it recovers the unknown target function. This applies to a large variety of learning frameworks including network models (Rumelhart and McClelland, 1986; Poggio and Girosi, 1989), PAC (Valiant, 1984) learning, and classical pattern recognition (Duda and Hart, 1973). *Active Learning* describes a different class of example-based learning paradigms where the learner explicitly seeks for new training examples of high utility, and can thus be viewed as a form of instance selection in the data-mining sense. By judiciously querying the teacher for new training examples from specific input-space locations instead of passively accepting new training data randomly chosen by the teacher, *active learning* techniques can conceivably yield better approximation results with less training data than classical passive learning methods.

This chapter presents a Bayesian formulation for active learning within a classical *function approximation learning* framework, and shows how the active learning formulation can be adapted to perform instance selection for a real-world learning task on object detection. Section 20.2

explains the active learning theory: given a classical function approximation based learning task and some prior information about the target
function, define a principled strategy for selecting useful training data
in some "optimal" fashion. Section 20.3 compares the training data requirements of this active example selection strategy with that of random
sampling for some simple target function classes. Finally, section 20.4
describes the instance selection procedure in our object detection application.

20.2.   THE THEORETICAL FORMULATION

We briefly review *function approximation* as a lead-in to our active
learning formulation. Let $\mathcal{D}_n = \{(\mathbf{x_i}, y_i) \in \Re^d \times \Re | i = 1, \ldots, n\}$ be
a set of n data points obtained by sampling an unknown function g,
possibly in the presence of noise, where d is the input dimensionality.
Given $\mathcal{D}_n$ and a set of admissible target functions $\mathcal{F}$, where each $f \in \mathcal{F}$
has known prior probability $\mathcal{P}_{\mathcal{F}}(f)$, *function approximation* attempts
to estimate the unknown g by means of a "best" approximator $\hat{g} \in$
$\mathcal{F}$. The *regularization* approach (Tikhonov and Arsenin, 1977; Bertero,
1986; Poggio and Girosi, 1989) to function approximation selects a Bayes
optimal approximator, $\hat{g}$, that maximizes the a-posteriori conditional
probability:

$$P(\hat{g}|\mathcal{D}_n) \propto P(\mathcal{D}_n|\hat{g})\mathcal{P}_{\mathcal{F}}(\hat{g}) \qquad (20.1)$$

In the case where output noise (i.e. noise in the y_i measurements) at
the n data points is *identically independently Gaussian distributed* with
variance σ^2, the following relationship applies:

$$P(\mathcal{D}_n|\hat{g}) \propto \exp\left(-\sum_{i=1}^{n} \frac{1}{2\sigma^2}(y_i - \hat{g}(\mathbf{x_i}))^2\right) \qquad (20.2)$$

20.2.1.   Active Learning

In general, one can obtain a better Bayes optimal estimate of the unknown function, g, by adding more data samples to the original training
set $\mathcal{D}_n$. Our *active learning* formulation addresses the following issue:
*At what input location, $\mathbf{x}_{(n+1)}$, should one sample the next data point,
$(\mathbf{x}_{(n+1)}, y_{(n+1)})$, in order to obtain the "best" possible Bayes optimal estimate of g from the enlarged data set $\mathcal{D}_n \cup (\mathbf{x}_{(n+1)}, y_{(n+1)})$?*
Using ideas from *optimal experiment design* (Federov, 72), one can
approach the active data sampling problem in two stages:

1 **Define mathematically the notion of a "best" possible Bayes optimal estimate of an unknown target function.** For this part, we propose an optimality criterion for evaluating the "goodness" of a solution with respect to an *unknown* target function (see Section 20.2.2).

2 **Formalize mathematically the task of determining the "best" possible next sample location.** We express the above mentioned optimality criterion as a cost function to be minimized, and the task of choosing the next sample location as one of minimizing the cost function with respect to the input location of the next data sample (see Section 20.2.3).

A number of authors (Cohn, 1991; MacKay, 1992; Sollich, 1994; Sung and Niyogi, 1996) have developed similar *optimal experiment design* based techniques for sampling data. As the differences between these techniques are minor and are not critical for appreciating the general active learning framework presented in this book chapter, we refer the interested reader to (Sung and Niyogi, 1996) for a detailed comparison.

20.2.2. An Optimality Criterion for Learning An Unknown Target

Let g be the target function that we want to estimate by means of an approximation function $\hat{g} \in \mathcal{F}$, where $\mathcal{F}$ is the set of admissible target functions. If the target function g were known, then one can quantify how closely $\hat{g}$ approximates g by computing their *Integrated Squared Difference* (ISD) over the input space, $\Re^d$, or over an appropriate region of interest in the input space:

$$\delta(\hat{g}, g) = \int_{\mathbf{x} \in \Re^d} (g(\mathbf{x}) - \hat{g}(\mathbf{x}))^2 d\mathbf{x}. \qquad (20.3)$$

In most function approximation tasks, the target g is unknown, so we clearly cannot express the quality of a learning result in terms of g. We can, however, compute an *expected* integrated squared difference (EISD) between $\hat{g}$ and its unknown target g as a closeness measure for the approximation, by treating the unknown target g as a random variable in the admissible concept class $\mathcal{F}$. Let $\mathcal{P}_{\mathcal{F}}(g)$ be the prior distribution of g in $\mathcal{F}$ as assumed by the function approximation task. Taking into account $\mathcal{D}_n$, the n data points seen so far, we have the following a-posteriori likelihood for g: $\mathcal{P}(g|\mathcal{D}_n) \propto \mathcal{P}(\mathcal{D}_n|g)\mathcal{P}_{\mathcal{F}}(g)$, where $\mathcal{P}(\mathcal{D}_n|g)$ has form similar to Equation 20.2. The *expected* integrated squared difference (EISD) between an unknown target, g, and its estimate, $\hat{g}$, is thus:

$$E_{\mathcal{F}}[\delta(\hat{g}, g)|\mathcal{D}_n] = \int_{g \in \mathcal{F}} \mathcal{P}(g|\mathcal{D}_n)\delta(\hat{g}, g)dg$$

$$= \int_{g \in \mathcal{F}} \mathcal{P}_{\mathcal{F}}(g)\mathcal{P}(\mathcal{D}_n|g)\delta(\hat{g}, g)dg. \qquad (20.4)$$

As an optimality criterion for evaluating solutions to an *unknown* target, the EISD is intuitively pleasing because it favors outcomes where the Bayes optimal estimate $\hat{g}$ is very likely to be close (in the classical ISD sense) to the actual unknown target g, even if the two functions are not identical.

20.2.3.   Selecting The Next Sample Location

Given the above learning goal, a natural sampling strategy would be to choose the next example from the input location $\mathbf{x}_{(n+1)} \in \Re^d$ that minimizes the EISD between g and its new Bayes optimal estimate $\hat{g}_{(n+1)}$. How then does one predict the new EISD that results from sampling the next data point at location $\mathbf{x}_{(n+1)}$?

Suppose we know the target output value (possibly noisy), $y_{(n+1)}$, at $\mathbf{x}_{(n+1)}$. The EISD between g and its new estimate $\hat{g}_{(n+1)}$ would then be $E_{\mathcal{F}}[\delta(\hat{g}_{(n+1)}, g)|\mathcal{D}_n \cup (\mathbf{x}_{(n+1)}, y_{(n+1)})]$, where $\hat{g}_{(n+1)}$ can be recovered from $\mathcal{D}_n \cup (\mathbf{x}_{(n+1)}, y_{(n+1)})$ via regularization. In reality, we do not know $y_{(n+1)}$, but we can derive its conditional probability distribution using $\mathcal{D}_n$, the n data samples seen so far, and other quantities available to the function approximation task:

$$\mathcal{P}(y_{(n+1)}|\mathbf{x}_{(n+1)}, \mathcal{D}_n) \propto \int_{f \in \mathcal{F}} \mathcal{P}(\mathcal{D}_n \cup (\mathbf{x}_{(n+1)}, y_{(n+1)})|f)\mathcal{P}_{\mathcal{F}}(f)df \quad (20.5)$$

Now treating $y_{(n+1)}$ as a random variable, we obtain the following *expected* value for the new EISD, if we sample our next data point at $\mathbf{x}_{(n+1)}$:

$$\mathcal{U}(\hat{g}_{(n+1)}|\mathcal{D}_n, \mathbf{x}_{(n+1)}) = \int_{-\infty}^{\infty} \mathcal{P}(y_{(n+1)}|\mathbf{x}_{(n+1)}, \mathcal{D}_n) \qquad (20.6)$$

$$E_{\mathcal{F}}[\delta(\hat{g}_{(n+1)}, g)|\mathcal{D}_n \cup (\mathbf{x}_{(n+1)}, y_{(n+1)})]dy_{(n+1)}$$

Notice from Equation 20.4 that $E_{\mathcal{F}}[\delta(\hat{g}_{(n+1)}, g)|\mathcal{D}_n \cup (\mathbf{x}_{(n+1)}, y_{(n+1)})]$ in the above expression is actually independent of the unknown target function g, and so $\mathcal{U}(\hat{g}_{(n+1)}|\mathcal{D}_n, \mathbf{x}_{(n+1)})$ (henceforth referred to as the *total output uncertainty*) is fully computable from information already available in the function approximation learning model. Clearly, the optimal input location to sample next is the location that minimizes $\mathcal{U}(\hat{g}_{(n+1)}|\mathcal{D}_n, \mathbf{x}_{(n+1)})$, i.e.:

$$\hat{\mathbf{x}}_{(n+1)} = \arg \min_{\mathbf{x}_{(n+1)}} \mathcal{U}(g_{(n+1)}|\mathcal{D}_n, \mathbf{x}_{(n+1)}). \qquad (20.7)$$

20.2.4.   Summary of The Active Learning Procedure

Given $\mathcal{D}_n = \{(\mathbf{x_i}, y_i) \in \Re^d \times \Re | i = 1, \ldots, n\}$, the n data points seen so far in a function approximation task, and $\mathcal{P}_{\mathcal{F}}(f)$, the prior distribution of approximation functions in the target class $\mathcal{F}$, we summarize the key steps involved for finding the optimal next sample location $\hat{\mathbf{x}}_{(n+1)}$:

1 Compute $\mathcal{P}(g|\mathcal{D}_n)$. This is the same a-posteriori distribution used by *regularization* methods for finding $\hat{g}_n$, the current Bayes optimal estimate of the unknown target function g.

2 Assume a new point $\mathbf{x}_{(n+1)}$ to sample.

3 Assume a value $y_{(n+1)}$ at this $\mathbf{x}_{(n+1)}$. One can compute $\mathcal{P}(g|\mathcal{D}_n \cup (\mathbf{x}_{(n+1)}, y_{(n+1)}))$ and hence the *expected* integrated squared difference (EISD) between the target and its new estimate $\hat{g}_{(n+1)}$ given by $E_{\mathcal{F}}[\delta(\hat{g}_{(n+1)}, g)|\mathcal{D}_n \cup (\mathbf{x}_{(n+1)}, y_{(n+1)})]$ in Equation 20.4.

4 At the assumed $\mathbf{x}_{(n+1)}$, $y_{(n+1)}$ has a probability distribution given by Equation 20.5. Averaging the resulting EISD over all $y_{(n+1)}$'s, we obtain the *total output uncertainty* for $\mathbf{x}_{(n+1)}$, given by $\mathcal{U}(\hat{g}_{(n+1)}|\mathcal{D}_n, \mathbf{x}_{(n+1)})$ in Equation 20.6.

5 Sample at the input location $\hat{\mathbf{x}}_{(n+1)}$ that minimizes the cost function $\mathcal{U}(\hat{g}_{(n+1)}|\mathcal{D}_n, \mathbf{x}_{(n+1)})$.

Some final remarks about our example selection strategy: Intuitively, a reasonable selection criterion should choose new examples that provide dense information about the target function g. Furthermore, the choice should also take into account the learner's current state, namely $\mathcal{D}_n$ and $\hat{g}_n$, so as to maximize the *net* amount of information gained from each new data sample. The scheme here treats an approximation function's *expected misfit* with respect to the unknown target (i.e. their EISD) as a measure of *uncertainty* in the current solution. It selects new examples, based on the data that it has already seen to minimize the *expected* value of the resulting EISD measure. In doing so, it essentially maximizes the net amount of information gained with each new example.

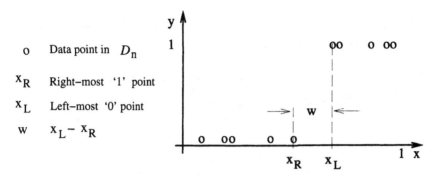

Figure 20.1 Diagram showing the notation used for our unit-step example.

20.3. COMPARING SAMPLE COMPLEXITY

To demonstrate the usefulness of the above active learning procedure, we show analytically and empirically that the active strategy learns target functions more accurately with fewer training examples than random sampling for three specific approximation function classes: (1) unit step functions, (2) polynomial approximators and (3) Gaussian radial basis function networks. For all three function classes, one can derive exact analytic data selection algorithms following the key steps outlined in Section 20.2.4.

20.3.1. Unit Step Functions

We first consider a target function class, $\mathcal{F}$, of one-dimensional unit-step functions described by a single parameter $0 \le a \le 1$:

$$u(x, a) = \begin{cases} 1 & \text{if } x \ge a \\ 0 & \text{otherwise} \end{cases}$$

Assume that $\mathcal{P}_{\mathcal{F}}(g(x) = u(x, a))$ is such that the parameter a has an a-priori uniform distribution in the range $0 \le a \le 1$. Suppose the data set, $\mathcal{D}_n = \{(x_i, y_i); i = 1, ..n\}$, is noiseless and consistent with some unknown target function, $g(x) = u(x, a)$, that the learner has to approximate. We want to find the best input location to sample next, $x \in [0, 1]$, that would return maximum information to better approximate the unknown target. Let x_R be the right most point in $\mathcal{D}_n$ whose y value is 0, i.e., $x_R = \max_{i=1,..n}\{x_i|y_i = 0\}$ (see Figure 20.1). Similarly, let $x_L = \min_{i=1,..n}\{x_i|y_i = 1\}$ and $w = (x_L - x_R)$. Following the general procedure outlined in Section 20.2.4, we go through the following steps:

1 Derive $\mathcal{P}(g|\mathcal{D}_n)$. One can show that:

$$P(g(x) = u(x, a)|\mathcal{D}_n) = \begin{cases} 1/w & \text{if } a \in [x_R, x_L] \\ 0 & \text{otherwise} \end{cases}$$

2 Suppose we sample next at a particular $x \in [0, 1]$, we would obtain y with the distribution:

$$P(y = 0 | \mathcal{D}_n, x) = \begin{cases} \frac{(x_L - x)}{x_L - x_R} = \frac{(x_L - x)}{w} & \text{if } x \in [x_R, x_L] \\ 1 & \text{if } x \leq x_R \\ 0 & \text{otherwise} \end{cases}$$

$$P(y = 1 | \mathcal{D}_n, x) = \begin{cases} \frac{(x - x_R)}{x_L - x_R} = \frac{(x - x_R)}{w} & \text{if } x \in [x_R, x_L] \\ 1 & \text{if } x \geq x_L \\ 0 & \text{otherwise} \end{cases}$$

3 For a particular y, the new data set would be $\mathcal{D}_{(n+1)} = \mathcal{D}_n \cup (x, y)$ and the corresponding EISD can be obtained using the distribution $\mathcal{P}(g | \mathcal{D}_{(n+1)})$. Averaging this over $\mathcal{P}(y | \mathcal{D}_n, x)$ as in **step 4** of the general procedure, we obtain:

$$\mathcal{U}(\hat{g}_{(n+1)} | \mathcal{D}_n, x) = \begin{cases} w^2/12 & \text{if } x \leq x_R \text{ or } x \geq x_L \\ \frac{(x_L - x)^3 + (x - x_R)^3}{12w} & \text{otherwise} \end{cases}$$

4 Clearly the new input location that minimizes $\mathcal{U}(\hat{g}_{(n+1)} | \mathcal{D}_n, x)$, the *total output uncertainty* measure, is:

$$\hat{x}_{(n+1)} = \arg \min_{x \in [0,1]} \mathcal{U}(\hat{g}_{(n+1)} | \mathcal{D}_n, x) = (x_L + x_R)/2$$

which is the midpoint between x_L and x_R.

Thus, by applying the general procedure to this trivial case of one-dimensional unit-step functions, we get the familiar binary search learning algorithm that queries the midpoint of x_R and x_L. For this function class, one can show analytically in PAC-style (Valiant, 1984) that this active data sampling strategy takes fewer examples to learn an unknown target function to a given level of *total output uncertainty* than randomly drawing examples according to a uniform distribution in x.

Theorem 1. Suppose we want to collect examples so that we are guaranteed with high probability (i.e. probability $> 1 - \delta$) that the *total output uncertainty* is less than ϵ. Then a passive learner would require at least $\frac{1}{\sqrt{48\epsilon}} \ln(1/\delta)$ examples while the active strategy described earlier would require at most $(1/2) \ln(1/12\epsilon)$ examples.

20.3.2. Polynomial Approximators

We consider next a univariate polynomial target and approximation function class with maximum degree K, i.e.:

$$\mathcal{F} = \left\{ p(x, \mathbf{a}) = p(x, a_0, \dots, a_K) = \sum_{i=0}^{K} a_i x^i \right\}$$

The model parameters to be learnt are $\mathbf{a} = [a_0 \ a_1 \ \dots \ a_K]^\mathsf{T}$ and x is the input variable. We assume that $\mathcal{F}$ has a zero-mean Gaussian distribution with covariance $\Sigma_{\mathcal{F}}$ on the model parameters $\mathbf{a}$:

$$\mathcal{P}_{\mathcal{F}}(p(x, \mathbf{a})) = \mathcal{P}_{\mathcal{F}}(\mathbf{a}) = \frac{1}{(2\pi)^{(K+1)/2} |\Sigma_{\mathcal{F}}|^{1/2}} \exp(-\frac{1}{2} \mathbf{a}^\mathsf{T} \Sigma_{\mathcal{F}}^{-1} \mathbf{a}) \quad (20.8)$$

Our task is to approximate an *unknown* target function $p \in \mathcal{F}$ within the input range $[x_{\mathrm{LO}}, x_{\mathrm{HI}}]$ on the basis of sampled data. Let $\mathcal{D}_n = \{(x_i, y_i = p(x_i, \mathbf{a}) + \eta) | i = 1, \dots, n\}$ be a *noisy* data sample from the unknown target in the input range $[x_{\mathrm{LO}}, x_{\mathrm{HI}}]$, where η is an additive zero-mean Gaussian noise term with variance σ_s^2. We compare two different ways of selecting the next data point: (1) sampling the function at a random point x according to a uniform distribution in $[x_{\mathrm{LO}}, x_{\mathrm{HI}}]$ (i.e. passive learning), and (2) using our active learning framework to derive an exact algorithm for determining the next sampled point.

20.3.2.1 The Active Strategy.

We go through the general active learning procedure outlined in Section 20.2.4 to derive an exact expression for $\hat{x}_{(n+1)}$, the next query point:

1 Let $\bar{\mathbf{x}}_\mathbf{i} = [1 \ x_i \ x_i^2 \ \dots \ x_i^K]^\mathsf{T}$ be a length $(K+1)$ power vector of the ith data sample's input value. The a-posteriori approximation function class distribution, $\mathcal{P}(\mathbf{a}|\mathcal{D}_n)$, is a multivariate Gaussian centered at $\hat{\mathbf{a}}$ with covariance Σ_n (see Appendix A.1.1 of (Sung and Niyogi, 1996)), where:

$$\hat{\mathbf{a}} = \Sigma_n \left(\frac{1}{\sigma_s^2} \sum_{i=1}^{n} \bar{\mathbf{x}}_\mathbf{i} y_i \right) \quad (20.9)$$

$$\Sigma_n^{-1} = \Sigma_{\mathcal{F}}^{-1} + \frac{1}{\sigma_s^2} \sum_{i=1}^{n} \left(\bar{\mathbf{x}}_\mathbf{i} \bar{\mathbf{x}}_\mathbf{i}^T \right) \quad (20.10)$$

2 Deriving the *total output uncertainty* expression requires several steps (see Appendix A.1.2 and A.1.3 of (Sung and Niyogi, 1996)). Taking advantage of the Gaussian distribution on both the parameters $\mathbf{a}$ and the noise term, we eventually get:

$$\mathcal{U}(\hat{p}_{(n+1)}|\mathcal{D}_n, x_{(n+1)}) \propto |\Sigma_{(n+1)}|, \qquad (20.11)$$

$\Sigma_{(n+1)}$ has the same form as Σ_n and depends on the previous data, the priors, noise and the next sample location $x_{(n+1)}$. When minimized over $x_{(n+1)}$, we get $\hat{x}_{(n+1)}$ as the maximum utility location where the active learner should next sample the unknown target.

20.3.2.2   Simulations — Error Rate versus Sample Size.

We perform simulations to compare the active strategy's sample complexity with that of a passive learner which receives uniformly distributed random training examples on the input domain: $[x_{\text{LO}}, x_{\text{HI}}] = [-5, 5]$.

Experiment 1: Learning with an Exact Prior Model.

We first show that our active example selection strategy learns an unknown target more accurately with less data than the passive strategy, if the *exact* prior distribution on the unknown target is known to the learning task. For this experiment, we set $K = 9$ for the target polynomial function class $\mathcal{F}$, and use a fixed multi-variate Gaussian prior model, $\mathcal{P}_{\mathcal{F}}(\mathbf{a})$, where $\Sigma_{\mathcal{F}}$ in Equation 20.10 is a $(K + 1) \times (K + 1)$ diagonal matrix whose i^{th} diagonal element is $s_{i,i} = 0.9^{2(i-1)}$. Qualitatively, this prior model favor smooth functions by assigning higher probabilities to polynomials with smaller coefficients, especially for higher powers of x.

We randomly generate 1000 target polynomial functions using the assumed prior, $\mathcal{P}_{\mathcal{F}}(\mathbf{a})$. For each target polynomial, we collect data sets with noisy y values ($\sigma^2 = 0.1$) ranging from 3 to 50 samples, using both the active and passive sampling strategies. We also use the same priors and Gaussian noise model to obtain the Bayes optimal estimate of the target polynomial for each data set. Because we know the actual target polynomial for each data set, one can compute the actual integrated squared difference between the target and its estimate as an approximation error measure. Figure 20.2(a) compares the two sampling strategies by separately averaging their approximation error rates for each data sample size over the 1000 different target polynomials.

Experiment 2: Learning with an Inexact Prior Model.

We next show that our active example selection strategy still learns an unknown target more accurately with less data than the passive strategy, even if the learning procedure assumes a prior distribution model that has a different form from that of the target class. Let $p(x, \mathbf{a}) \in \mathcal{F}$ be a polynomial in the approximation function class. One can quantify the overall "smoothness" of p by integrating its squared first derivative over the input domain $[x_{\text{LO}}, x_{\text{HI}}]$:

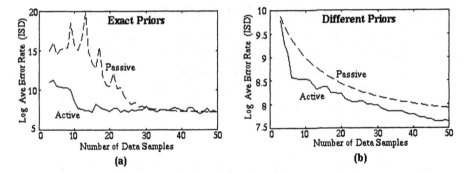

Figure 20.2 Comparing active and passive learning average error rates at different data sample sizes for polynomial approximators of maximum degree $K = 9$. In **(a)**, we use the same priors for generating and learning target functions. In **(b)**, we use slightly different priors.

$$Q(p(x, \mathbf{a})) = \int_{x_{LO}}^{x_{HI}} \left[\frac{dp(x, \mathbf{a})}{dx} \right]^2 dx. \qquad (20.12)$$

Using the "smoothness" measure above, one can define a prior distribution on $\mathcal{F}$ that also favors smoothly varying functions with small constant (a_0) terms (see Appendix A.1.4 of (Sung and Niyogi, 1996) for a detailed derivation):

$$
\begin{aligned}
\mathcal{P}'_{\mathcal{F}}(\mathbf{a}) \quad &\propto \quad \exp(-Q(p(x, \mathbf{a}))) \exp(-\frac{a_0^2}{2\sigma_0^2}) \\
&= \quad \frac{1}{(2\pi)^{(K+1)/2}|\Sigma_{\mathcal{F}}|^{1/2}} \exp(-\frac{1}{2}\mathbf{a}^T \Sigma_{\mathcal{F}}^{-1} \mathbf{a}) \qquad (20.13)
\end{aligned}
$$

$$
\Sigma_{\mathcal{F}}^{-1}(i, j) =
\begin{cases}
1/\sigma_0^2 & \text{if } i = j = 1 \\
2\frac{(i-1)(j-1)}{i+j-3}(x_{HI}^{i+j-3} - x_{LO}^{i+j-3}) & \text{if } 2 \le \{i, j\} \le (K+1) \\
0 & \text{otherwise}
\end{cases}
$$

We repeat the active versus passive function learning simulations by generating 1000 target polynomials according to the original distribution $\mathcal{P}_{\mathcal{F}}(\mathbf{a})$ from Experiment 1, but sampling (for the active case) and learning the unknown target from each data set according to the new "smoothness prior" $\mathcal{P}'_{\mathcal{F}}(\mathbf{a})$ with $\sigma_0 = 0.8$. We then average and compare their approximation errors in the same way as before. Figure 20.2(b) suggests that despite the inexact priors, the active learner still outperforms the passive strategy.

20.3.3.   Gaussian Radial Basis Functions

Our final example looks at an approximation function class $\mathcal{F}$ of d-dimensional Gaussian radial basis functions with K fixed centers. Let $\mathcal{G}_i$ be the ith basis function with a fixed center $\mathbf{c_i}$ and a fixed covariance $\mathcal{S}_i$. The model parameters to be learnt are the weight coefficients denoted by $\mathbf{a} = [a_1\, a_2\, \cdots\, a_K]^{\mathsf{T}}$. An arbitrary function $r \in \mathcal{F}$ in this class can thus be represented as:

$$r(\mathbf{x}, \mathbf{a}) = \sum_{i=1}^{K} a_i \mathcal{G}_i(\mathbf{x}) = \sum_{i=1}^{K} \frac{a_i}{(2\pi)^{d/2}|\mathcal{S}_i|^{1/2}} \exp(-\frac{1}{2}(\mathbf{x} - \mathbf{c}_i)^{\mathsf{T}} \mathcal{S}_i^{-1}(\mathbf{x} - \mathbf{c}_i))$$

We impose a prior $\mathcal{P}_{\mathcal{F}}(\mathbf{a})$ on the approximation function class by putting a zero-centered Gaussian distribution with covariance $\Sigma_{\mathcal{F}}$ on the model parameters $\mathbf{a}$. Thus, for an arbitrary function $r(\mathbf{x}, \mathbf{a})$:

$$\mathcal{P}_{\mathcal{F}}(r(\mathbf{x}, \mathbf{a})) = \mathcal{P}_{\mathcal{F}}(\mathbf{a}) = \frac{1}{(2\pi)^{K/2}|\Sigma_{\mathcal{F}}|^{1/2}} \exp(-\frac{1}{2}\mathbf{a}^{\mathsf{T}}\Sigma_{\mathcal{F}}^{-1}\mathbf{a})$$

Lastly, the learner has access to noisy data, $\mathcal{D}_n = \{(\mathbf{x_i}, y_i = r(\mathbf{x_i}, \mathbf{a}) + \eta) : i = 1, \ldots, n\}$, where g is an unknown target function and η is a zero-mean additive Gaussian noise term with variance σ_s^2. This leads to an expression for $\mathcal{P}(\mathcal{D}_n | r(\mathbf{x}, \mathbf{a}))$ for every $r(\mathbf{x}, \mathbf{a}) \in \mathcal{F}$ that is of the form:

$$\mathcal{P}(\mathcal{D}_n | r(\mathbf{x}, \mathbf{a})) \propto \exp\left(-\frac{1}{2\sigma_s^2}\sum_{i=1}^{n}(y_i - r(\mathbf{x_i}, \mathbf{a}))^2\right)$$

Learning a Bayes optimal estimate of the unknown target from $\mathcal{D}_n$ involves finding a set of model parameters $\hat{\mathbf{a}}$ that maximizes $\mathcal{P}(\mathbf{a}|\mathcal{D}_n) = \mathcal{P}_{\mathcal{F}}(\mathbf{a})\mathcal{P}(\mathcal{D}_n | r(\mathbf{x}, \mathbf{a}))$. Let $\bar{\mathbf{z_i}} = [\mathcal{G}_1(\mathbf{x_i})\, \mathcal{G}_2(\mathbf{x_i})\, \ldots\, \mathcal{G}_K(\mathbf{x_i})]^{\mathsf{T}}$ be a vector of RBF kernel output values for the ith input value. Regardless of how the n data points are selected, one can show (see Appendix A.2.1 of (Sung and Niyogi, 1996)) that:

$$\hat{\mathbf{a}} = \Sigma_n \left(\frac{1}{\sigma_s^2}\sum_{i=1}^{n}\bar{\mathbf{z}}_i y_i\right) \tag{20.14}$$

$$\Sigma_n^{-1} = \Sigma_{\mathcal{F}}^{-1} + \frac{1}{\sigma_s^2}\sum_{i=1}^{n}(\bar{\mathbf{z_i}}\bar{\mathbf{z_i}}^{\mathsf{T}}) \tag{20.15}$$

20.3.3.1   The Active Strategy.

As in the polynomial example, we go through the general active learning procedure outlined in Section 20.2.4 to derive an exact expression for $\hat{x}_{(n+1)}$, the optimal next query point.

1 Find an analytical expression for $\mathcal{P}(\mathbf{a}|\mathcal{D}_n)$. One can show that this a-posteriori distribution is a multivariate Gaussian density with mean $\hat{\mathbf{a}}$ and covariance Σ_n in Equation 20.15.

2 Deriving the RBF *total output uncertainty* cost function. This requires several steps as detailed in Appendix A.2.2 and A.2.3 of (Sung and Niyogi, 1996). We eventually get:

$$\mathcal{U}(\hat{r}_{(n+1)} | \mathcal{D}_n, \mathbf{x}_{(n+1)}) \propto |\Sigma_{(n+1)}|. \tag{20.16}$$

$\Sigma_{(n+1)}$ has the same form as Σ_n in Equation 20.15 and depends on the previous data sample locations $\{\mathbf{x}_i : i = 1, \ldots, n\}$, the model priors $\Sigma_{\mathcal{F}}$, the data noise variance σ_s^2, and the next sample location $\mathbf{x}_{(n+1)}$. When minimized over $\mathbf{x}_{(n+1)}$, we get $\hat{\mathbf{x}}_{(n+1)}$ as the maximum utility location where the active learner should next sample the unknown target function.

20.3.3.2 Simulations — Error Rate versus Sample Size.

As in the case for polynomial approximators, we perform simulations to compare the active strategy's sample complexity with that of a passive learner under the following two conditions:

Experiment 1: Learning with an Exact Prior Model.

For simplicity, we work in a one-dimensional input domain $[x_{\mathrm{LO}}, x_{\mathrm{HI}}] = [-5, 5]$. The approximation and target function classes are RBF networks with $K = 8$ fixed centers, arbitrarily located within the input domain. Each RBF kernel has a fixed 1-dimensional Gaussian "covariance" of $\mathcal{S}_i = 1.0$. Finally, we assume a prior model, $\mathcal{P}_{\mathcal{F}}(\mathbf{a})$, with an identical independent Gaussian distribution on the model parameters $\mathbf{a}$, i.e. $\Sigma_{\mathcal{F}} = \mathbf{I}_K = \mathbf{I}_8$, where $\mathbf{I}_K$ stands for a $K \times K$ *identity* covariance matrix.

The simulation proceeds as follows: We randomly generate 5000 target RBF functions according to $\mathcal{P}_{\mathcal{F}}(\mathbf{a})$. For each target function, we collect data sets with noisy y values ($\sigma_s = 0.1$) ranging from 3 to 50 samples in size, using both the active and passive sampling strategies. We then obtain a Bayes optimal estimate of the target function for each data set using Equation 20.14, and finally, we compute the actual integrated squared difference between the target and its estimate as an approximation error measure. Figure 20.3(a) shows that the active example selection strategy learns an unknown target more accurately with less data than the passive strategy.

Experiment 2: Learning with an Inexact Prior Model.

We use the same prior model above, $\mathcal{P}_{\mathcal{F}}(\mathbf{a})$, to generate the 5000 target RBF functions, but sample (for the active case) and learn the unknown target from each data set using a slightly different model class: (i) Each center is slightly displaced from its true location (i.e. its location in

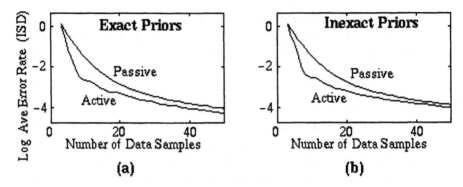

Figure 20.3 Comparing active and passive learning average error rates at different data sample sizes for Gaussian RBF approximators with $K = 8$ centers. In (**a**), we use the same priors for generating and learning target functions. In (**b**), we use slightly different priors.

the target function class) by a random distance with Gaussian standard deviation $\sigma = 0.1$. (ii) The learner's priors on model parameters, $\mathcal{P}'_{\mathcal{F}}(\mathbf{a})$, has a slightly different multivariate Gaussian covariance matrix: $\Sigma_{\mathcal{F}} = 0.9\mathbf{I}_8$. Figure 20.3(b) shows that despite slightly incorrect priors, the active example selection strategy still learns the unknown target more accurately with less data than the passive strategy.

20.4. INSTANCE SELECTION IN AN OBJECT DETECTION SCENARIO

Our active learning strategy chooses its next sample location by minimizing $\mathcal{U}(\hat{g}_{(n+1)}|\mathcal{D}_n, \mathbf{x}_{(n+1)})$ in Equation 20.6, which is an *expected misfit* cost function between the unknown target, g, and its new Bayes optimal estimate, $\hat{g}_{(n+1)}$. In general, this cost function may not have an analytical form for arbitrary target function classes, $\mathcal{F}$, and so the resulting active learning procedure can be computationally intractable. This section adapts the active learning formulation into a practical instance selection strategy for a real-world learning task, without departing too much from its original goal of selecting optimal utility data.

20.4.1. Sampling with A Simpler Example Utility Measure

We present a different example selection *heuristic* that uses a simpler but less comprehensive example utility measure, to efficiently reduce the *expected misfit* between g and $\hat{g}_{(n+1)}$ with each new data sample. In

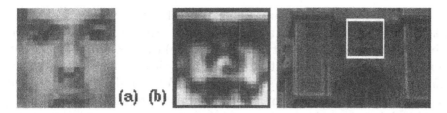

Figure 20.4 **(a):** A standard frontal face pattern. The face detector looks for square image patches with this general structure. **(b):** An example of a *high utility* non-face pattern that resembles a human face when viewed in isolation.

terms of this high-level learning objective, both this new heuristic and the original active learning formulation are very similar in spirit.

Consider the following *local* "uncertainty" measure for the current Bayes optimal estimate $\hat{g}_n$ at $\mathbf{x}$:

$$\mathcal{L}(\hat{g}_n|\mathcal{D}_n, \mathbf{x}) = \int_{g \in \mathcal{F}} \mathcal{P}(g|\mathcal{D}_n)(\hat{g}_n(\mathbf{x}) - g(\mathbf{x}))^2 dg \qquad (20.17)$$

Notice that unlike the EISD (see Equation 20.4) which measures *global* misfit between g and $\hat{g}_n$ over the entire input space, $\mathcal{L}(\hat{g}_n|\mathcal{D}_n, \mathbf{x})$ is just a *local* "error bar" measure between g and $\hat{g}_n$ only at $\mathbf{x}$. In a loose sense, this *local* "error bar" is information that the learner *lacks* at input location $\mathbf{x}$ for an exact solution. Given such an interpretation, one can also regard $\mathcal{L}(\hat{g}_n|\mathcal{D}_n, \mathbf{x})$ as an *example utility* measure, because the learner essentially gains this amount of information at $\mathbf{x}$ by sampling there. The new example selection *heuristic* can thus be formulated as minimizing misfit between g and $\hat{g}_{(n+1)}$ by choosing the next data sample where the new example's *utility value* is *currently* greatest, i.e., by sampling next where the learner *currently* lacks most information:

$$\mathbf{x}_{(n+1)} = \arg \max_{\mathbf{x} \in \Re^d} \mathcal{L}(\hat{g}_n|\mathcal{D}_n, \mathbf{x}) = \arg \max_{\mathbf{x} \in \Re^d} \int_{g \in \mathcal{F}} \mathcal{P}(g|\mathcal{D}_n)(\hat{g}_n(\mathbf{x}) - g(\mathbf{x}))^2 dg$$

$$(20.18)$$

20.4.2.   Human Face Detection

We now describe a real-world learning scenario where instance selection plays a critical role during training. Given a digitized image, we want a computer vision algorithm that locates all upright frontal views of human faces in the image. Computational efficiency aside, one can perform the task by searching the image exhaustively (over all locations and across multiple scales) for square patches having a standard "face-like" structure similar to Figure 20.4(a).

The learning part involves training a classifier to recognize such "face-like" image patterns using a comprehensive but tractably small set of positive (i.e. face) and negative (i.e. non-face) examples. For "face" patterns, one can simply collect all frontal face views from image archives, and still have a manageably small data set. For "non-face" patterns, any image window that does not tightly contain a frontal human face is a valid training example, so our negative example set can grow intractably large without performing instance selection.

20.4.3.   The "Boot-Strap" Paradigm

To identify high utility "non-face" pattern for our training database, we outline a variant of the example selection heuristic in Equation 20.18, which we shall henceforth refer to as the "boot-strap" paradigm (Sung and Poggio, 1998):

1 Start with a small and possibly highly non-representative set of "non-face" examples in the training database.

2 Train a classifier to identify face patterns from non-face patterns, using examples from the current database of labeled face and non-face patterns.

3 Run the trained face detector on images with no faces. Collect one or more "non-face" patterns that the current system mis-classifies as "faces" (see for example Figure 20.4(b)).

4 Add these "non-face" patterns to the training database as new negative examples and return to Step 2.

Note that one can also use the "boot-strap" paradigm to select useful training examples of face patterns, if necessary.

In the same spirit as the previously discussed simplified example selection heuristic, the "boot-strap" paradigm selects high utility data for further training, where utility is also measured in terms of *local* information gained. Instead of directly computing $\mathcal{L}(\hat{g}_n|\mathcal{D}_n, \mathbf{x})$, the *local* "error bar" utility measure in Equation 20.17 (which may still not have an analytical form for complex target function classes), "boot-strapping" takes advantage of the already available output value (i.e., class label) at each candidate location to implement a coarse but very simple *local* "error bar" measure for selecting new data. Points that are classified correctly have low actual output errors, and are assumed to also have low *local* uncertainty "error bars". They are discarded as examples containing little new information. Conversely, points that are wrongly classified have high actual output errors. They are conveniently assumed to also have

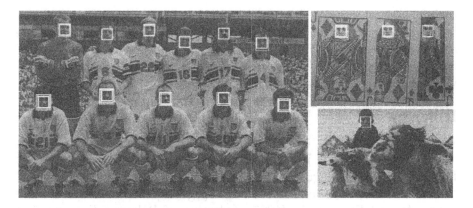

Figure 20.5 Some face detection results by the frontal view system, trained using a "boot-strap" strategy for choosing useful non-face examples.

high *local* uncertainty "error bars", and are therefore selected as high utility examples for the learning task.

20.4.4. Training Results

Using a Radial Basis Function-like network architecture for pattern classification, we trained a frontal view face detection system (see (Sung and Poggio, 1998) for details) in two "boot-strap" cycles. The final training database contains 4150 "face" patterns and over 40000 "non-face" patterns, of which about 6000 were selected during the second "boot-strap" cycle. As a control, we trained a second system without "boot-strapping". The second training database contains the same 4150 "face" patterns and another 40000 randomly selected "non-face" patterns.

We ran both systems on a test database of 23 cluttered images with 149 frontal faces. The first system detected 126 faces (84.6%) and produced 13 false detects, while the second system found only 83 faces (55.7%) but produced 56 false positives. The results suggest that despite possibly being "sub-optimal" at choosing new training patterns, the "boot-strap" strategy is still a very simple and effective technique for sifting through unmanageably large data sets for useful examples.

20.5. CONCLUSION

The active learning formulation presented in this book chapter is based on sampling new data according to a theoretical optimality criterion for learning an unknown target function. It is desirable that such frameworks eventually translate into practical algorithms that can be used in complex learning scenarios. As an attempt to bridge the gap between theory and practice, we have argued how the "boot-strap" paradigm can be derived from our active learning framework and applied with some success within a sophisticated face detection system.

References

Aha, D. W., Kibler, D., and Albert, M. K. (1991). Instance-based Learning Algorithms. *Machine Learning*, 6:37-66.

Bertero, M. (1986). Regularization Methods for Linear Inverse Problems. In C. Talenti, ed, *Inverse Problems*. Sprg-Vrlg.

Blum, A. and Langley, P. (1997). Selection of Relevant Features and Examples in Machine Learning. *Artificial Intelligence*, 97:245-271.

Cohn, D. (1991). A Local Approach to Optimal Queries. In D. Touretzky, editor, In *Proc. Connectionist Summer School*.

Duda, R. and Hart, P. (1973). *Pattern Classification and Scene Analysis*. John Wiley and Sons Inc., New York.

Fayyad, U., Piatetsky-Shapiro, G. and Smyth, P. (1996). From Data Mining to Knowledge Discovery: An Overview. In *Proc. Advances in KDD*.

Fedorov, V. (1972). *Theory of Optimal Experiments*. Academic Press.

MacKay, D. (1992). *Bayesian Methods for Adaptive Models*. PhD thesis, California Institute of Technology, USA.

Musick, R., Catlett, J. and Russel, S. (1993). Decision Theoretic Subsampling for Induction on Large Databases. In *Proc. International Conference on Machine Learning*.

Poggio, T. and Girosi, F. (1989). A Theory of Networks for Approximation and Learning. AIM–1140, MIT AI Lab.

Rumelhart, D. and McClelland, J. (1986). *Parallel Distributed Processing*, volume 1. MIT Press.

Sollich, P. (1994). Query Construction, Entropy, Generalization in Neural Network Models. *Physical Review E*, 49:4637–4651.

Sung, K. and Niyogi, P. (1996). A Formulation for Active Learning with Appl. to Obj. Detection. AIM-1438, MIT-AI.

Sung, K. and Poggio, T. (1998). Example-based Learning for View-based Human Face Detection. *IEEE. Trans. Pattern Analysis and Machine Intelligence*, 20:1.

Syed, N., Liu, H. and Sung, K. (1999). A Study of SVMs on Model Independent Example Selection. In *Proc. SIGKDD*.

Tikhonov, A. and Arsenin, V. (1977). *Solutions of Ill-Posed Problems*. W. H. Winston, Washington, DC.

Valiant, L. (1984). A Theory of Learnable. In *Pr. STOC*.

Chapter 21

FILTERING NOISY INSTANCES AND OUTLIERS

Dragan Gamberger
Rudjer Bošković Institute
Bijenička 54, 10000 Zagreb, Croatia
Dragan.Gamberger@irb.hr

Nada Lavrač
Jožef Stefan Institute
Jamova 39, 1000 Ljubljana, Slovenia
Nada.Lavrac@ijs.si

Abstract   Instance selection methods are aimed at finding a representative data subset that can replace the original dataset but still provide enough information to solve a given data mining task. If instance selection is done by sampling, the sample should preferably exclude noisy instances and outliers. This chapter presents methods for noise and outlier detection that can be incorporated into sampling as filters for data cleaning. The chapter presents the following filtering algorithms: a saturation filter, a classification filter, a combined classification-saturation filter, and a consensus saturation filter. The distinguishing feature of the novel consensus saturation filter is its high reliability which is due to the multiple detection of outliers and/or noisy instances. Medical evaluation in the problem of coronary artery disease diagnosis shows that the detected instances are indeed noisy or non-typical class representatives.

Keywords: Noise, outliers, saturation filter, classification filter, literals.

21.1.   INTRODUCTION

Instance selection is an important part of the KDD process. It is aimed at finding a representative data subset that can replace the original dataset, still solving a data mining task as if the whole dataset were used. For most data mining tasks, including classification tasks, the selected dataset should preferably exclude noisy instances.

Effective noise handling is an important task in inductive machine learning. Prediction accuracy and applicability of induced knowledge significantly depend on appropriate noise handling procedures. Noise usually means random errors in training examples (erroneous attribute values and/or erroneous classification). In this work we use the term noise in a broader sense of outliers (Weisberg, 1985) denoting all examples that do not follow the same model as the rest of the data. This definition "includes not only erroneous data but also *surprising* veridical data" (John, 1995), including non-typical class representatives.

A *target concept* of a given problem domain is pre-defined as the source of all possible correct examples of the concept. An inductive learning task is to find a good representation of the target concept in a selected hypothesis language. This representation is called a *target theory*. The main property of a target theory is that it should be correct for all the correct domain examples. A domain may consist of instances of a single concept, a set of subconcepts, a main concept and some subconcepts, etc. These domain characteristics together with the restrictions of the used hypothesis language may cause that a target theory may either have the form of a single concept description, a set of subconcept descriptions, or a main concept description and its exceptions (descriptions of small subconcepts or outlier descriptions).

In an ideal inductive learning problem, the induced hypothesis H is indeed a target theory that will 'agree' with the classifications of training examples and will thus perform as a perfect classifier on yet unseen instances. In practice, however, it frequently happens that available data contain various kinds of errors, either random or systematic. Random errors are usually referred to as noise. As mentioned above, the term noise is used here in a broader sense that includes errors (incorrect examples) as well as exceptions (correct examples representing some relatively rare subconcept of the target concept). The reasoning behind this decision is that, from the point of view of induction, exceptions have the same effect on the induction process as erroneous examples themselves.

Detection and filtering of noisy examples from the training set helps in the induction of the target hypothesis - a hypothesis induced from noiseless data will be less complex and more accurate when classifying unseen cases. For small datasets, noise filtering can be used as an important step in the preprocessing of training data; for large datasets, on the other hand, the sampling process that selects a subset of the training set should include or should be followed by noise filtering. Notice, however, that if noise filtering is combined with random sampling, the following phenomenon has to be taken into account. In domains with uneven concept (or class) distributions random sampling may result in only few instances of some subconcepts to be present in the sample (John and Langley, 1995). Such instances, despite the fact that they are important representatives of a rare subconcept, are outliers in the sample. This observation is similar to the side-effect noticed in instance selection approaches based on critical and boundary points, like instance-based learning (Aha et al., 1991) and support vector machines (Burges, 1998), which tend to increase the relative number of noisy instances in the selected subset.

The above observations support the development of cautious procedures for explicit detection and filtering of noisy instances, as presented in this chapter. This work upgrades the earlier work of the authors on Occam's razor applicability for noise detection and elimination (Gamberger and Lavrač, 1997). The chapter first elaborates on the background and related work, and then presents various noise filtering algorithms, that are used for noise handling in our rule learning system ILLM (Inductive Learning by Logic Minimization) (Gamberger, 1995).

The chapter presents a series of noise detection experiments performed in the problem of diagnosis of coronary artery disease (Grošelj et al., 1997), using ILLM without noise handling, using either the saturation filter or the classification filter with ILLM, and by upgrading ILLM with two novel combined approaches: the combined classification-saturation filter and the consensus saturation filter. The best results were achieved by the latter, employing various levels of consensus achieved by different ways of n-fold saturation filtering, aimed at increasing the reliably of the detection of noisy instances.

21.2.   BACKGROUND AND RELATED WORK

In machine learning, numerous approaches to noise handling have been developed. The most popular noise handling approaches, such as rule truncation and tree pruning (Quinlan, 1987; Mingers, 1989; Quinlan, 1993) learn from a noisy dataset and try to avoid overfitting the noisy instances. These 'noise-tolerant' approaches treat erroneous instances and exceptions as being mislabeled. They do not distinguish between noisy instances and exceptions. The problem of distinguishing exceptions from noise, and hypothesis generation including the representation of exceptions has been studied by other authors (Srinivasan et al., 1992; Dimopulos and Kakas, 1995).

Our work follows the Occam's razor principle, commonly attributed to William of Occam (early 14th century), that states: "Entities should not be multiplied beyond necessity." This principle is generally interpreted as: "Among the theories that are consistent with the observed phenomena, one should select the simplest theory" (Li and Vitányi, (1993)). The Occam's razor principle suggests that among all the hypotheses that are correct for all (or for most of) the training examples one should select the simplest hypothesis; it can be expected that this hypothesis is most likely to capture the structure inherent in the problem and that its high prediction accuracy can be expected on objects outside the training set (Rissanen, 1978). This principle is frequently used by noise handling algorithms (including the above-mentioned rule truncation and tree pruning algorithms) because noise handling aims at simplifying the generated rules or decision trees in order to avoid overfitting a noisy training set.

Despite the successful use of the Occam's razor principle as the basis for hypothesis construction, several problems arise in practice. First is the problem of the definition of the most appropriate complexity measure that will be used to identify the simplest hypothesis, since different measures can select different simplest hypotheses for the same training set. This holds for any Kolmogorov complexity based measure (Li and Vitányi, (1993)), including MDL (Minimal Description Length) (Rissanen, 1978), that use approximations of an ideal measure of complexity. Second, recent experimental work has undoubtly shown that applications of the Occam's razor may not always lead to best prediction accuracy. A systematic way of improving the results generated by the application of the Occam's razor principle has been presented by Webb (Webb, 1996). Further empirical evidence against the use of Occam's razor principle is provided by boosting and bagging approaches (Quinlan1996) in which an ensemble of classifiers (a more complex hypothesis) typically achieves better accuracy than any single classifier. In addition to experimental evidence, much disorientation was caused by the so-called "conservation law of generalization performance" (Schaffer, 1994). Although it is rather clear (Rao et al., 1995) that real-word learning tasks are different from the set of all theoretically possible learning tasks as defined in (Schaffer, 1994), there remains the so-called "selective superiority problem" that each algorithm performs best in some but not all domains (Brodley, 1995); the latter can be in a way explained by the bias-variance tradeoff (Kohavi and Wolpert, 1996). Some of the above contradictory observations have been recently explained by Domingos (Domingos, 1999).

Our work, reported in (Gamberger and Lavrač, 1997), contributes to the understanding of the abovementioned phenomena by studying for which domains (real or theoretically possible) an inductive learning algorithm based on the Occam's razor principle has a theoretical chance for successful induction. The paper (Gamberger and Lavrač, 1997) theoretically elaborates the conditions for Occam's razor applicability. For practical applications, the paper indicates the properties that need to be satisfied for the effective induction using the Occam's razor principle. This theoretical work is the basis for the *saturation filter* used for explicit noise and outlier detection presented in this chapter. The approach to explicit detection and filtering of noise and outliers is based on the observation that the elimination of noisy examples, in contrast to the elimination of examples for which the target theory is correct, reduces the CLCH value of the training set (CLCH stands for the Complexity of the Least Complex correct Hypothesis). This noise detection algorithm is called the saturation filter since it employs the CLCH measure to test whether the training set is saturated, i.e., whether, given a selected hypothesis language, the dataset contains a sufficient number of examples to induce a stable and reliable target theory.

An alternative approach to explicit noise detection and elimination has been recently suggested by Brodley and Friedl (Brodley and Friedl, 1999). The basic idea of their noise elimination algorithm, called in this chapter the *classification filter*, is to use one or more learning algorithms (that may but do not have to include explicit noise handling) to create classifiers that serve as filters for the training data. The experimental results published in (Brodley and Friedl, 1999) provide evidence that classification filters can successfully deal with noise. It has also been shown that consensus filters are conservative at eliminating good data while majority filters are better at detecting bad data that makes them more appropriate for situations with an abundance of data.

The reader interested in a comprehensive list of references to related work can find it in the related work section of the Brodley and Friedl paper.

21.3. NOISE FILTERING ALGORITHMS

Four noise filtering algorithms are presented, aimed at the detection and elimination of noisy instances: our noise filtering algorithm (here called the saturation filter (Gamberger and Lavrač, 1997; Gamberger et al., 2000)), the noise filtering algorithm by Brodley and Friedl (here called the classification filter (Brodley and Friedl, 1999)) and two new algorithms that combine the two approaches to noise filtering.

21.3.1.   Saturation Filter

Our approach to noise filtering has its theoretical foundation in the saturation property of training data (Gamberger and Lavrač, 1997). The algorithm is outlined here for the sake of completeness. For a more detailed description of the saturation filter and the results of experiments in a number of UCI domains, the reader is referred to (Gamberger et al., 2000).

Suppose that a hypothesis complexity measure c is defined and that for any hypothesis H its complexity $c(H)$ can be determined. Based on this complexity measure, for a training set E one can determine the complexity of the least complex hypothesis correct for all the examples in E; this complexity, denoted by $g(E)$ is called the CLCH value (Complexity of the Least Complex Hypothesis, correct for all the examples in E).

In (Gamberger and Lavrač, 1997) we have shown that if E is noiseless and saturated (containing enough training examples to find a correct target hypothesis), then $g(E) < g(E_n)$, where $E_n = E \cup \{e_n\}$ and e_n is a noisy example for which the target hypothesis is not correct. The property $g(E) < g(E_n)$ means that noisy examples can be detected as those that enable CLCH value reduction. The approach in an iterative form is applicable also when E_n includes more than one noisy example.

It must be noted that the saturation property of a training set is the main theoretical condition for the presented filter. In practice many domains have a restricted number of training examples and hence we may assume that these domains do not satisfy the saturation condition. Notice, however, that the described algorithm is applicable without changes also in this case. The reason is that for some subconcept of the domain there may still be enough training examples so that this subpart of the domain is saturated; hence, a subconcept description is induced by the learner whereas all other examples are eliminated since the learner will treat them as being erroneous or exceptions of the subtheory being learned.

The greatest practical problem of the saturation filter is the computation of the CLCH value $g(E)$ for a training set E. In rule-based induction, the hypothesis complexity measure $c(H)$ can be defined as the number of attribute value tests (literals) used in the hypothesis H. In this case, the corresponding $g(E)$ value can be defined as the minimal number of literals that are necessary to build a hypothesis that is correct for all the examples in E.

Suppose that the training set is contradiction free (there are no examples that differ only in their class value), and that the set of literals L defined in the hypothesis language is sufficient for finding a hypothesis H that is correct for all examples in E. Also, suppose that a set of all possible pairs of examples U is defined such that the first example in the pair is a positive example from E and the second one is negative. The importance of such pairs follows from the fact that only differences among positive and negative examples determine properties of the induced hypothesis H. A literal *covers* a pair of examples if it is *true* for the positive and *false* for the negative example in the pair (Lavrač et all., 1998). A literal covering many example pairscan be recognized as more important than another literal covering only a few examples pairs while literals that do not cover any pair of examples can be directly detected as useless. The necessary and sufficient condition for a subset $L' \subseteq L$ to be sufficient for finding a hypothesis H that is correct for all examples in E is that for every possible pair of examples there must be at least one literal in L' that covers the pair. The proof of the theorem is in (Gamberger and Lavrač, 1997) and (Gamberger et al., 2000). This fact enables that the $g(E)$ value defined by the minimal number of literals can be computed by any minimal covering algorithm over the set of example pairs. In this work, the ILLM heuristic minimal covering algorithm, presented here as Procedure 1, is used for the computation of the minimal L'. The advantages of this approach are: $g(E) = \min |L'|$ computation does not require the actual construction of a hypothesis and the $g(E)$ value can be determined relatively fast. This approach presents the heart of the saturation noise filtering method.

The procedure starts with the empty set of selected literals L' (step 1) and the set U' of yet uncovered example pairs equal to all possible pairs of one positive and one negative example from the training set (step 3), for which in (step 2) weights $v(e_i, e_j)$ have been computed. The weight of a pair is high if the pair is covered by a small number of distinct literals from L (small z value in step 2). The meaning of this measure is that for a 'heavy' example pair there are only a few literals covering the pair and it will be more difficult to find an appropriate literal that will cover this pair than for a pair with a small weight.

Each iteration of the main algorithm loop (steps 4 – 11) adds one literal to the minimal set L' (step 9). At the same time, all example pairs covered by the selected literal are eliminated from U' (step 10). The algorithm terminates when U' remains empty. In each iteration we try to select the literal that covers a maximal number of 'heavy' example pairs (pairs with high weight).

This is achieved by detecting a pair (e_a, e_b) that is covered by the least number of literals (step 5). At least one of the literals from the set L_{ab} with literals that cover this pair (step 6), must be included into the minimal set L'. To determine this literal, for each of them weight $w(l)$ is computed (step 7) and the literal with the maximal weight is selected (step 8). The weight of a literal is the sum of the weights of example pairs that are covered by the literal.

Procedure 1: *MinimalCover(U)*
Input: U (set of example pairs), L (set of literals)
Output: L' (minimal set of literals)

$L' \leftarrow \emptyset$ (1)

for every $(e_i, e_j) \in U$ compute weights   (2)
    $v(e_i, e_j) = 1/z$, where z is the number of
    literals $l \in L$ that cover (e_i, e_j)

$U' \leftarrow U$ (3)

while $U' \neq \emptyset$ **do**   (4)
  select (e_a, e_b) (5)
    $(e_a, e_b) \in U'$: $(e_a, e_b) = arg \quad max \quad v(e_i, e_j)$,
    where *max* is over all $(e_i, e_j) \in U'$
  $L_{ab} \leftarrow \{l \mid l \in L \text{ covering } (e_a, e_b)\}$ (6)
  for every $l \in L_{ab}$ compute   (7)
    $w(l) = \sum v(e_i, e_j)$, where *sum* is over all
    $(e_i, e_j) \in U'$ covered by l
  select literal l_s: $l_s = arg \quad max \quad w(l)$,   (8)
    where *max* is over all $l \in L_{ab}$
  $L' \leftarrow L' \cup \{l_s\}$ (9)
  $U' \leftarrow U' \setminus \{\text{all } (e_i, e_j) \text{ covered by } l_s\}$ (10)
end while   (11)

Algorithm 1 presents the saturation filter. It begins with the reduced training set E' equal the input training set E and an empty set of detected noisy examples A (step 1). The algorithm supposes that the set of all appropriate literals L for the domain is defined. U represents a set of all possible example pairs where the first example in the pair is from the set of all positive training examples P' in the reduced set E', and the second example is from the set N' of all negative examples in the reduced training set E'. The algorithm detects one noisy example per iteration. The base for noise detection are weights $w(e)$ that are computed for each example e from E'. Initially all $w(e)$ values are initialized to 0 (step 5). At the end, the example with maximum weight $w(e)$ is selected (step 16). If the maximum $w(e)$ value is greater than the parameter ε_h predefined value then the corresponding training example is included into the set A and eliminated from the reduced training set E'

(step 18). The new iteration of noise detection begins with this reduced training set (steps 2–20). The algorithm terminates when in the last iteration no example has $w(e)$ greater than ε_h. Noisy examples in A and the noiseless E' are the output of the algorithm.

Algorithm 1: *SaturationFilter(E)*
Input: E (training set), L (set of literals)
Parameter: ε_h (noise sensitivity parameter)
Output: A (detected noisy subset of E)

$E' \leftarrow E$ and $A \leftarrow \emptyset$	(1)
while $E' \neq \emptyset$ **do**	(2)
determine set U of all possible example pairs	(3)
for examples in E'	
call **Procedure 1** to find minimal L', $L' \subseteq L$ so that	(4)
$\forall (e_i, e_j) \in U \; \exists\, l \in L'$ with the property l covers (e_i, e_j)	
initialize $w(e) \leftarrow 0$ for all $e \in E'$	(5)
for every $l \in L'$ **do**	(6)
$P^* \leftarrow \emptyset,\ N^* \leftarrow \emptyset$	(7)
for every $(e_i, e_j) \in U$	(8)
if (e_i, e_j) covered by l and no other literal from L'	(9)
then $P^* \leftarrow P^* \cup \{e_i\},\ N^* \leftarrow N^* \cup \{e_j\}$	(10)
end for	(11)
if $P^* = \emptyset$ **then** $L' \leftarrow L' \setminus \{l\}$ and **goto step 5**	(12)
for every $e \in P^*$ **do** $w(e) \leftarrow w(e) + \frac{1}{|P^*|}$	(13)
for every $e \in N^*$ **do** $w(e) \leftarrow w(e) + \frac{1}{|N^*|}$	(14)
end for	(15)
select example e_s: $e_s = arg\ max\ w(e)$,	(16)
where max is computed over all $e \in E'$	
if $w(e_s) > \varepsilon_h$ **then**	(17)
$A \leftarrow A \cup \{e_s\}$ and $E' \leftarrow E' \setminus \{e_s\}$	(18)
else exit with generated sets A and E'	(19)
end while	(20)

Computations in each iteration begin with the search for the minimal set of literals L' that cover all example pairs in U (calling Procedure 1 in step 4). A pair of examples is covered by a literal l if the literal is evaluated *true* for the positive example and evaluated *false* for the negative example in the pair. This step represents the computation of the $g(E')$ value. Next, a heuristic approach is used to compute weights $w(e)$ that measure the possibility that the elimination of an example e would enable $g(E')$ reduction. Weights $w(e)$ are computed so that for every literal l from L', minimal sets of positive (P^*) and negative examples (N^*) are determined, such that if P^* or N^* are eliminated from E' then l becomes unnecessary in L'. This is done in a loop (steps 8–11) in which every example pair is tested if it is covered by a single literal l. If such a pair is detected (step 9) then its positive example is included into the set P^* and its negative example into the set N^* (step 10). Literal elimination from L' presents the reduction of the $g(E')$ value. If a literal can be made unnecessary by the elimination of a very small subset of training examples, then this indicates that these examples might be noisy. In steps 13 and 14, the $w(e)$ weights are incremented only for the examples that are the members of the P^* and N^* sets. The weights are incremented by the inverse of the total number of examples in these sets. Weights are summed over all literals in L'. Step 12 is necessary because of the imperfectness of the heuristic minimal cover algorithm. Namely, if some $l \in L'$ exists for which there is no example pair that is covered only by this literal (i.e., for which either $P^* = \emptyset$ or $N^* = \emptyset$), this means that L' is actually not the minimal set because $L' \setminus \{l\}$ also covers all example pairs in U. In such case L' is substituted by $L' \setminus \{l\}$.

The presented saturation filter uses the parameter ε_h that determines noise sensitivity of the algorithm. The parameter can be adjusted by the user in order to tune the algorithm to the domain characteristics. Reasonable values are between 0.25 and 2. For instance, the value 1.0 guarantees the elimination of every such example by whose elimination the set L' will be reduced for at least one literal. Lower ε_h values mean greater sensitivity of the algorithm (i.e., elimination of more examples): lower ε_h values should be used when the domain noise is not completely random, and when dealing with large training sets (since statistical properties of noise distribution in large training sets can have similar effects). In ILLM the default values of ε_h are between 0.5 and 1.5, depending on the number of training examples in the smaller of the two subsets of E: the set of positive examples P or the set of negative examples N. Default values for the saturation filter's noise sensitivity parameter ε_h are: 1.5 for training sets with 2–50 examples, 1.0 for 51–100 examples, 0.75 for 101–200 examples, and 0.5 for more than 200 examples.

21.3.2. Classification Filter

Algorithm 2: $ClassificationFilter(E)$
Input: E (training set)
Parameter: n (number of subsets, typically 10)
Output: A (detected noisy subset of E)
form n disjoint almost equally sized   (1)
   subsets E_i, where $\cup_i E_i = E$
  $A \leftarrow \emptyset$ (2)
  for $i = 1, \ldots, n$ do   (3)
    form $E_y \leftarrow E \setminus E_i$ (4)
    induce H_y based on examples in E_y (5)
     (using some inductive learning system)

for every $e \in E_i$ **do**	(6)
if H_y incorrectly classifies e	(7)
then $A \leftarrow A \cup \{e\}$	
end for	(8)
end for	(9)

The n-fold cross-validation is a substantial part of this algorithm. The classification filter begins with n equal-sized disjoint subsets of the training set E (step 1) and the empty output set A of detected noisy examples (step 2). The main loop (steps 3–9) is repeated for each training subset E_i. In step 4, subset E_y is formed that includes all examples from E except those in E_i. Set E_y is used as the input for an arbitrary inductive learning algorithm that induces a hypothesis (a classifier) H_y (step 5). Examples from E_i for which the hypothesis H_y does not give the correct classification, are added to A as potentially noisy (step 7).

21.3.3.   Combined Classification-Saturation Filter

A combined classification-saturation algorithm is very similar to the original classification filtering approach. The only difference is that saturation filtering is used for every subset E_y in order to eliminate noise from E_y. The intention of this modification is the induction of more appropriate hypotheses H_y and more reliable noise detection based on these. The combined classification-saturation filter algorithm is the same as the classification filter with added saturation-based filtering between its steps 4 and 5.

21.3.4.   Consensus Saturation Filter

Algorithm 3: *ConsensusSaturationFilter(E)*
Input: E (training set)
Parameter: n (number of subsets, default $= 10$)
Parameter: v (consensus level, default $= n - 1$)
Output: A (detected noisy subset of E)

form n disjoint almost equally sized	(1)
subsets E_i, where $\cup_i E_i = E$	
$A \leftarrow \emptyset$	(2)
for $i = 1, \ldots, n$ **do**	(3)
form $E_y \leftarrow E \setminus E_i$	(4)
$A_i \leftarrow SaturationFilter(E_y)$	(5)
end for	(6)
for every $e \in E$ **do**	(7)
$c \leftarrow 0$	(8)
for $i = 1, \ldots, n$ **if** $e \in A_i$, $c \leftarrow c + 1$	(9)
if $c \geq v$, $A \leftarrow A \cup \{e\}$	(10)

end for   (11)

The n-fold cross-validation method is also a substantial part of the consensus saturation filtering algorithm. Like in the combined classification–saturation filter, subset E_y is constructed (step 4) and noise eliminated from it by a saturation filter. In contrast to the combined classification-saturation filter, this reduced E_y is not used for hypothesis induction. Instead, detected noise is simply saved in the set A_i (step 5). When the procedure is performed for all E_i subsets, then n generated A_i sets are used to determine the elements of the output set A (steps 7–11). Since the same example e occurs in $n - 1$ of E_y subsets, in an ideal case the same noisy example may occur in $n - 1$ sets A_i. In this situation, detected by counter c (steps 9–11), example e is added to the output noisy set A (step 10), since the consensus of all A_i sets is reached. In practice we may allow that an example is detected as noisy and added to A also in cases when it is included in less than $n - 1$ sets A_i. Parameter v with values less than $n - 1$ (e.g., $n - 2$, $n - 3$) enables this possibility.

21.4.   EXPERIMENTAL EVALUATION

A problem of diagnosis of coronary artery disease (Grošelj et al., 1997) is used to perform a series experiments with noise filtering methods. Medical evaluation in this section shows that the detected instances are indeed noisy or non-typical class representatives.

21.4.1.   Domain Description: Diagnosis of Coronary Artery Disease

Coronary artery disease (CAD) is a result of diminished blood flow through coronary arteries due to stenosis or occlusion. The consequence of CAD is an impaired function of the heart and possible necrosis of the myocardium (myocardial infarction).

The dataset, collected at the University Medical Center, Ljubljana, Slovenia, includes 327 patients (250 men and 77 women, mean age 55 years). Each patient had performed clinical and laboratory examinations including ECG during rest and exercise, myocardial perfusion scintigraphy and finally the coronary angiography that provides for the actual diagnosis of coronary artery disease. In 229 patients, CAD was confirmed by angiography, and for 98 patients it was not confirmed. The patients' clinical and laboratory data are described by 77 attributes. This dataset was previously used for inducing diagnostic rules by a number of machine learning algorithms (Grošelj et al., 1997; Kukar et al., 1997).

21.4.2. Experimental Setting

According to the standard 10-fold cross-validation procedure, the original data set was partitioned into 10 folds with 32 or 33 examples each. Training sets are built from 9 folds, with the remaining fold as a test set. Let G denote the entire set of training examples, T_i is an individual test set (consisting of one fold), and G_i the corresponding training set ($G_i \leftarrow G \setminus T_i$, composed on nine folds). In this way, 10 training sets G_0 - G_9, and 10 corresponding test sets, T_0 - T_9, were constructed. Every example occurs exactly once in a test set, and 9 times in training sets.

Different noise detection procedures were used on the training sets and after the elimination of potentially noisy examples and hypothesis generation the prediction accuracy was measured on the test sets. The hypothesis was always constructed with the same algorithm so that the differences in the obtained prediction accuracy reflect only the differences in noise detection. The used rule construction algorithm was the ILLM (Inductive Learning by Logic Minimization) system used without its noise handling capability. The ILLM rule learning algorithm is similar to the standard covering algorithms for rule construction (Clark and Niblett, 1989; Fürnkranz, 1999).

In the first test we used ILLM (without its noise handling capability) to induce rules on complete G_0 - G_9 training sets. On the test sets T_0 - T_9 we measured the number of prediction errors: on the average, there were 5.5 prediction errors per test set. This corresponds to the 83.1% average prediction accuracy. This result, presented in the first row (A) of Table 21.1, presents the baseline for comparing the quality of different noise elimination algorithms. The average number of eliminated training examples is 0 because no noise elimination has been used. Generated rules are complete and consistent with all training examples in corresponding training sets.

In all other experiments instead of G_0 - G_9, reduced training sets G'_0 - G'_9 were used as the input to inductive learning. Reduced training sets G'_i were obtained by the elimination of noisy examples, $G'_i \leftarrow G_i \setminus A_i$, where A_i represent noisy set outputs of different filtering algorithms for G_i as input training sets.

21.4.3. Results

21.4.3.1 Saturation Filter.
By using the saturation filter presented in Section 21.3.1 with the ε_h parameter set to 1.0, the average prediction error was 3.6 examples per test set, which corresponds to the 89.0% average prediction accuracy. The result is presented in the second row (S) of Table 21.1. Compared to the results in row A this

is a substantial improvement. This result was obtained by an average elimination of 30.2 examples per iteration, which represents about 10% of the training sets. A comparison to prediction results obtained by other authors, using very different machine learning methods, that are all between 86.6% and 89.7% (Kukar et al., 1997), indicates that 10% of detected and eliminated noise seems to be realistic. Both the prediction accuracy and the number of eliminated examples demonstrate that saturation filtering can be used as an effective noise handling mechanism.

Table 21.1 Average number of prediction errors (with standard deviation in parentheses), average prediction accuracy, and average number of eliminated examples (with standard deviation in parentheses) for different noise detection algorithms: A - ILLM without noise handling, S - saturation filter, C - classification filter, CS - combined classification-saturation filter, S9,S8,S7 - consensus saturation filters with varied levels of consensus.

Algorithm	Prediction error	Accuracy	Eliminated examples
A	5.5 (3.03)	83.1 %	0
S	3.6 (1.96)	89.0 %	30.2 (2.78)
C	4.7 (1.64)	85.6 %	52.6 (4.58)
CS	3.6 (1.71)	89.0 %	36.5 (4.34)
S9	4.2 (1.48)	87.2 %	12.5 (2.46)
S8	3.6 (1.17)	89.0 %	17.5 (2.12)
S7	3.6 (1.35)	89.0 %	19.8 (2.20)

21.4.3.2   Classification Filter.   Using the classification filter, described in Section 21.3.2, with parameter $n = 10$ (10-fold cross-validation), in average 4.7 prediction errors per test set were made. This corresponds to the 85.6% average prediction accuracy. This result, presented in the third row (C) of Table 21.1, represents an improvement in accuracy when compared to the result of row A although it is not as good as the result obtained by the saturation filter (row S). Nevertheless, this result is important because the classification filter is computationally much simpler than the saturation filter. Additionally, according to Brodley and Friedl (Brodley and Friedl, 1999), the use of simple voting mechanisms based on the results of filtering by different learning approaches (not necessarily only inductive learning approaches) can further improve the reliability of this noise detection process. This possibility was not investigated in this work.

The average number of eliminated examples per training set was 52.6. This means that the classification filter practically detected every fifth training example as potentially noisy. This shows a weakness of the classification filter: it is non-selective. Too many examples are detected as being potentially noisy.

21.4.3.3   Combined Classification-Saturation Filter.

Results obtained with noise filtering performed by the combined classification-saturation filter (see Section 21.3.3, $n=10$, $\varepsilon_h = 1.0$) are presented in CS row of Table 21.1. The measured average error was 3.6 examples representing the 89.0% average accuracy. It can be concluded that the combination of the two different noise detection algorithms is advantageous only in comparison with the classification filtering approach. The combined approach namely resulted in the average of 36.5 detected and eliminated potentially noisy examples per training set. Although this number is smaller than the one for the classification filter (row C with 52.6 examples) it seems that non-selectiveness is an inherent characteristic of classification filtering when a single classifier is used.

21.4.3.4   Consensus Saturation Filter.

The relative success of the classification filter, enabling significant prediction accuracy improvement using a learning algorithm without noise handling, stimulated a series of experiments in which we tested if a mechanism similar to the one used in the classification filter could improve the results of saturation filtering. In all the experiments parameter n was set to 10 and $\varepsilon_h=1.0$, while changing the consensus level v. With a default value $v = 9$ the average number of eliminated examples per training set was only 12.5 and the average number of prediction errors on test sets was 4.2 (presented in S9 row of Table 21.1).

This result is important since a satisfactory prediction accuracy was achieved by the elimination of a very small number of training examples. When compared with saturation filtering itself (row S) we see that by requiring the consensus of filtering in 10-fold validation, lower prediction accuracy (87.2% instead of 89.0%) was achieved by a significantly smaller average number of eliminated examples (12.5 instead of 30.2).

The described consensus filtering approach seems to be a reliable noise detection algorithm. A further proof of this claim is the following: from the total of 125 examples eliminated by the consensus filter (row S9), 124 of them were detected also by the combined classification-saturation filter (row CS) that, however, eliminated 365 examples in total.

Requiring consensus in 10-fold validation results in noise detection of high specificity. However, the decreased prediction accuracy indicates that the algorithm's sensitivity is too low. This can be the consequence of a too high consensus level (in order to declare an example to be potentially noisy, the example had to be tagged as potentially noisy in all subsets in which it occurs, i.e., in 9 subsets).

The algorithm's noise sensitivity can easily be increased by decreasing the consensus level v. In this way, relaxed consensus filters can be constructed. Rows S8 and S7 present results obtained by setting the consensus levels to 8 and 7, respectively. In both cases the increase in sensitivity resulted in the increase of the achieved prediction accuracy. Good prediction accuracy of 89.0%, as in cases S and CS, is obtained by relaxed consensus filtering but with significantly fewer potentially noisy examples eliminated.

21.4.4. Mediacal Evaluation

The results of described experiments suggest that consensus saturation filtering (with consensus level 9, row S9 of Table 21.1) presents a reliable tool for the detection of a minimal number of examples that are indeed noisy. In order to test the practical usefulness of noise detection by the consensus saturation filter, we applied the described approach to the entire coronary artery disease dataset of 327 examples. In this case the training sets G_0 - G_9 were used as 10 training subsets for the consensus filter. In total 15 potentially noisy examples were detected. This is in accordance with the result obtained for G_0 - G_9 sets for which the average value of eliminated examples was 12.5.

The detected examples were shown to a domain expert for evaluation. In order to make the task more difficult to the expert, we formed a set of 20 examples consisting of 15 examples detected as noisy by the consensus saturation filter, and additional 5 randomly selected non-noisy examples. The expert analyzed the group of these 20 examples as a whole. Out of these 20 examples, 13 were of class *CAD non-confirmed* (11 detected potentially noisy and 2 other examples of class non-confirmed), and 7 were of class *CAD confirmed* (4 detected potentially noisy examples and 3 other examples of class confirmed).

In the class *non-confirmed* the expert recognized 11 cases as being outliers, and these examples were exactly those selected by the noise detection algorithm. For 8 of these 11 cases the problem was a high grade of stenosis of coronary arteries, very close to the predefined value that distinguishes between patients with confirmed and non-confirmed coronary artery disease. Other 3 cases represented patients who did not have actual coronary heart disease problems but rather problems due to recent miocardial infarction, functionally malfunctioning by-pass, and valve disease.

For the class *confirmed* the expert detected 4 potentially noisy cases three of which were selected also by the noise detection algorithm and one of the randomly selected cases of this class. For all 4 cases the detected grade of stenosis was very close to the border line between the two classes. Additionally, for one patient the values of measured parameters were detected as non-typical due to the improper level of stress during measurements.

Further analysis showed that most of cases detected as noise represented a special group of patients whose coronary angiography measurements (measuring the percentage of stenosis or occlusion in the main coronary artery, left anterior descendens, left circumflex and right coronary artery) are very close to the border line between the two classes CAD confirmed and not-confirmed. By a slight change of the definition of class *confirmed* these cases can change their class value, therefore it is clear that they are non-typical training cases. Their elimination from the training set, as well as the elimination of patients who did not have actual coronary heart disease problems and patients with improperly measured parameters, is reasonable if one wants to induce diagnostic rules uncovering characteristic properties of patients with CAD.

For a more detailed medical evaluation of results of the experiments in the coronary artery disease diagnosis problem the reader is referred to (Gamberger et al., 1999), which served as a basis for this chapter.

21.5.   SUMMARY AND FURTHER WORK

The introductory section of this volume identifies three prominent functions of instance selection: enabling, focusing, and cleaning. Enabling and focusing use complexity reduction and transformation techniques to enable handling of the given, potentially large and sparsely defined data mining problem by the available data mining resources and to increase data mining efficiency by concentrating on relevant parts of the data.On the other hand, data cleaning denotes the elimination of irrelevant, erroneous, and misleading instances leading to increased data quality and resulting in improved quality of results and reduced data mining costs. Our work contributes to data cleaning, in particular towards explicit detection and elimination of potentially noisy instances, including outliers.

Sensitivity and reliability are the main characteristics of the different noise detection approaches. Only an appropriate combination of both properties can ensure high prediction quality of the induced knowledge. In this work we have presented two novel approaches to noise filtering and showed their performance on a real word medical domain in comparison with the saturation and classification noise filtering algorithms. Standard deviations, presented as parenthesized numbers in Table 21.1, show that the differences of achieved prediction accuracies by different noise filtering algorithms are not significant. On the other hand, most of the differences in the numbers of eliminated noisy examples are significant and represent genuine characteristics of the presented noise filtering algorithms.

It is interesting that a simple classification filtering approach, based on an iterative application of a machine learning algorithm (without noise handling) by itself enables an improvement of the prediction accuracy. However, in the coronary artery disease diagnostic domain, best prediction accuracies were achieved by the approaches that include saturation filtering. The proposed consensus saturation filter is an interesting combination: it eliminates only a small number of potentially noisy examples, with a very high probability of actually being noisy. This property may turn out to be decisive for a broader applicability of noise detection algorithms. The experiments suggest that a relaxed consensus saturation filter (see row S8 in Table 21.1) represents a very good solution to the problem of noise filtering.

From the point of view of practical data mining we need to recognize that all presented noise filtering algorithms are time and space consuming. Their practical applicability is today limited to domains with a few thousand instances. In this sense classification based approaches are more flexible than saturation based systems, but in large domains both can be applied in combination with instance selection algorithms. Explicit noise detection in large domains seems only possible based on a modified combined classification-saturation filter. In this approach a reasonably large subset $E' \in E$ should first be selected, the saturation based filter is then used to eliminate noise from E' and then reliable classification rules based on examples in the reduced E' set are constructed. Finally, these classification rules are used to test examples in the remaining space $E \setminus E'$ and those examples that contradict the rules are detected as noise. Induction of highly reliable and sensitive rules for noise detection is the topic of our further research. The selection of a representative subset of E' using some algorithms presented in this book is the first step in this process.

Acknowledgement

We are grateful to Ciril Grošelj from the University Medical Centre Ljubljana for providing the data and the medical interpretation and evaluation of the results of this study, and to Matjaž Kukar from the

Faculty of Computer and Information Sciences, University of Ljubljana, for his help in the preparation of data as well as the evaluation of results.

References

Aha, D. W., Kibler, D., and Albert, M. K. (1991). Instance-based learning algorithms. *Machine Learning*, 6:37–66.

Brodley, M. (1995). Recursive automatic bias selection for classifier construction. *Machine Learning*, 20:63–94.

Brodley, C.E. and Friedl, M.A. (1999). Identifying mislabeled training data. *Journal of Artificial Intelligence Research*, 11:131–167.

Burges, C. (1998). A tutorial on support vector machines. *Journal of data Mining and Knowledge Discovery*, 2(2).

Clark, P. and Niblett, T. (1989). The CN2 induction algorithm. *Machine Learning*, 3:261–283.

Dimopoulos, Y. and Kakas, A. (1995). Learning non-monotonic logic programs: Learning exceptions. In *Proc. of the Eighth European Conference on Machine Learning*, pp. 122–137. Springer-Verlag.

Domingos, P. (1999). The role of Occam's razor in knowledge discovery. *Data Mining and Knowledge Discovery*, 3(4):409–425.

Fürnkranz, J. (1999). Separate-and-conquer rule learning. *Artificial Intelligence Review* 13(1):3–54.

Gamberger, D. (1995). A minimization approach to propositional inductive learning. In *Proc. of the Eighth European Conference on Machine Learning*, pp. 151–160. Springer-Verlag.

Gamberger, D. and Lavrač, N. (1997). Conditions for Occam's razor applicability and noise elimination. In *Proc. of the Ninth European Conference on Machine Learning*, pp. 108–123. Springer-Verlag.

Gamberger, D., Lavrač, N., and Grošelj, C. (1999). Experiments with noise filtering in a medical domain. In *Proc. of the Sixteenth International Conference of Machine Learning*, pp. 143–151. Morgan Kaufmann.

Gamberger, D., Lavrač, N., and Džeroski, S. (2000). Noise detection and elimination in data preprocessing: experiments in medical domains. *Applied Artificial Intelligence*, 14:205–223.

Grošelj, C., Kukar, M., Fetich, J., and Kononenko, I. (1997). Machine learning improves the accuracy of coronary artery disease diagnostic methods. *Computers in Cardiology*, 24:57–60.

John, G.H. (1995). Robust decision trees: Removing outliers from data. In *Proc. of the First International Conference on Knowledge Discovery and Data Mining*, pp. 174–179. AAI Press.

John, G.H. and Langley, P.(1996). Static versus dynamic sampling for data mining. In *Proc. of the Second International Conference on Knowledge Discovery and Data Mining*, pp. 367–370. AAI Press.

Kohavi, R. and Wolpert, D.H. (1996). Bias plus variance decomposition for zero-one loss functions. In *Proc. of the Thirteenth International Conference on Machine Learning*, pp. 275–283. Morgan Kaufmann.

Kukar, M., Grošelj, C., Kononenko, I., and Fetich, J. (1997). An application of machine learning in the diagnosis of ischaemic heart disease. In *Proc. of the Sixth Conference on Artificial Intelligence in Medicine Europe*, pp. 461–464. Springer-Verlag.

Lavrač, N., Gamberger, D., and Turney, P. (1998). A relevancy filter for constructive induction. *IEEE Intelligent Systems* 13:50–56.

Li, M. and Vitányi, P. (1993). An Introduction to Kolmogorov Complexity and its Applications. Springer-Verlag.

Mingers, J. (1989). An empirical comparison of pruning methods for decision tree induction. *Machine Learning*, 4(2):227–243.

J.R. Quinlan (1987). Simplifying decision trees. *International Journal of Man-Machine Studies*, 27(3):221–234.

Quinlan, J.R. (1993). C4.5: Programs for Machine Learning. Morgan Kaufmann.

Quinlan, J.R. (1996). Bagging, boosting and C4.5. In *Proc. of the Thirteenth National Conference on Artificial Intelligence*, pp. 725–730. AAAI Press.

Rao, R., Gordon, D., and Spears, W. (1995). For every generalization action, is there really an equal or opposite reaction? Analysis of conservation law. In *Proc. of the Twelveth International Conference on Machine Learning*, pp. 471–479. Morgan Kaufmann.

Rissanen, J. (1978). Modeling by the shortest data description. *Automatica*, 14:465–471.

Schaffer, C. (1994). A conservation law for generalization performance. In *Proc. of the Eleventh International Conference on Machine Learning*, pp. 259–265. Morgan Kaufmann.

Srinivasan, A., Muggleton, S., and Bain, M. (1992). Distinguishing exceptions from noise in non-monotonic learning. In *Proc. of the Second International Workshop on Inductive Logic Programming*. Tokyo, ICOT TM-1182.

Webb, G.I. (1996). Further experimental evidence against the utility of Occam's razor. *Journal of Artificial Intelligence Research*, 4:397–417.

Weisberg, S. (1985). Applied Linear Regression. John Wiley & Sons.

Chapter 22

INSTANCE SELECTION BASED ON SUPPORT VECTOR MACHINE FOR KNOWLEDGE DISCOVERY IN MEDICAL DATABASE

Shinsuke Sugaya and Einoshin Suzuki

Yokohama National University, Japan

shinsuke@slab.dnj.ynu.ac.jp, suzuki@dnj.ynu.ac.jp

Shusaku Tsumoto

Shimane Medical University, Japan

tsumoto@computer.org

Abstract   This chapter presents an instance-discovery method based on support vector machines (SVMs). Applied to a binary classification problem which is linearly separable, a SVM generates, as a classifier, a hyperplane with the largest margin to the nearest instances. SVMs have achieved higher accuracies than conventional methods especially in a high-dimensional problem in which a noise-free training data set is available. In this chapter, we try to exploit SVMs for knowledge discovery, especially for discovering typical instances for each class, boundary instances for discrimination, and misclassified instances in classification. In order to improve the readability of the results, we also propose an attribute-selection method based on the classifier learned by a SVM. Our method has been applied to the meningoencephalitis data set, which is a benchmark problem in knowledge discovery. Comparison by a physician with a method based on Fisher's linear discriminant (FLD) demonstrated the effectiveness of our method in instance discovery. Moreover, our results led the physician to interesting discoveries.

Keywords: Support vector machines (SVMs), typical instance discovery, boundary instance discovery, attribute selection, medical data set.

22.1. INTRODUCTION

In conventional knowledge discovery, the target of discovery is represented by a pattern, which includes predictive models, rules, and clusters. Since a pattern is an abstracted representation of given data, it allows deeper comprehension of the problem. A drawback of discovering patterns, however, is that they are sometimes difficult to be interpreted. On the other hand, an instance can be often easily interpreted, especially when redundant attributes are omitted. In this chapter, we settle instances as the target of discovery, and call such discovery instance discovery. Alternatively, instance discovery can be regarded as discovery of interesting subsets of the given data set. We strongly believe that instance discovery will attract much attention of the knowledge discovery community.

Support vector machines (SVMs) (Burges, 1998; Cortes and Vapnik, 1995; Platt, 1999; Vapnik, 1995) resolve a binary classification problem by obtaining, when the classes are linearly separable, a hyperplane with the largest margins to the nearest instances. SVMs have achieved higher accuracies in classification problems in which the number of attributes is huge and a noise-free training data set is available. Examples of such problems include text classification (Joachims, 1998), face detection (Osuna, et al., 1997), and object recognition (Pontil and Verri, 1998). These empirical results suggest that SVMs produce highly accurate models for such kinds of problems. Although SVMs are mainly employed for classification, this intuition has led us believe that SVMs would be also effective for other problems in knowledge discovery.

In this chapter, we propose a novel method which employs SVMs for instance discovery, and evaluate its effectiveness with a medical data set. Our method discovers, typical instances for each class, boundary instances for discrimination, and misclassified instances in classification. Since this method also detects relevant attributes in discrimination, and thus deletes irrelevant attributes, the discovered instances show high readability. In the experiments, this method is compared with conventional methods such as a method based on Fisher's linear discriminant (FLD). The evaluation is subjective: it is done in medical context with the aid of a domain expert. The results were above our expectations: our method not only showed higher performance than the other methods, but also produced highly interesting patterns which led the expert to interesting discoveries.

22.2.   SUPPORT VECTOR MACHINES

Support vector machines (SVMs) (Burges, 1998; Cortes and Vapnik, 1995; Platt, 1999; Vapnik, 1995) resolve a binary classification problem by obtaining, when the classes are linearly separable, a hyperplane with the largest margins to the nearest instances. We here call this hyperplane an optimal separating hyperplane (OSH), each of the nearest instances a support vector (SV), and each of the instances which is not a SV a non-support vector (NSV). We show several concepts related to SVMs in Figure 22.1.

The distance between an OSH and a SV is called a margin, and as explained above, an OSH and SVs are determined so that the margin is the maximum. The problem of obtaining an OSH can be formalized as a quadratic problem (Bazaraa and Shetfy, 1979), which represents maximization of a quadratic function under linear constraints. Efficient algorithms exist, and we employed the method which was proposed by Platt (Platt, 1999).

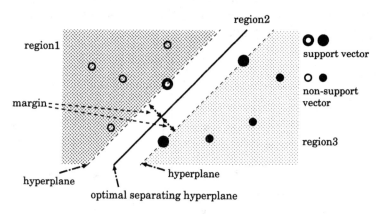

Figure 22.1 Concepts related to Support vector machines (SVMs).

For a binary classification problem in which the classes are not linearly separable, two approaches are frequently used. One is to project the instance space to a different space of higher dimensionality with the aid of a Kernel function. The other is to assign a cost for a misclassified instance. In this chapter, we employ the latter method (Burges, 1998).

22.3.   INSTANCE DISCOVERY BASED ON SUPPORT VECTOR MACHINES

22.3.1.   Attribute Selection

Based on SVMs presented in the previous section, we here propose a novel method for attribute selection. Our method can identify attributes each of which is relevant in discrimination.

Note that a normal vector of an OSH consists of coefficients for attributes in the instance space. For example, in Figure 22.2 (a), an OSH represents a straight line, and its normal vector consists of a coefficient for x axis and another for y axis. Our basic idea is to regard the unsigned magnitude of each coefficient as discrimination power. For example, in Figure 22.2 (a), the unsigned coefficient for x axis is greater than the one for y axis. As a result, in this example, the attribute x can discriminate black instances and white instances with a single threshold, while the attribute y requires multiples thresholds for the same purpose. Here, some readers would suggest a counterexample such as presented in Figure 22.2 (b) in which they attribute the dotted line as an OSH, and claim that an attribute with a small unsigned magnitude can be relevant in the discrimination. However, this case never happens since SVM produces a classification model that is represented by a straight line, which has the largest margin.

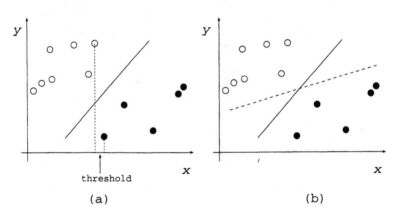

Figure 22.2   (a) A projection of the instances on the axis of an attribute of which coefficient in the normal vector has a large unsigned magnitude. (b) A counterexample for a discriminative attribute of which coefficient has a small unsigned magnitude. Actually, (b) never happens since the straight line, not the dotted line, is produced as the OSH.

Note that relevance of an attribute here is related to geometric positions of unweighted instances rather than accuracy. In Figure 22.3, the

x axis can almost discriminate black instances and white instances with a single threshold, but will be judged irrelevant in our method since the unsigned magnitude of its coefficient is small. This is caused by the small distance between the circled instances, which determined the OSH in this example. Our definition for relevance is subject to discussions, and we will empirically evaluate both definitions by comparison in section 22.4.3.

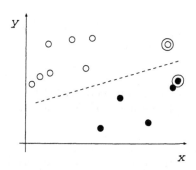

Figure 22.3 Illustration which shows that relevance of an attribute here is related to geometric positions of unweighted instances rather than accuracy for each attribute. The x axis can almost discriminate black instances and white instances with a single threshold, but will be judged irrelevant due to the small distance between the circled instances.

Note that the above discussions hold true when instances have similar values. This requires that the scales of the attributes are approximately equivalent since an unsigned magnitude of a coefficient in the normal vector also depends on the scale of the attribute. In order to understand this, consider the example in Figure 22.4. The left-hand side (a) and the right-hand side (b) represent expressing an attribute "HEIGHT" in inch or in centimeter respectively. Although these represent the same attribute, the line in the right-hand side is flatter due to the narrower scale for "HEIGHT". Hence, if scales of each attribute are totally different, it is difficult to determine relevant attributes for discrimination.

In order to circumvent this problem, we employ normalization of each attribute based on z-scores, which is frequently used in statistics. Normalization based on z-scores obtains a relative position of the data with respect to an attribute. A value of z-scores z_i represents the number of standard deviation s between an attribute value x_i and an average value x_m. Attribute values are transformed by

$$z_i = \frac{x_i - x_m}{s} \tag{22.1}$$

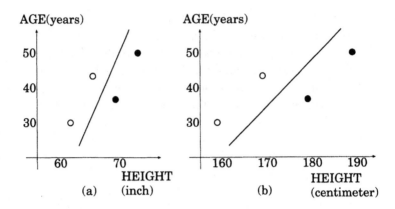

Figure 22.4   Effect of expressing an attribute "HEIGHT" in inch (a) or in centimeter (b) respectively. Line (b) is flatter due to its narrower scale for values.

Relevant attributes for discrimination are determined by this transformation, and this corresponds to adjusting attributes to approximately equivalent scales.

In summary, our method finds the n most relevant attributes for discrimination by inspecting the coefficients in a normal vector after having ordered the coefficients by their unsigned magnitude.

22.3.2.   Instance Discovery

We also exploited SVMs presented in section 22.2 for instance discovery. We begin by describing our motivations in three steps. First, a SV, which is the nearest instance to an OSH, is also nearest to the region of the other class and thus can be regarded as a boundary instance in discrimination. Second, the farthest instances among NSVs can be considered as typical instances of their class since they are farthest to the region of the other class[1]. Third, a misclassified instance, which exists when the classes are not linearly separable, can suggest important information since they cannot be explained by the OSH classifier.

Our method discovers three types of instances as follows. SVs are outputted as boundary instances in discrimination. The H % farthest instances from the OSH are outputted as typical instances for each class. The misclassified instances in classification, if they exist, are also outputted.

Note that an instance has low readability when the classification problem has a large number of attributes. A user is likely to be interested only in relevant attributes in discrimination when he inspects a discovered instance. In our method, each discovered instance is represented with respect to the K most relevant attributes based on the attribute selection procedure in the previous section.

22.4. APPLICATION TO THE MENINGOENCEPHALITIS DATA SET

22.4.1. KDD Contest

The main goal of a KDD (Knowledge Discovery in Databases) contest (Suzuki, 2000; Tsumoto, et al.) can be considered as comparing various knowledge discovery methods under the supervision of a domain expert. Several contests, including KDD Challenge 2000 (Suzuki, 2000) employed a common medical data set on meningoencephalitis. The results of these contests enable us to characterize KD (Knowledge Discovery) methods more concretely, and show the importance of interaction between KD researchers and domain experts in KDD.

The meningoencephalitis data set (Suzuki, 2000; Tsumoto, et al.; Tsumoto, 1999c) consists of the data of patients who suffered from meningitis and were admitted to the department of emergency and neurology in several hospitals. Tsumoto worked as a domain expert for these hospitals and collected those data from the past patient records (1979 to 1989) and the cases in which he made a diagnosis (1990 to 1993). Since this domain is relatively well understood and there is little disagreement between experts in the field, Tsumoto can evaluate discovered results by himself.

The database consists of 140 cases and all the data are described by 38 attributes, including present and past history, laboratory examinations, final diagnosis, therapy, clinical courses and final status after the therapy. More information about this data set is given in Appendix A, and also available on the WWW (Suzuki, 2000).

22.4.2. Conditions of The Experiments

Six tasks were settled by selecting Diag2 (grouped attribute of diagnosis), EEG_WAVE (electroencephalography), CT_FIND (CT finding), CULT_FIND (whether bacteria or virus is specified or not), COURSE (Grouped) (grouped attribute of clinical course at discharge) and RISK (Grouped) (grouped attribute of risk factor) as a target class. These tasks are the main interest of medical experts. Since SVMs deal with a binary classification problem, an attribute with more than three values cannot be regarded as a class.

Since SVMs can be applied to only continuous attributes, nominal attributes with more than two values should be either converted to several binary attributes or ignored. Since the first approach gave relatively poor results, we took the latter approach. We consider that this choice depends on the problem, and our method would each time require careful preprocessing of a data set[2]. Note that this preprocessing problem typically applies to all KDD endeavors, and does not suggest limitations of our approach.

Here, attributes concerning therapy and courses, such as COURSE, can be measured only after differential diagnosis process. Thus, they were ignored in four tasks with respect to Diag2, EEG_WAVE, CT_FIND and CULT_FIND since they are not available in determining these classes. Moreover, RISK was removed for data analysis because it was selected as a decision attribute. We hereafter abbreviate COURSE (Grouped) and RISK (Grouped) to COURSE and RISK respectively, if there is no risk of confusion.

After the experiments shown in (Tsumoto, 1999c), we generalized several attributes, which have too many values and seem to lose essential information for data analysis. Our method was applied after these preprocessing procedures, and the parameters were settled as $K = 10$, $H = 10$.

Fisher's linear discriminant (FLD) (Hand, 1981) outputs a linear model and is widely used in statistics. For comparison, we applied the same procedure to the model produced by FLD, and discovered relevant attributes and three types of instances.

22.4.3.   Results of Attribute Selection

We first evaluated the effect of using z-scores presented in section 22.3.1, and briefly describe the results. For this purpose, we have chosen Diag2 as the class, since the domain expert (Tsumoto) possesses rich experience and knowledge for this problem. Attributes with narrow scales were never selected when z-scores were not used, but some of them were selected when z-scores were used. Especially, several attributes which are relevant according to domain knowledge were selected only using z-scores. We can therefore justify our use of z-scores from these results.

Next, we evaluated the effectiveness of the proposed method in attribute selection. Tsumoto ranked, from one to five, selected attributes with respect to validness, unexpectedness and usefulness. Here, validness indicates that discovered results agree with the medical context, and unexpectedness represents that discovered results can be partially explained by the medical context but are not accepted as common sense. Usefulness indicates that discovered results are useful in medical context.

According to Tsumoto, it is relatively difficult, for a discovery method, to select ten relevant attributes in this problem.

We employed, for comparison, the method based on Fisher's linear discriminant (FLD) and a simple method (naive method). The naive method is employed to validate the use of a classifier in this procedure and our definition of attribute relevance presented in section 22.3.1. The naive method first projects each instance in the instance space to the axis of each attribute, and obtains a boundary with the maximum accuracy for each attribute. It then ranks the attributes according to their accuracies. The results obtained by applying the three methods to six tasks are summarized in Table 22.1.

Table 22.1 Average scores of the proposed method, the naive method, and the FLD-based method with respect to attribute selection. The largest values for validness and usefulness as well as the smallest values for unexpectedness are boldfaced.

| Task | Attribute selection | | |
	Validness	Unexpectedness	Usefulness
Diag2	**4.8**/4.6/4.7	**1.0**/1.4/**1.0**	**4.8**/4.5/**4.8**
EEG_WAVE	**4.7**/4.0/4.2	**1.2**/2.2/1.3	4.1/4.1/**4.5**
CT_FIND	**4.9**/4.8/4.8	**1.0**/**1.0**/1.5	**5.0**/4.5/**5.0**
CULT_FIND	**4.8**/4.7/4.7	1.8/**1.4**/1.6	**4.9**/4.8/4.8
COURSE	4.9/**5.0**/4.5	1.2/**1.0**/1.5	4.9/**5.0**/4.7
RISK	**4.8**/4.1/4.5	**1.1**/1.9/1.5	**4.9**/4.4/4.6

For attribute selection, we see that our method often outperforms the other two. For validness, our method outperformed the other approaches in five tasks among six. For usefulness, our method shows the best results in four tasks among six. For unexpectedness, all three approaches show relatively low scores, and can be judged as conservative in knowledge discovery. The poor performance of the naive method in the experiments can be attributed to its lack of consideration on the dependencies among attributes. The FLD-based method was also frequently outperformed by our method, which can be attributed to the nature of the problem: the number of attributes is huge and a noise-free training data set is available. In this kind of problem, a classifier which is based on the boundary instances can be considered as more appropriate than a classifier which is based on the variance of training instances. From these results, we can conclude that our method shows the best performance among the three approaches for these tasks, and often discovers valid and useful results in medical context.

22.4.4.   Results of Instance Discovery

We also evaluated the effectiveness of the proposed approach in instance discovery by the degree that discovered instances match medical knowledge. We show, in Tables 22.2 and 22.3, results on typical and boundary instance discovery. In the tables, each row represents a discovery task, and rows correspond to, from left to right, Diag2, EEG_WAVE, CT_FIND, CULT_FIND, COURSE, and RISK.

Table 22.2 Results for discovery of typical instances for each class. Rows correspond to, from left to right, Diag2, EEG_WAVE, CT_FIND, CULT_FIND, COURSE, and RISK. The column "expert" represents numbers of instances discovered by a domain expert. The other two columns, "SVM" and "FLD", represent (number of discovered instances which match the instances of the expert)/ (number of discovered instances) for the proposed method and the FLD-based method respectively.

Task	Di	E	CT	CU	CO	R
expert	31	35	28	20	25	29
SVM	13/14	11/15	10/12	10/10	9/12	9/13
FLD	11/15	11/16	10/12	8/12	11/13	9/12

Table 22.3 Results for discovery of boundary instances in discrimination.

Task	Di	E	CT	CU	CO	R
expert	34	35	49	58	47	39
SVM	23/24	21/29	22/26	28/31	19/32	26/32
FLD	9/15	6/13	7/12	7/12	11/12	10/14

These tables show that the proposed method outperforms the FLD-based method except for COURSE. This superiority is especially clear for boundary instance discovery. Interestingly, our method and the FLD-based method share few instances, and this fact corresponds to an empirical evidence that the discovery results of the two approaches are definitely different.

Table 22.4 Results for discovery of misclassified instances in classification.

Task	Di	E	CT	CU	CO	R
expert	0	9	15	36	27	22
SVM	-/0	-/0	15/20	36/58	19/32	8/10
FLD	0/7	8/12	10/18	29/35	11/31	10/16

Table 22.4 shows results on misclassified instance discovery. From the table, we see that several tasks are difficult even for an expert. For matching rates, our method outperforms the FLD-based method except for CULT_FIND, and discovers results which are similar to judgment of the expert. Our method and the FLD-based method share more instances than in typical instance discovery and boundary instance discovery, but this is partially due to the fact that a larger number of instances were discovered.

These results can be summarized as follows. Our method outperforms the FLD-based method in typical instance discovery and in boundary instance discovery, and discovers similar instances in misclassified instance discovery. As stated in the previous section, the reason can be attributed to the appropriateness of SVMs for a high-dimensional classification problem in which a noise-free training data set is available.

22.4.5. Discovery by An Expert

Tsumoto analyzed the discovered instances according to the selected attributes, and discovered several pieces of interesting knowledge.

Discovery from the task of COURSE (Grouped) is the most interesting one: the comparison between typical instances and boundary instances suggests that the combination of LOC_DAT and FOCAL is much more important than other attributes. While the values of LOC_DAT and FOCAL in 9/11 instances are both negative in the typical negative instances, 17/28 instances have positive values at least in one of the two attributes in SVs. This is also true in 41 positive instances and 99 negative instances. In these positive instances, 35/41 instances have positive values of LOC_DAT or FOCAL. On the other hand, only 29(/99) negative instances have positive ones: 70 instances have negative values for both attributes. Although these results are not statistically significant, these observations are reasonable to medical knowledge: LOC_DAT usually measures global brain function and FOCAL usually measures local brain functions. Thus, if both attributes are negative, a brain does not have serious damage. In other words, both of negative values are required for good prognosis.

Although these results have been expected by experiences of neurologists, they have never been verified by using a clinical data set before. It is notable that the classification of instances into typical and boundary ones together with the attribute selection enabled a medical expert to compare them and to discover such important knowledge.

22.5.   DISCUSSION

In this chapter, SVMs were compared with FLD from the viewpoint of attribute selection and instance discovery rather than discrimination. The experimental results show that SVMs-based method is much similar to experts' knowledge and comparison of discovered instances (typical instances and boundary instances) led an expert to interesting discoveries.

One of the important characteristics of this comparison is that it was based on experts' knowledge, not on some objective criteria. This style of comparison can be called "subjective evaluation" because knowledge of experts includes subjective belief. Its advantage is that extracted rules or instances are triggered to enable a novel discovery guided by experts' sophisticated reasoning. On the other hand, its disadvantage is that it is a little difficult to check whether these results are objective and can be generalized.

However, from the viewpoint of discovery process, "subjective evaluation" is as important as "hypothesis generation". It is well known that a novel scientific discovery is frequently initiated by a hypothesis inspired by unexpected or interesting results. Thus, when we consider that interactions between computer-based discovery systems and human beings are indispensable to a novel discovery, the most important point is whether induced results inspire an interesting hypothesis. In this experiment, discovered instances have actually contributed to generation of a novel hypothesis.

The experimental results also suggest that instance discovery will be a novel research direction for knowledge discovery. Although rule induction methods have often been used in KDD research, instance discovery has several advantages. First, instances have richer information than rules: rules have a fixed representation, mainly the conjunction of attribute-value pairs, which is typically a projection of instances (of high dimensionality) to pieces of knowledge of low dimensionality. Thus, substantial amount of information is lost through rule induction process. On the other hand, instances are free from such a representation style and experts can take a closer look at discovered instances. Second, classification of instances into typical and boundary instances or into other categories will make comparison between these instances easier. Rules usually classify them into several classes, but may not consider the degree of contribution of each instance to classification. Typical and boundary instances roughly capture the degree that domain experts usually use in a natural way and comparison of this degree can generate novel information, which cannot be extracted from rules. Thus, this degree of membership to a class can be viewed as a novel axis for knowledge discovery and can be extracted by instance discovery.

However, instance discovery methods have also several disadvantages. First, if the data have a very high dimensionality, comparison of instances is difficult even for domain experts. Second, discovery by a hyperplane may not be effective to some domain in which a hyperplane may not be a good classifier.

It is notable that our method also achieves a simple solution to the first problem: our method outputs typical and boundary instances by using the attributes judged relevant in discrimination. Thus, discovered instances are described by reduced attributes, which make the interpretation by experts easier.

The second problem will be an important problem which should be solved in the near future. The generated hyperplane has global information about classification, but in some domains, local information is more important. Incorporating local information may extend the capabilities of our method, which is a future work of our research.

22.6.   CONCLUSIONS

This chapter has explored the capability of support vector machines (SVMs) in instance discovery. Our method discovers, typical instances for each class, boundary instances in discrimination, and misclassified instances in classification. These instances are outputted according to the relevant attributes since our method possesses attribute-selection capability. We employed a medical data set, which has been provided for KDD contests as a common problem, in order to evaluate the effectiveness of the proposed method from the viewpoint of domain knowledge. The proposed method outperformed a method based on Fisher's linear discriminant (FLD) and a projection-based naive method in selection of relevant attributes. As an instance-discovery method, it outperformed the FLD-based method in discovery of typical and boundary instances, and showed similar results to a domain expert in misclassified instance discovery. Our work can lead to a convenient scheme for interpreting support vectors and non-support vectors within an instance-discovery framework.

Knowledge discovery has long employed rule discovery as its basic method, but will certainly incorporate instance discovery. A rule has relatively a small amount of information since it contains only a small number of attributes in its premise and predicts only a small number of attribute-value pairs in its conclusion. On the other hand, an instance is rich in information with a relatively larger number of attributes. The advantage of our approach in the medical context is that it discovers typical instances and boundary instances with a reduced attribute set. This information is found highly useful for hypothesis generation, and has actually led an expert to discoveries of interesting knowledge. This knowledge is judged to be difficult to be discovered using statistical methods and/or rule-based methods. We have demonstrated empirically, in this chapter, that classification of instances into typical ones and boundary ones with relevant attributes selection represents a promising approach for discovering interesting knowledge.

Notes

1. Note that the term "typical" is used to represent "dissimilar to the other class". Discovery of "average" instances in a class requires a different method, but we will not discuss this issue in this chapter.

2. The use of z-scores also represents a preprocessing method, and its effectiveness thus depends on the problem.

References

Bazaraa, M. and Shetfy, C. (1979) *Nonlinear Programming*, Wiley, New York.

Burges, C. (1998) "A Tutorial on Support Vector Machines for Pattern Recognition", *Data Mining and Knowledge Discovery*, 2(2):121–167.

Cortes, C and Vapnik, V. (1995). "Support Vector Network", *Machine Learning*, 20(3):1–25.

Hand, D. J. (1981). *Discrimination and Classification*, Wiley, New York.

Joachims, T. (1998). "Text Categorization with Support Vector Machines: Learning with Many Relevant Features", *Proc. Tenth European Conf. Machine Learning (ECML)*, pages 137-142.

Osuna, E., Freund, R. and Girosi, F. (1997). "Training Support Vector Machines: an Application to Face Detection", *Proc. Computer Vision and Pattern Recognition (CVPR)*, pages 130-136.

Platt, J. (1999). "Fast Training of Support Vector Machines Using Sequential Minimal Optimization", *Advances in Kernel Methods: Sup-*

port Vector Learning, Schölkopf, B., Burges, C. and Smola, A. (eds.), pages 185-208, MIT Press, Cambridge, Mass.

Pontil, M. and Verri, A. (1998) "Support Vector Machines for 3D Object Recognition", *IEEE Trans. Pattern Analysis and Machine Intelligence*, 20(6):637–646.

Sugaya, S., Suzuki, E. and Tsumoto, S. (1999). "Support Vector Machines for Knowledge Discovery", *Principles of Data Mining and Knowledge Discovery, LNAI 1704 (PKDD)*, pages 561–567.

Suzuki, E (ed.) (2000). *Proc. Int'l Workshop of KDD Challenge on Real-world Data (KDD Challenge)*, 2000 (data sets are available from http://challenge2000.slab.dnj.ynu.ac.jp).

Tsumoto, S. et al. (1999) "Comparison of Data Mining Methods using Common Medical Datasets", *ISM Symposium: Data Mining and Knowledge Discovery in Data Science*, pages 63–72.

Tsumoto, S. (1999). "Knowledge Discovery in Clinical Databases: an Experiment with Rule Induction and Statistics", *Proc. Eleventh Int'l Symp. Methodologies for Intelligent Systems (ISMIS), LNAI 1609*, pages 349-357, Springer.

Vapnik, V. (1995). *The Nature of Statistical Learning Theory*, Springer, New York, 1995.

Appendix: Meningoencepalitis Data Set

Tables A.1 to A.3 show information about attributes which describe the meningoencephalitis data set.

Table A.1 Personal information, diagnosis and present history.

1.	AGE:	age
2.	SEX:	gender
3.	DIAG:	Original Diagnosis
4.	Diag2:	Grouped Attribute of DIAG
		BACTERIA
		VIRUS
5.	COLD:	Since when the patient has symptoms like common cold.
6.	HEADACHE:	Since when he/she has a headache.
7.	FEVER:	Since when he/she has a fever.
8.	NAUSEA:	when nausea starts
9.	LOC:	when loss of conscious starts
10.	SEIZURE:	when convulsion is observed
11.	ONSET:	ACUTE
		SUBACUTE
		CHRONIC
		RECURR: recurrent

Table A.2 Physical and laboratory examinations.

Physical Examinations when admitted to the hospital

12.	BT:	Body Temperature
13.	STIFF:	Neck Stiffness
14.	KERNIG:	Kernig sign
15.	LASEGUE:	Lasegue sign
16.	GCS:	Glasgow Coma Scale
17.	LOC_DAT:	loss of consciousness
18.	FOCAL:	Focal sign (Grouped)

Laboratory Examinations when admitted to the hospital

19.	WBC:	White Blood Cell Count
20.	CRP:	C-Reactive Protein
21.	ESR:	Blood Sedimentation Test
22.	CT_FIND:	CT Findings (Grouped)
23.	EEG_WAVE:	Electroencephalography(EEG) Wave Findings
24.	EEG_FOCUS:	Focal Sign in EEG
25.	CSF_CELL:	Cell Count in Cerebulospinal Fluid(CSF)
26.	Cell_Poly:	Cell Count (Polynuclear cell) in CSF
27.	Cell_Mono:	Cell Count (Mononuclear cell) in CSF
28.	CSF_PRO:	Protein in CSF
29.	CSF_GLU:	Glucose in CSF
30.	CULT_FIND:	Whether bacteria or virus is specified or not. (Grouped)
31.	CULTURE:	The name of Bacteria or Virus

Table A.3   Therapy and clinical course.

Therapy and Clinical Courses		
32.	THERAPY2:	Therapy
33.	CSF_CELL3:	CSF_CELL, 3 days after the admission
34.	CSF_CELL7:	CSF_CELL, 7 days after the admission
35.	C_COURSE:	Clinical Course at discharge
36.	COURSE(Grouped):	Grouped attribute of C_COURSE (n: negative, p: positive)
37.	RISK:	Risk Factor
38.	RISK(Grouped):	Grouped attribute of RISK (n: negative, p: positive)

Index